137 Biological metal extraction
138 Recycling metals
139 Life-cycle assessments
140 Extended response – Reactivity of metals
141 The Haber process
142 More about equilibria

CHEMISTRY PAPER 4

143 The alkali metals
144 The halogens
145 Reactions of halogens
146 Halogen displacement reactions
147 The noble gases
148 Extended response – Groups
149 Rates of reaction
150 Investigating rates
151 Exam skills – Rates of reaction
152 Heat energy changes
153 Reaction profiles
154 Calculating energy changes
155 Crude oil
156 Fractional distillation
157 Alkanes
158 Incomplete combustion
159 Acid rain
160 Choosing fuels
161 Cracking
162 Extended response – Fuels
163 The early atmosphere
164 Greenhouse effect
165 Extended response – Atmospheric science

PHYSICS PAPERS 5 & 6

166 Key concepts

PHYSICS PAPER 5

167 Scalars and vectors
168 Speed, distance and time
169 Equations of motion
170 Velocity/time graphs
171 Determining speed
172 Newton's first law
173 Newton's second law
174 Weight and mass
175 Force and acceleration
176 Circular motion
177 Momentum and force
178 Newton's third law
179 Human reaction time
180 Stopping distances
181 Extended response – Motion and forces
182 Energy stores and transfers
183 Efficient heat transfer
184 Energy resources
185 Patterns of energy use
186 Potential and kinetic energy
187 Extended response – Conservation of energy
188 Waves
189 Wave equations
190 Measuring wave velocity
191 Waves and boundaries
192 Waves in fluids
193 Extended response – Waves
194 Electromagnetic spectrum
195 Investigating refraction
196 Wave behaviour
197 Dangers and uses
198 Changes and radiation
199 Extended response – Light and the electromagnetic spectrum
200 Structure of the atom
201 Atoms and isotopes
202 Atoms, electrons and ions
203 Ionising radiation
204 Background radiation
205 Measuring radioactivity
206 Models of the atom
207 Beta decay
208 Radioactive decay
209 Half-life
210 Dangers of radiation
211 Contamination and irradiation
212 Extended response – Radioactivity

PHYSICS PAPER 6

213 Work, energy and power
214 Extended response – Energy and forces
215 Interacting forces
216 Free-body force diagrams
217 Resultant forces
218 Extended response – Forces and their effects
219 Circuit symbols
220 Series and parallel circuits
221 Current and charge
222 Energy and charge
223 Ohm's law
224 Resistors
225 I–V graphs
226 Electrical circuits
227 The LDR and the thermistor
228 Current heating effect
229 Energy and power
230 a.c. and d.c. circuits
231 Mains electricity and the plug
232 Extended response – Electricity and circuits
233 Magnets and magnetic fields
234 Current and magnetism
235 Current, magnetism and force
236 Extended response – Magnetism and the motor effect
237 Electromagnetic induction and transformers
238 Transmitting electricity
239 Extended response – Electromagnetic induction
240 Changes of state
241 Density
242 Investigating density
243 Energy and changes of state
244 Thermal properties of water
245 Pressure and temperature
246 Extended response – Density
247 Elastic and inelastic distortion
248 Springs
249 Forces and springs
250 Extended response – Forces and matter
251 Biology Answers
256 Chemistry Answers
262 Physics Answers
268 The Periodic Table of the Elements
269 Physics Equations List

A small bit of small print:
Edexcel publishes Sample Assessment Material and the Specification on its website. This is the official content and this book should be used in conjunction with it. The questions in Now try this have been written to help you practise every topic in the book. Remember: the real exam questions may not look like this.

REVISE EDEXCEL GCSE (9–1)

Combined Science

REVISION GUIDE

Higher

Series Consultant: Harry Smith

Authors: Pauline Lowrie, Mike O'Neill and Nigel Saunders

A note from the publisher

In order to ensure that this resource offers high-quality support for the associated Pearson qualification, it has been through a review process by the awarding body. This process confirms that this resource fully covers the teaching and learning content of the specification or part of a specification at which it is aimed. It also confirms that it demonstrates an appropriate balance between the development of subject skills, knowledge and understanding, in addition to preparation for assessment.

Endorsement does not cover any guidance on assessment activities or processes (e.g. practice questions or advice on how to answer assessment questions) included in the resource, nor does it prescribe any particular approach to the teaching or delivery of a related course.

While the publishers have made every attempt to ensure that advice on the qualification and its assessment is accurate, the official specification and associated assessment guidance materials are the only authoritative source of information and should always be referred to for definitive guidance.

Pearson examiners have not contributed to any sections in this resource relevant to examination papers for which they have responsibility.

Examiners will not use endorsed resources as a source of material for any assessment set by Pearson.

Endorsement of a resource does not mean that the resource is required to achieve this Pearson qualification, nor does it mean that it is the only suitable material available to support the qualification, and any resource lists produced by the awarding body shall include this and other appropriate resources.

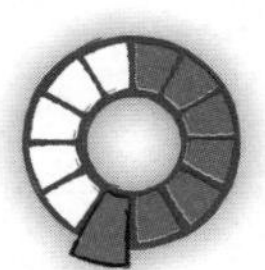

Question difficulty

Look at this scale next to each exam-style question. It tells you how difficult the question is.

For the full range of Pearson revision titles across KS2, KS3, GCSE, Functional Skills, AS/A Level and BTEC visit:
www.pearsonschools.co.uk/revise

Pearson

Conte nts

BIOLOGY PAPERS 1 & 2
1 Plant and animal cells
2 Different kinds of cell
3 Microscopes and magnification
4 Dealing with numbers
5 Using a light microscope
6 Drawing labelled diagrams
7 Enzymes
8 pH and enzyme activity
9 The importance of enzymes
10 Getting in and out of cells
11 Osmosis in potatoes
12 Extended response – Key concepts

BIOLOGY PAPER 1
13 Mitosis
14 Cell growth and differentiation
15 Growth and percentile charts
16 Stem cells
17 Neurones
18 Responding to stimuli
19 Extended response – Cells and control
20 Meiosis
21 DNA
22 Genetic terms
23 Monohybrid inheritance
24 Family pedigrees
25 Sex determination
26 Variation and mutation
27 The Human Genome Project
28 Extended response – Genetics
29 Evolution
30 Human evolution
31 Classification
32 Selective breeding
33 Genetic engineering
34 Stages in genetic engineering
35 Extended response – Genetic engineering
36 Health and disease
37 Common infections
38 How pathogens spread
39 STIs
40 Human defences
41 The immune system
42 Immunisation
43 Treating infections
44 New medicines
45 Non-communicable diseases
46 Alcohol and smoking
47 Malnutrition and obesity
48 Cardiovascular disease
49 Extended response – Health and disease

BIOLOGY PAPER 2
50 Photosynthesis
51 Limiting factors
52 Light intensity
53 Specialised plant cells
54 Transpiration
55 Translocation
56 Water uptake in plants
57 Extended response – Plant structures and functions
58 Hormones
59 Adrenalin and thyroxine
60 The menstrual cycle
61 Control of the menstrual cycle
62 Assisted Reproductive Therapy
63 Blood glucose regulation
64 Diabetes
65 Extended response – Control and coordination
66 Exchanging materials
67 Alveoli
68 Blood
69 Blood vessels
70 The heart
71 Aerobic respiration
72 Anaerobic respiration
73 Rate of respiration
74 Changes in heart rate
75 Extended response – Exchange
76 Ecosystems and abiotic factors
77 Biotic factors
78 Parasitism and mutualism
79 Fieldwork techniques
80 Organisms and their environment
81 Human effects on ecosystems
82 Biodiversity
83 The carbon cycle
84 The water cycle
85 The nitrogen cycle
86 Extended response – Ecosystems and material cycles

CHEMISTRY PAPERS 3 & 4
87 Formulae
88 Equations
89 Ionic equations
90 Hazards, risks and precautions
91 Atomic structure
92 Isotopes
93 Mendeleev's table
94 The periodic table
95 Electronic configurations
96 Ions
97 Formulae of ionic compounds
98 Properties of ionic compounds
99 Covalent bonds
100 Simple molecular substances
101 Giant molecular substances
102 Other large molecules
103 Metals
104 Limitations of models
105 Relative formula mass
106 Empirical formulae
107 Conservation of mass
108 Reacting mass calculations
109 Concentration of solution
110 Avogadro's constant and moles
111 Extended response – Types of substance

CHEMISTRY PAPER 3
112 States of matter
113 Pure substances and mixtures
114 Distillation
115 Filtration and crystallisation
116 Paper chromatography
117 Investigating inks
118 Drinking water
119 Extended response – Separating mixtures
120 Acids and alkalis
121 Strong and weak acids
122 Bases and alkalis
123 Neutralisation
124 Salts from insoluble bases
125 Salts from soluble bases
126 Making insoluble salts
127 Extended response – Making salts
128 Electrolysis
129 Electrolysing solutions
130 Investigating electrolysis
131 Extended response – Electrolysis
132 The reactivity series
133 Metal displacement reactions
134 Explaining metal reactivity
135 Metal ores
136 Iron and aluminium

Plant and animal cells

Animals and plants are formed from **cells**. Animal cells and plant cells have some parts in common. These parts have particular functions in a cell.

Generalised structures

Generalised animal cell

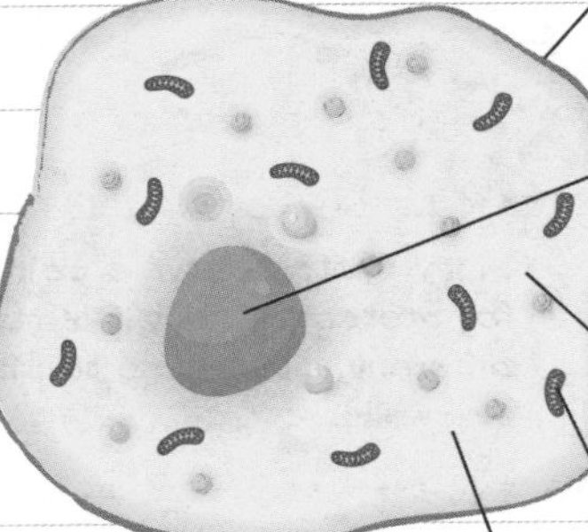

cell membrane: controls what enters and leaves the cell, e.g. oxygen, carbon dioxide, glucose

nucleus: a large structure that contains genes that control the activities of the cell

cytoplasm: jelly-like substance that fills the cell – many reactions take place here

mitochondria (single: mitochondrion): tiny structures where respiration takes place, releasing energy for cell processes

ribosomes (present in the cytoplasm but not visible at this size): where proteins are made (protein synthesis)

Generalised plant cell

cell wall

central vacuole

chloroplasts

Worked example

Name the three structures that are found in most plant cells but not animal cells, and describe their functions. **(4 marks)**

Chloroplasts are the structures where photosynthesis takes place to make food for the plant cell.

The cell wall is made of cellulose, and is tough so that it helps support the cell and helps it keep its shape.

The large central vacuole contains cell sap, which helps to keep the plant cell rigid.

1 mark is for naming the three structures and there is 1 mark for each function.

Watch out! The cell membrane and cell wall are different and separate structures.

Now try this

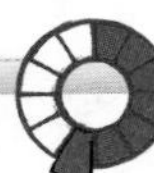

1 Muscle cells contain more mitochondria than skin cells. Suggest why. **(3 marks)**

2 Plants don't have skeletons. Explain how they stand upright. **(2 marks)**

3 Explain why not all plant cells have chloroplasts. **(2 marks)**

Had a look ☐ Nearly there ☐ Nailed it! ☐

Different kinds of cell

Some plant and animal cells are **specialised** for different functions. Bacteria have a different kind of **cell structure** from plant and animal cells.

Bacterial cells

Bacteria have a simple cell structure. Like animal and plant cells, they have a cell membrane surrounding the cytoplasm. But they do not have a nucleus.

A single loop of **chromosomal DNA** lies free in the cytoplasm. This carries most of the bacterial genes.

cell membrane

Some bacteria have a **flagellum** to help them move.

Ribosomes are tiny structures that make proteins.

Some bacteria have extra circles of DNA called **plasmid DNA.** Plasmids contain additional genes that are not found in chromosomes.

Many bacteria have a **cell wall** for protection, but it is made of different substances to plant cell walls.

Worked example

Many cells are specialised to carry out a particular function. The diagrams show three specialised human cells. Explain how the specialisation of each cell is related to its function. **(6 marks)**

An egg cell contains nutrients in the cytoplasm to supply the growing embryo. It has a haploid nucleus that can fuse with another haploid nucleus from the sperm to form a diploid zygote. After fertilisation, the membrane changes so that no more sperm cells can enter.

A sperm cell has a tail for swimming to the egg cell for fertilisation. Many mitochondria around the base of the tail release the energy needed to propel the sperm. The sperm cell has a haploid nucleus that fuses with the egg nucleus to form a diploid zygote. The acrosome contains enzymes to digest a way through the egg cell membrane.

Epithelial cells line tubes, such as the trachea. Cilia move things along the tube, such as mucus. This cell has a lot of cilia to move mucus, containing dirt and bacteria, away from the lungs.

Egg cell

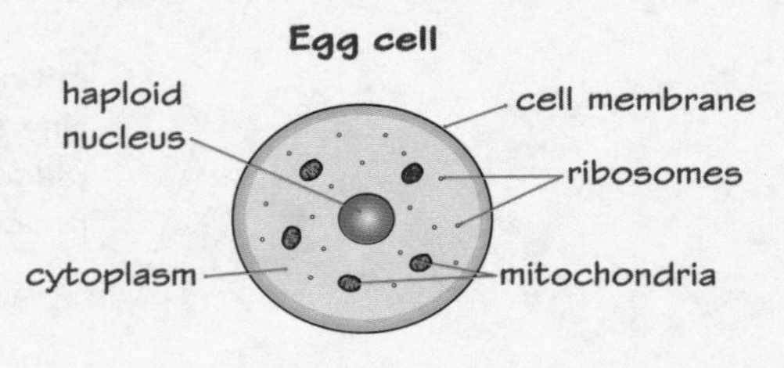

Sperm cell

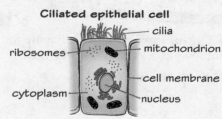

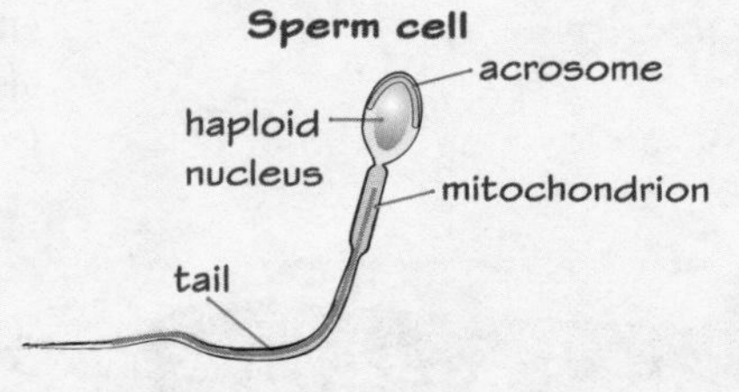

Ciliated epithelial cell

cilia
mitochondrion
ribosomes
cell membrane
cytoplasm
nucleus

Make sure that you can recognise an unusual feature of a cell, which may be a specialisation to allow the cell to carry out a particular function.

Now try this

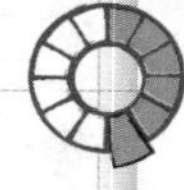

1 Give **one** similarity and **one** difference between a bacterial cell and an animal cell. **(2 marks)**

2 The diagram shows a root hair cell. Explain how the shape of this cell is related to its function. **(2 marks)**

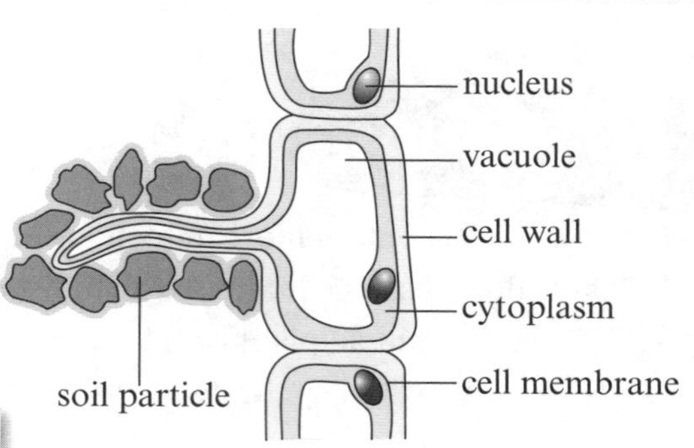

Describe how the shape of the cell is different from other cells, and then give a reason why this is an advantage.

Microscopes and magnification

As microscope technology has developed over time, we have been able to see more of the **structures of cells** and to know more about the **role** of sub-cellular structures.

Using microscopes

Before microscopes were invented about 350 years ago, people could not see the cells in organisms. Magnification enables you to see plant cells, animal cells and bacterial cells, and the structures inside them.

A **light microscope** uses light to magnify objects. The greatest possible magnification using a light microscope is about ×2000.

An **electron microscope** uses electrons to view an object. This makes it possible to magnify objects up to about ×10 million. You can see objects in cells more clearly and in far more detail with an electron microscope than with a light microscope.

Calculating magnification

$$\text{magnification (M)} = \frac{\text{image size (I)}}{\text{real size (R)}}$$

- Measure image in millimetres (mm).
- Multiply by 1000 to get measurement in micrometres (µm).
- If the image has a scale bar, use that to find the answer.

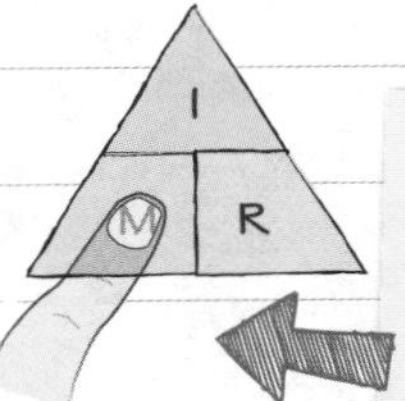

Cover up the term you are trying to find to show the expression required to calculate it.

Worked example

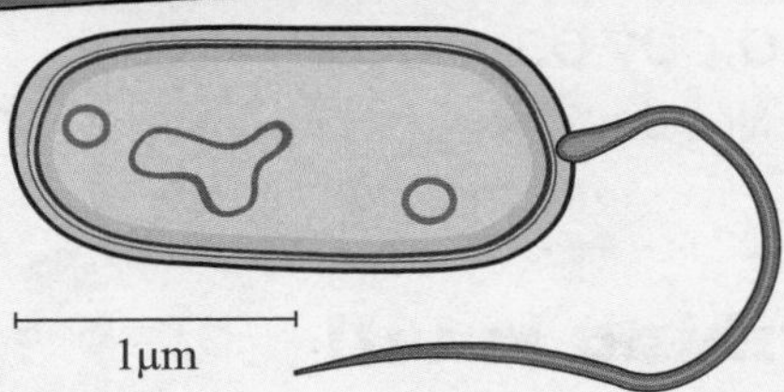

Calculate the magnification of this image. **(2 marks)**

scale bar (image size, I) = 20 mm

$20 \times 10^3 = 20\,000$ µm

so I = 20 000 µm

real size (R) = 1 µm (from diagram)

$\text{magnification} = \frac{I}{R} = \frac{20\,000}{1}$

magnification = ×20 000

Worked example

Some cells were viewed by microscope using a ×4 eyepiece and a ×20 objective. Calculate the magnification of the cells seen through the microscope. **(1 mark)**

magnification of object

= magnification of eyepiece × magnification of objective

= 4 × 20 = 80

The cells will be magnified 80 times by the microscope.

Always show your working in a calculation. Even if you get the answer wrong you may be able to show that you understand the method.

Now try this

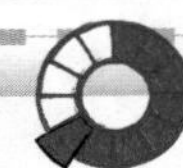

A bacterium is viewed under a light microscope using a ×40 objective and a ×10 eyepiece. The image is 1.2 mm long. Calculate the actual length of the cell. **(2 marks)**

Had a look ☐ Nearly there ☐ Nailed it! ☐

Dealing with numbers

Many structures in biology are very small so you will use very small **units of measurement.**

Standard form

Numbers in standard form have two parts.

$$7.3 \times 10^{-6}$$

This part is a number greater than or equal to 1 and less than 10.

This part is a power of 10.

You can use standard form to write very large or very small numbers.

$920000 = 9.2 \times 10^{5}$

Numbers greater than 10 have a positive power of 10.

$0.00703 = 7.03 \times 10^{-3}$

Numbers less than 1 have a negative power of 10.

Maths skills: Counting decimal places

You can count decimal places to convert between numbers in standard form and ordinary numbers.

3 jumps

$7900 = 7.9 \times 10^{3}$

7900 > 10
So the power is positive.

4 jumps

$0.00035 = 3.5 \times 10^{-4}$

0.00035 < 1
So the power is negative.

Be careful!

Don't just count zeros to work out the power.

Describing small structures

Smaller structures need smaller units:

unit	milli-	micro-	nano-	pico-
	10^{-3}	10^{-6}	10^{-9}	10^{-12}

The diameter of a human red blood cell is 9×10^{-6} m.

The diameter of DNA is about 4×10^{-9} m.

A millimetre is one thousandth of a metre, or 10^{-3} metres.

A micrometre is one millionth of a metre, or 10^{-6} metres.

Worked example

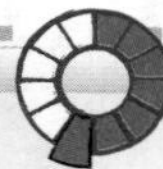

(a) State how many picograms there are in 1 g. **(1 mark)**

1 000 000 000 000 or 10^{12}

A mitochondrion measures 0.000 002 m.

(b) (i) Write this in standard form. **(1 mark)**

2×10^{-6} m

(ii) Write this in micrometres. **(1 mark)**

2 micrometres or 2 μm

(a) 'Pico' means 10^{-12}.

Remember to count the number of decimal places.

Remember that 'micro' means 10^{-6}.

Now try this

1 Match the structure to its correct average size.

protein molecule	chloroplast	cell
2 micrometres	10 nanometres	0.1 millimetres

(2 marks)

2 A human egg is 0.000 13 m long.
(a) State the length of the egg in standard form. **(1 mark)**
(b) State the length of the egg in micrometres. **(1 mark)**

3 The bacterial cell on page **3** is 2 μm long. Use this information to estimate the width of the cell. **(1 mark)**

Using a light microscope

You can use a **microscope** to **observe cells**. You can then produce accurate labelled diagrams from your observations. See page 6 for how to produce a labelled diagram.

Core practical

Using a light microscope safely

- Always start with the lowest **power objective** under the eyepiece.
- Clip the **slide** securely on the stage.
- Adjust the **light source (mirror)** so that light goes up through the slide.

If you use the Sun as a source of light, make sure the microscope mirror does not point directly at the Sun as this could permanently damage your eyesight.

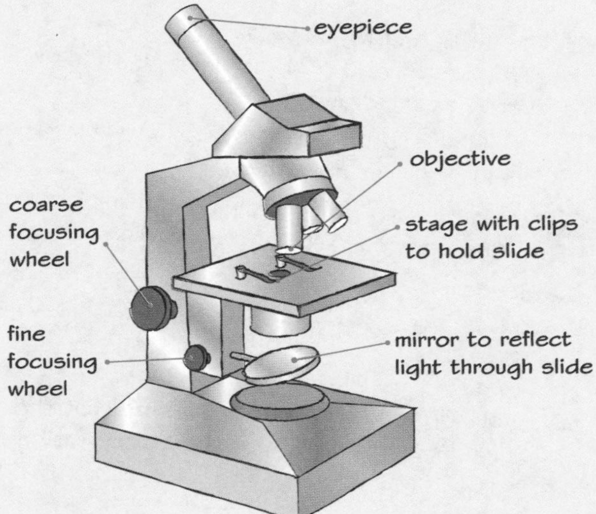

- Use the **coarse focusing wheel** to focus on the slide.
- Move the slide so that the cell you will draw is in the middle of the view.
- If needed, move a higher power objective into position above the slide.
- Use the **fine focusing wheel** to bring the cell back into focus.

Never use the coarse focusing wheel with a higher power objective, as it may crash into the slide.

If you cannot see the part of the slide you need with a higher power objective, go back to using the lower power objective to bring it back to centre view and to focus it before returning to the higher power objective.

Now try this

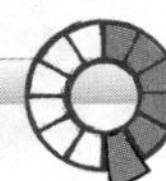

A student is given a microscope slide of a section through a leaf.

(a) Describe how the student should use a light microscope to study xylem cells in the leaf section. **(4 marks)**

(b) The lens closest to the specimen is called the

☐ **A** subjective

☐ **B** eyepiece

☐ **C** objective

☐ **D** stage **(1 mark)**

Drawing labelled diagrams

You can use a microscope to produce **accurate labelled diagrams** of cells and their structures. See page 5 for how to set up and use a light microscope.

Core practical

Aim

To produce a labelled diagram of a plant cell.

Apparatus

- light microscope
- unlined paper
- sharp HB pencil
- rubber
- ruler

If a line is in the wrong place, rub it out cleanly before drawing it in the correct position. Your drawing should be as clear as possible.

Method

Focus the microscope on a single cell.

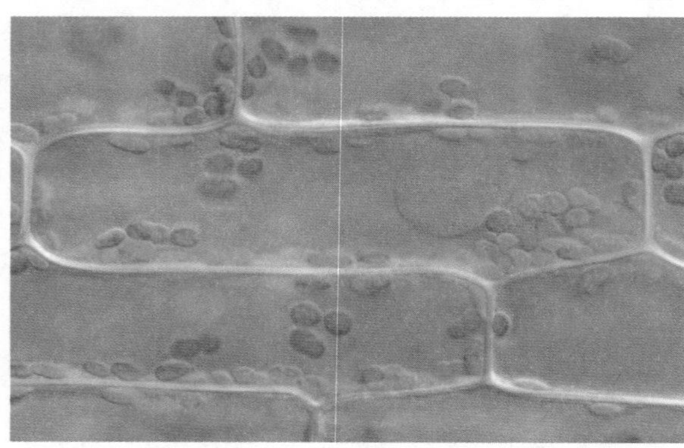

Keep the relative sizes of structures approximately correct and label all the key parts of your drawing.

Labels should be outside the drawing and label lines must not cross.

Use a ruler to draw label lines to link labels to the correct parts of the drawing.

Never use shading.

Carefully draw details of the parts that are important to your study. Other parts can be drawn just in outline.

Keep looking back at the specimen as you make the drawing, and only draw what you see, not what you think you ought to see.

Give your drawing a clear title saying what the specimen is, and always give the magnification, e.g. ×100. (See page 3 to help you calculate the magnification.)

Results

Pondweed cell ×650

chloroplast

cell walls of two neighbouring cells

cytoplasm and vacuole (not obvious)

nucleus

Now try this

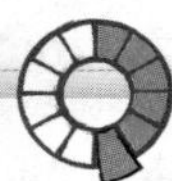

The photo shows some cells in a blood smear seen under a light microscope at ×1000.
Draw and label one example of each cell in the smear. **(3 marks)**

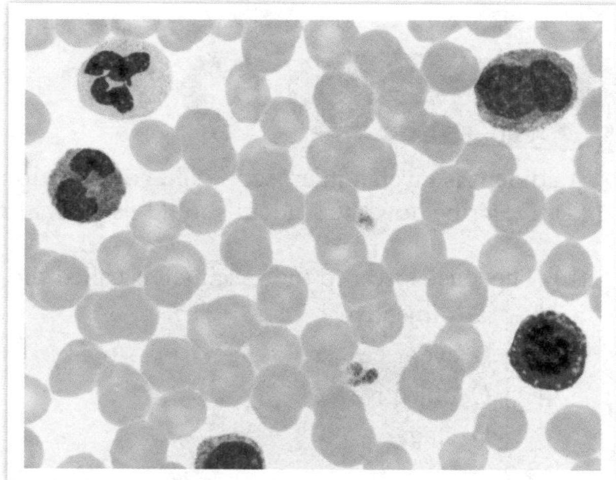

Enzymes

Enzymes are biological **catalysts** that control reactions in the body. Each enzyme is **specific** (only works with one substrate) to its substrate and the **activity** of enzymes is affected by temperature, substrate concentration and pH.

Effect of temperature and substrate concentration

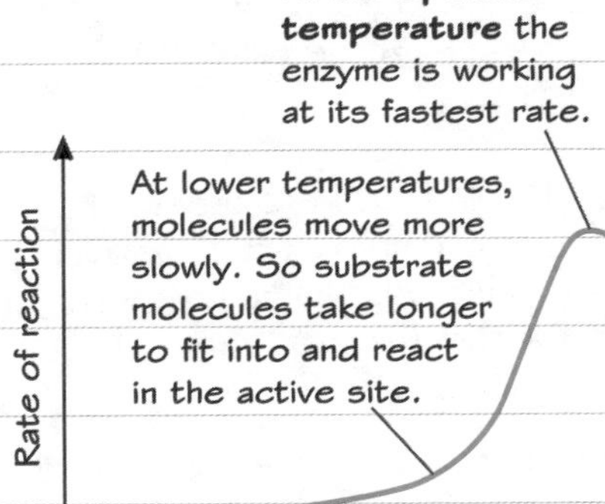
At the **optimum temperature** the enzyme is working at its fastest rate.
Higher temperatures cause the active site to change shape, so it can't hold the substrate as tightly and the reaction goes more slowly.
At lower temperatures, molecules move more slowly. So substrate molecules take longer to fit into and react in the active site.
At very high temperatures the active site breaks up and the enzyme is **denatured**.
Rate of reaction
10 20 30 40 50 60
Temperature (°C)

Adding more substrate at this point has little effect because the active site of every enzyme molecule is busy.

At this point not every active site of each enzyme molecule is busy, so adding more substrate increases the rate of reaction.

Rate of reaction

Substrate concentration

Worked example

The diagram shows an enzyme-controlled reaction.

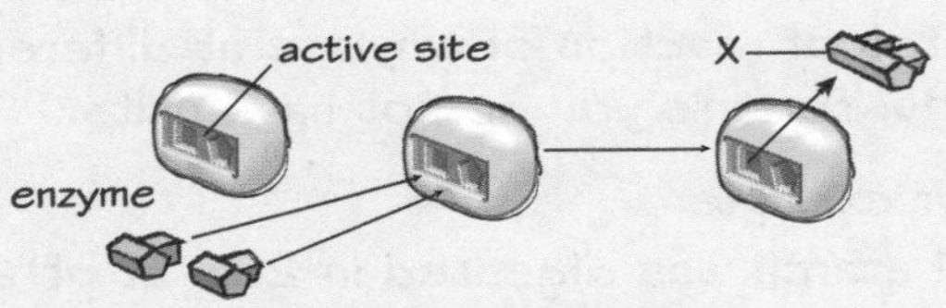
active site
X
enzyme
two different substrate molecules

(a) State the name of the part labelled X. **(1 mark)**

product molecule

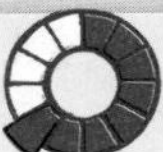

(b) Explain the role of the active site in an enzyme-controlled reaction. **(4 marks)**

The shape of the active site matches the shape of the substrate molecules and holds them close together so bonds can form between them to make the product. The product molecule doesn't fit the active site well so it is released from the enzyme.

The **substrate** fits like a 'key' into the **active site** (the 'lock'). Enzymes are specific because only a substrate with the right shape can fit the active site.

Worked example

Draw a graph of rate of reaction against pH for an enzyme with an optimum pH of 6. Label your graph to explain what you have drawn. **(2 marks)**

Changing the pH can change the shape of the enzyme's active site, and so change its ability to bond with the substrate.

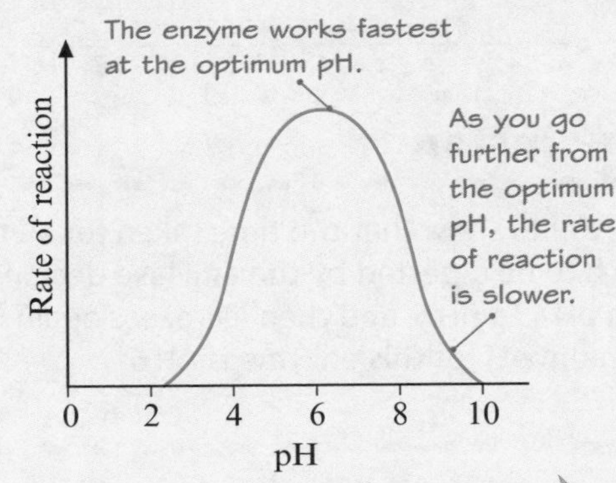
The enzyme works fastest at the optimum pH.
As you go further from the optimum pH, the rate of reaction is slower.
Rate of reaction
0 2 4 6 8 10
pH

Now try this

1 Explain what is meant by the active site of an enzyme. **(2 marks)**

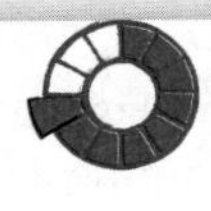

2 Enzymes in the human liver have an optimum temperature of about 37 °C. Explain why our body temperature is controlled to stay at about 37 °C. **(2 marks)**

pH and enzyme activity

You can **investigate the effect of pH on the rate of enzyme activity** by measuring the rate of an enzyme-controlled reaction at one pH, and comparing it with the rate at other pH values.

You can revise enzymes on page 7.

Core practical

Aim

To investigate the effect of pH on amylase activity.

Method

- Add amylase to buffered starch solution in a test tube.
- Place the tube in a water bath heated by a Bunsen burner for a constant temperature.
- Take samples of the mixture at regular intervals (e.g. every 10 s) and mix them with a fresh drop of iodine solution on a dimple tile.
- Repeat the test until the iodine solution stops changing colour when the starch/amylase mixture is added. Record the time taken for this to happen.
- Repeat the procedure at different pH values.

Results

Record the values for each pH in a table. Then, you can draw a graph like the one given below.

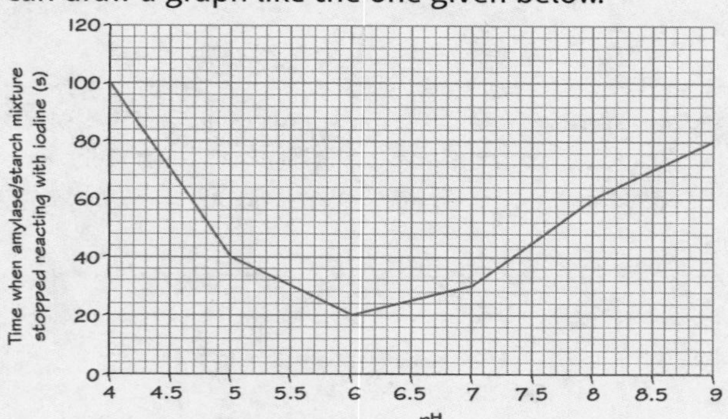

Conclusion

The results show that the time taken for all the starch to be digested by the amylase decreases from pH 4 to pH 6, and then increases again. The optimum pH for this enzyme is pH 6.

Starch is broken down by amylase to sugars.

Buffering the solution makes sure the pH doesn't change during the experiment.

When controlling the temperature, bear in mind the effect of temperature on enzyme action. Keeping the solution at the optimum temperature for the enzyme is best.

Iodine solution is usually yellow/orange. In the presence of starch, it turns blue/black.

Maths skills: Reaction rate

You can calculate the relative rate of reaction for a particular pH, as $\frac{1}{\text{time}}$, because the mass of starch used at each pH is the same. You can then compare the rates of reaction of amylase at different pHs. For this you do not have units.

For example:

All starch was digested in 20 s at pH 6.
The rate of reaction at pH 6 is $\frac{1}{20} = 0.05$. However, this is not a true rate. For a true rate of reaction you would need to have the mass of starch digested by the enzyme divided by the time taken. The units for this would be g/s.

See page 73 for more about rates.

Improvements

This method can be improved by:

- using more accurate measuring apparatus
- taking the mean of several repeats at each pH to help reduce the effect of random variation
- taking measurements over a narrower range of pH.

Now try this

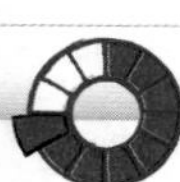

1 (a) Calculate the relative rate of reaction of amylase at pH 9 in the core practical. **(1 mark)**
(b) Explain why the graph shows that the optimum pH for this enzyme was pH 6. **(2 marks)**

2 A student repeated the experiment using solutions of pH 5.2, 5.6, 6.0, 6.4 and 6.8. Explain why the results of this experiment could produce a better answer for the optimum pH of the enzyme. **(2 marks)**

The importance of enzymes

Enzymes are biological **catalysts** that control reactions in the body. They catalyse reactions that **synthesise** large molecules from smaller ones and **break down** large molecules into smaller ones.

Enzymes as catalysts

Enzymes **speed up** the rate of a chemical reaction but are not used up in the reaction. This means they can be used over and over again.

Make sure you remember at least some examples of enzymes and the reactions they catalyse.

Digestion and synthesis

Some enzymes **digest** large molecules into smaller molecules. This happens in the gut where large food molecules are broken down so that they can be absorbed into the blood.

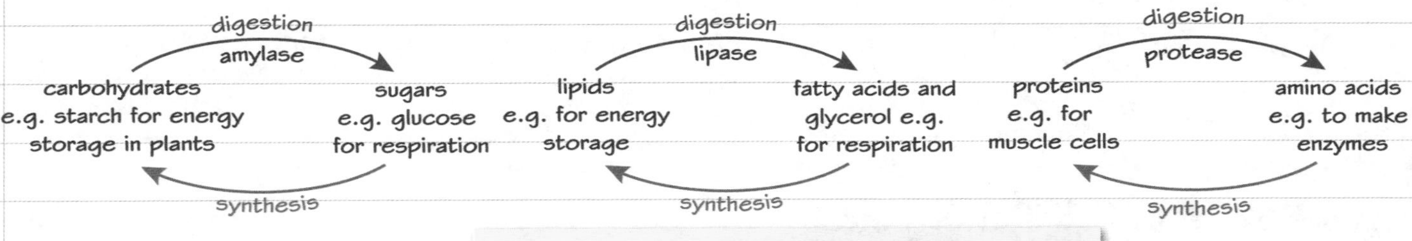

Some enzymes **synthesise** larger molecules from smaller molecules. This is important inside cells for supporting life processes and growth.

Worked example

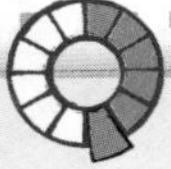

Complete the table. **(3 marks)**

Large molecule	Smaller molecules it is broken down into
carbohydrates	sugars
proteins	amino acids
lipids	fatty acids and glycerol

Remember you need to mention **both** fatty acids **and** glycerol to get the mark.

Worked example

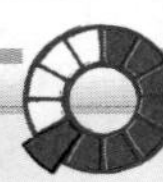

People with cystic fibrosis have thick mucus lining their gut. They have to take capsules containing enzymes before meals. Explain why. **(3 marks)**

The thick mucus prevents enzymes being secreted into the gut. If they didn't take enzyme capsules before meals, their food would not be digested into smaller, soluble molecules. As a result, they would not absorb enough nutrients.

Now try this

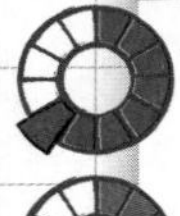

1 The enzyme that digests lipids will not digest proteins. Explain why. **(2 marks)**

2 Where in a cell would you expect to find enzymes that synthesise amino acids into proteins? **(1 mark)**

Think about what you know about cell structure (see page 1 for the generalised structures of plant and animal cells).

Had a look ☐ Nearly there ☐ Nailed it! ☐

Getting in and out of cells

Dissolved substances **(solutes)** move into and out of cells by **diffusion** and **active transport.** Water moves into and out of cells by **osmosis.**

Diffusion

high concentration of dissolved molecules (concentrated solution)

partially permeable membrane

More molecules move from the high concentration to the lower concentration than vice versa, so the net movement is **down** the concentration gradient.

low concentration of dissolved molecules (dilute solution)

Diffusion is important in the body, for example, to move oxygen into cells and to remove carbon dioxide.

Active transport

partially permeable membrane

Active transport needs energy from respiration.

low concentration of dissolved molecules (dilute solution)

high concentration of dissolved molecules (concentrated solution)

There is net movement against the concentration gradient.

Active transport makes it possible for cells to absorb ions from very dilute solutions, e.g. root cells absorb minerals from soil water, and small intestine cells absorb glucose from digested food in the gut into the body.

Worked example

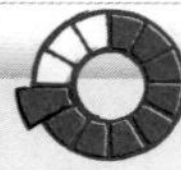

The diagram shows the results of an experiment. At the start, the level of solution inside the tube and the level of water in the beaker were the same. Name the process that caused the change and explain what happened. **(3 marks)**

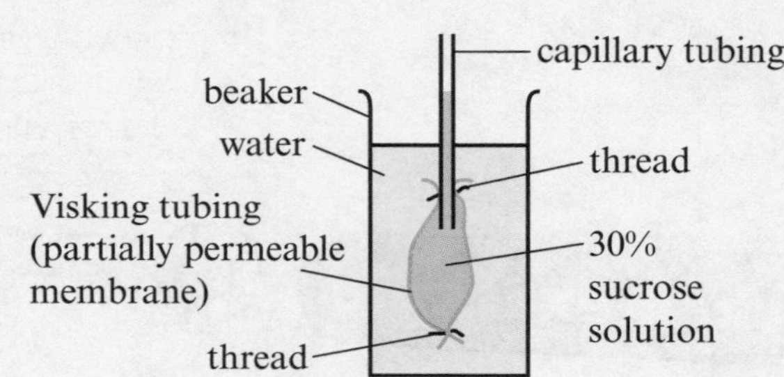

The process is called osmosis. To start with there were more water molecules in the water than in the same volume of sucrose solution, so more water molecules crossed the membrane into the tubing than going the other way. So the level of solution in the capillary rose.

Osmosis is the name given to a special case of diffusion. Osmosis is the net movement of water molecules across a partially permeable membrane.

Now try this

1 Define **osmosis**. **(2 marks)**

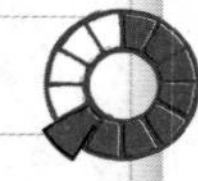

2 (a) Give **one** similarity between diffusion and osmosis. **(1 mark)**
(b) Give **one** difference between diffusion and active transport. **(1 mark)**

3 A plant root is treated with a poison that prevents respiration. Explain whether the root cells will still be able to absorb water and mineral ions from a dilute solution. **(4 marks)**

Remember that respiration provides the energy for active transport but there are other ways that substances can enter cells which do not require energy from respiration.

Osmosis in potatoes

You can investigate **osmosis** by calculating the change in mass of pieces of potato that have been placed in solutions of different solute concentration.

Core practical

Aim

To investigate osmosis using potatoes.

Apparatus

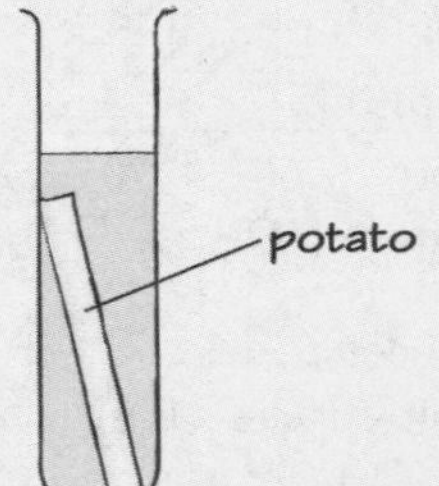

- pieces of potato about 3 × 1 × 1 cm
- boiling tubes
- accurate balance
- paper towels
- forceps
- solutions of different solute concentration (0, 0.2, 0.4, 0.6, 0.8, 1.0 mol dm^{-3})
- marker pen

Method

1 Mark the value of one solute concentration on one tube and repeat using a different tube for each concentration. Fill each tube two-thirds full of the appropriate solution.

2 Blot a piece of potato dry on a paper towel, then measure and record its mass. Use the forceps to place it into one of the tubes, and record the tube. Repeat for all tubes.

3 After 20 minutes, use the forceps to remove each piece of potato, blot it dry and measure its mass again. Record all final masses.

Results

The percentage change in mass of each potato slice is calculated and recorded, indicating whether mass was gained or lost.

Conclusion

The results show that when the solution concentration is very dilute, water enters the potato cells. This is due to osmosis because the solute concentration of the potato cells is greater than the surrounding solution. As the solute concentration of the solution increases above that inside the potato cells, osmosis causes water to be lost from the potato.

Worked example

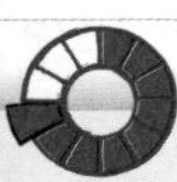

The initial mass of a potato slice was 13.54 g. After soaking in a solution the final mass was 14.66 g. Calculate the percentage change in mass. **(4 marks)**

% change in mass =

$$\frac{\text{final mass} - \text{initial mass}}{\text{initial mass}} \times 100\% =$$

$$\frac{14.66 - 13.54}{13.54} \times 100 = 8.27\% \text{ gained}$$

(or +8.27%)

The change in mass may be very small. You should measure to 2 d.p.

The solute in the solutions must be something that is too large to diffuse across the cell membrane. Sucrose is often used.

Blotting removes surface water, and can help to increase the accuracy and repeatability of measurements.

Repeating the test at each solute concentration, and calculating the mean of the results, can help to reduce the effect of random variation.

Now try this

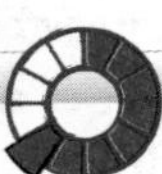

The table shows the results of an osmosis experiment using potato pieces.

(a) Calculate the percentage change in mass of potato at each concentration. **(2 marks)**

(b) Use your answers to draw a conclusion about the solute concentration of the potato cells. Explain your conclusion. **(2 marks)**

Solution concentration (mol dm^{-3})	Initial mass (g)	Final mass (g)
0.0	16.52	20.15
0.2	15.90	16.70
0.4	17.06	15.69
0.6	16.88	14.36
0.8	16.23	12.32

Extended response – Key concepts

There will be at least one 6-mark question on your exam paper. For these questions, you will need to think scientifically and structure your answer logically, showing how the points you make are related to each other. For the questions on this page, you can revise the topics on **enzyme activity** on pages 7–9.

Worked example

The graph shows the results of an experiment into the effect of temperature on the activity of the enzyme amylase on starch. The enzyme had been isolated from a bacterium. Explain the results shown in the graph. **(6 marks)**

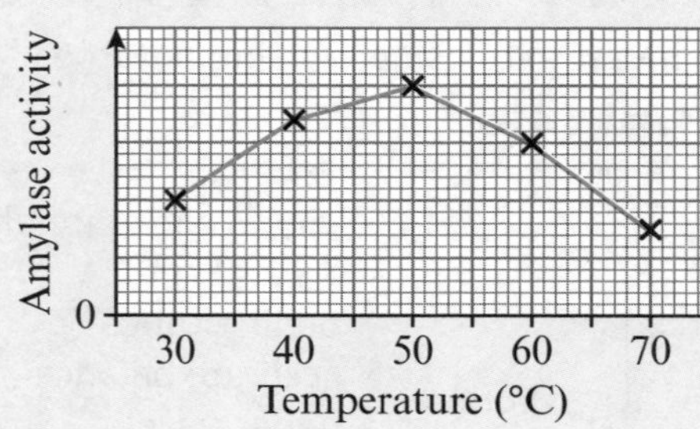

The graph shows that activity of the bacterial amylase increased from 30°C up to a maximum at 50°C, and then decreased as the temperature increased further to 70°C. The maximum activity is at the optimum temperature, which is about 50°C for this enzyme.

Before you explain the results, describe what the graph shows. Use values on the graph to add details to your answer. Remember to include important science words that are relevant in your answer.

Enzyme activity increases with increasing temperature up to the optimum because particles move faster. Starch molecules fit into the active site of enzyme molecules more quickly and are broken down more quickly, releasing the active site for another starch molecule.

Enzyme activity should be linked to how the active site of the enzyme and the substrate interact.

Beyond the optimum temperature, the shape of the active site starts to change, making it more difficult for the substrate to fit into it. So activity slows down. The active site shape changes more as temperature increases further, until the enzyme becomes denatured and stops being able to catalyse the reaction.

Make sure you explain each part of the graph clearly, linking your knowledge and understanding to give a good explanation of what is happening.

Command word: Explain

An **explain** question requires a reason for what is happening. Use a word such as 'because' to clearly link a description with its reason.

Now try this

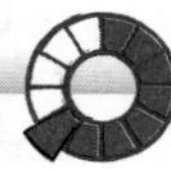

The graph on the right shows the effect of starch concentration on the activity of a bacterial amylase. Explain the results shown in the graph. **(6 marks)**

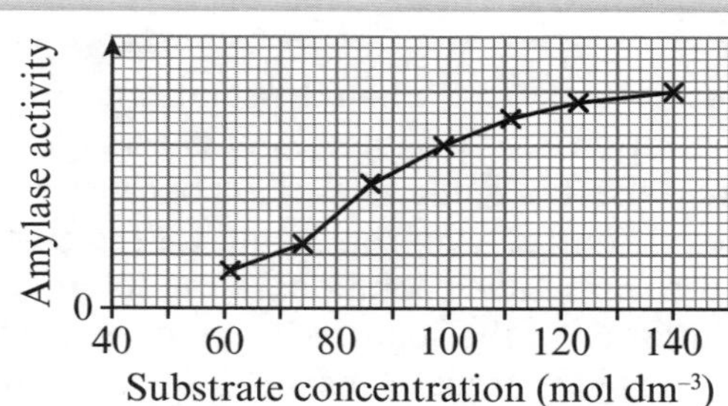

Mitosis

There are two types of cell division, **mitosis** and meiosis. Mitosis is covered here; meiosis is covered on page 20.

Mitosis is a type of **cell division** that takes place in several stages. There are two types of **cell division**. **Mitosis** is the cell division that happens in body cells. A **body cell** is any cell except those that produce **gametes** (sex cells).

The cell that is dividing is called the **parent cell** and the two new cells that are formed are called **daughter cells.** The daughter cells are identical to the parent cell, so if the parent cell is diploid then the daughter cells will be diploid too. You should be clear about the differences between meiosis and mitosis.

Three things to remember

- ✓ Mi-to-sis makes two cells.
- ✓ MiTosis makes genetically idenTical cells.
- ✓ Diploid means Double (two sets of) chromosomes.

Stages of mitosis

Each chromosome consists of two chromatids.

The chromatids separate and one chromatid from each pair is pulled to each pole of the cell. The chromatids can now be called chromosomes.

The cell splits into two. This is called **cytokinesis**.

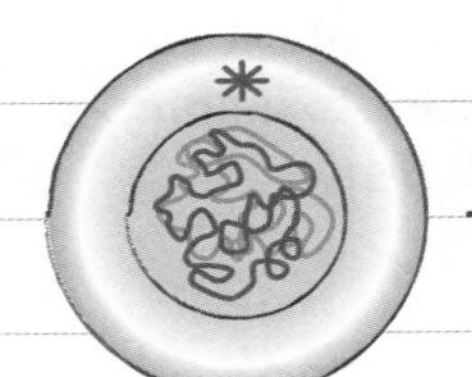

Interphase

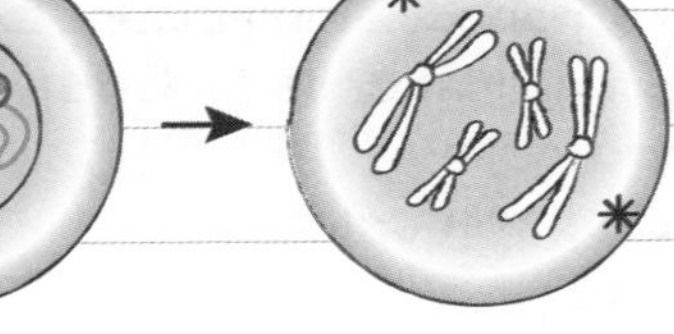

Prophase

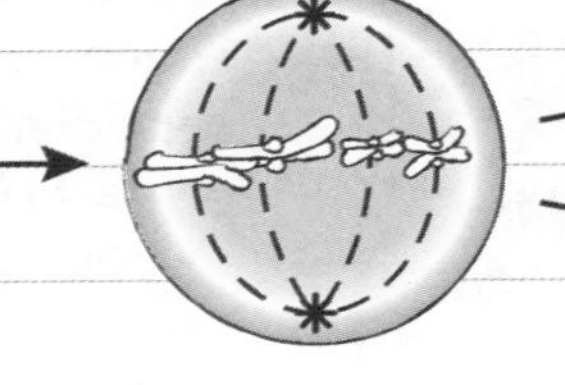

Metaphase

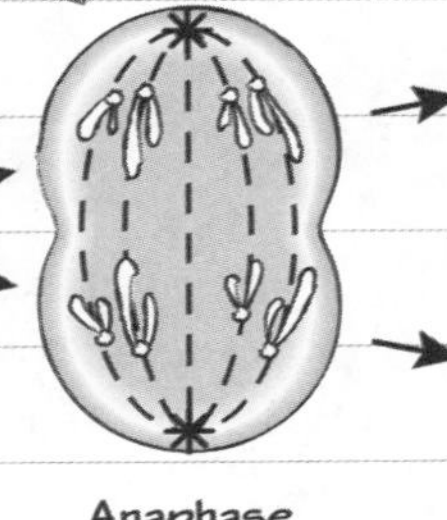

Anaphase

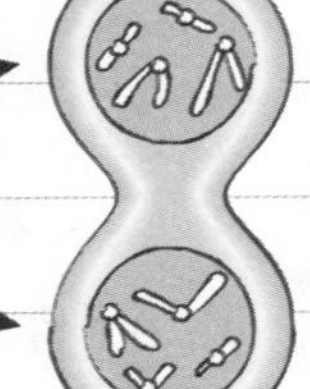

Telophase

At the end of interphase, chromosomes start to become visible. The DNA has already been copied.

The nuclear membrane breaks down. Chromosomes line up along the middle of the cell.

Spindle fibres disappear and a new nuclear membrane forms round each group of chromosomes.

Remember the stages of mitosis using the mnemonic IPMAT:
Interphase
Prophase
Metaphase
Anaphase
Telophase

Worked example

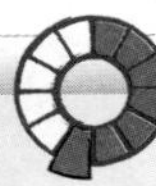

(a) Body cells have two sets of chromosomes. Explain how this happens. **(2 marks)**

One set of chromosomes comes from the mother and one set comes from the father.

(b) Explain how the daughter cells are genetically identical. **(2 marks)**

Each chromosome is copied; one copy of each chromosome goes into each daughter cell.

Interphase

A cell is in interphase most of the time.

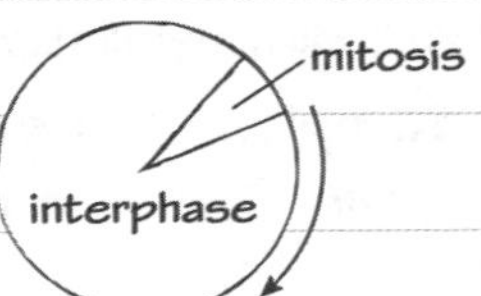

Now try this

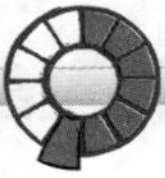

A cell divides by mitosis. Describe the daughter cells. **(2 marks)**

Had a look ☐ Nearly there ☐ Nailed it! ☐

Cell growth and differentiation

Mitosis is used for growth, repair and asexual reproduction. After growth, cells can then **differentiate** into specialised cells. Growth and differentiation happen in different ways in plants and animals.

When does mitosis happen?

Normally cells only divide by mitosis when new diploid cells are needed for:

- growth
- repair (replacement of damaged cells)
- asexual reproduction.

Uncontrolled cell division

Cancer cells are abnormal cells that divide uncontrollably by mitosis to form a **tumour**.

Cells usually stop dividing when growth has finished, except when repair is needed. Sometimes the controls that tell cells when to stop dividing go wrong which leads to cancer.

Worked example

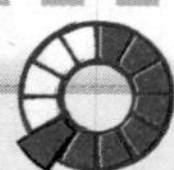

Aphids are insects that are pests of crop plants. In the summer they reproduce rapidly by asexual reproduction. Will the offspring of a single aphid be varied? Explain your answer. **(3 marks)**

The offspring will be produced by mitosis, so they will all be genetically identical to the parent aphid. They will all be diploid like the parent aphid.

Mitosis is used for asexual reproduction. This also occurs in plants, for example, when plants reproduce using bulbs or runners.

Growth in animals

In **animals**, a fertilised egg, or zygote, divides by mitosis to produce genetically identical daughter cells. These cells grow and divide by mitosis, and eventually **differentiate** into different types of cells to make up a whole organism. Differentiation creates **specialised cells** adapted to carry out a particular function. Examples of specialised animal cells include:

- red blood cells
- egg and sperm cells
- nerve cells
- bone cells
- smooth muscle cells.

Growth in plants

Plant cells divide by mitosis, just behind the tips of shoots and roots. After this the cells grow by enlarging. Young cells have small vacuoles which take in water by osmosis and enlarge, causing the cells to **elongate**. These cells can differentiate into specialised cell types. Most plant cells can **continue** to grow and differentiate throughout life.

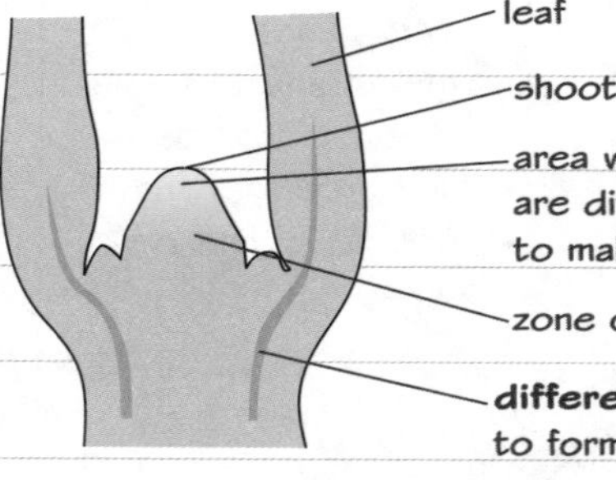

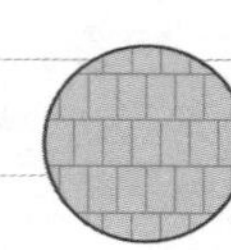

Examples of specialised plant cells include:

- xylem
- phloem
- mesophyll cells
- root hair cells
- stoma cells.

Now try this

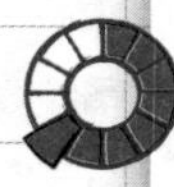

1 One drug that is used to treat cancer stops spindle formation in mitosis. Suggest how this drug helps to prevent the growth of a cancer. **(2 marks)**

2 State what is meant by a **specialised cell**. **(1 mark)**

3 Explain why cell differentiation is important. **(2 marks)**

Growth and percentile charts

When organisms **grow** they get bigger. Growth can be measured in different ways.

Estimating growth

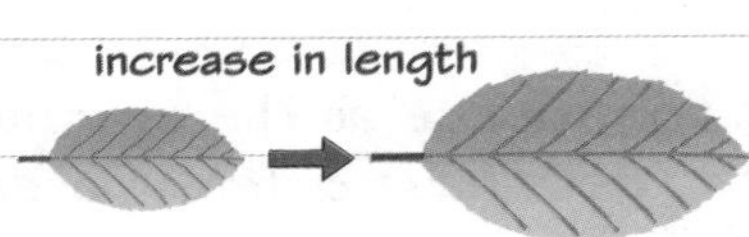

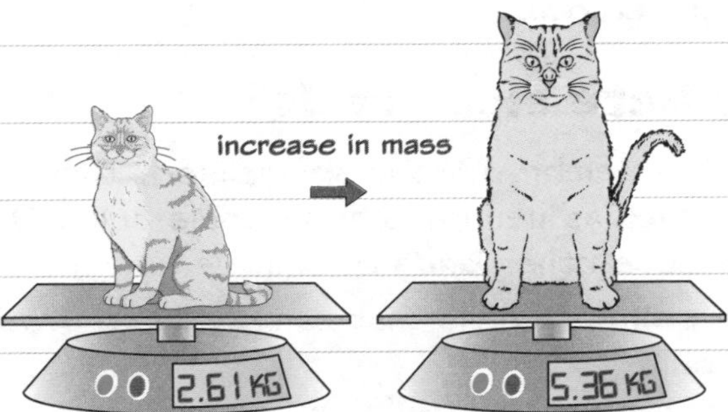

Growth is a **permanent** increase in size. For example, a balloon that is blown up a little more has not 'grown' in size.

Percentile charts

Percentile charts can help to show if a child is growing faster or more slowly than is normal for their age.

Babies with a mass above the 95th line or below the 5th line may not be growing properly.

A baby whose mass decreases by two or more percentiles over their first year may not be growing normally.

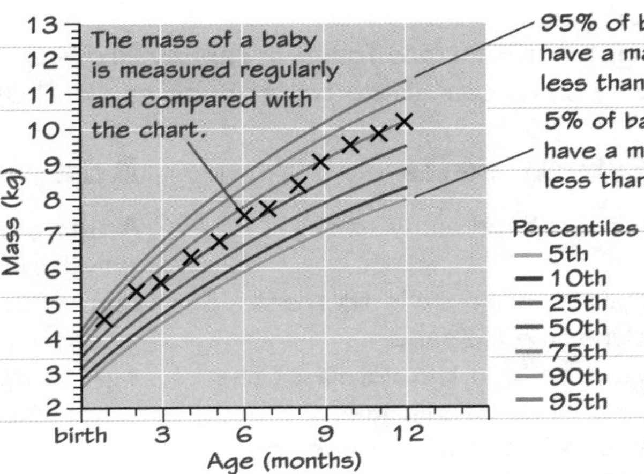

This chart is for baby girls. The chart for baby boys is similar, but they grow at a slightly different rate to girls.

Worked example

Look at the percentile chart. The crosses mark the mass of a baby girl measured each month.

(a) State the percentile that she belonged to in her first month. **(1 mark)**

75th percentile

(b) Do you think there was concern about this baby's mass increase in her first year? Explain your answer. **(2 marks)**

No, because her mass varied only between the 75th and 50th percentiles, and this amount of variation is normal.

Worked example

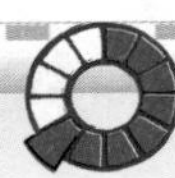

The government has produced percentile charts for BMI in children. A child is said to have a clinically healthy weight if their BMI is between the second and ninety-first percentile. A parent has been told that her child is on the 70th percentile for BMI. Explain to her what this means. **(2 marks)**

This means the child has a higher BMI than 70% of children and a lower BMI than 30% of children. Because this falls between the 2nd and 91st percentile, the child has a clinically healthy weight.

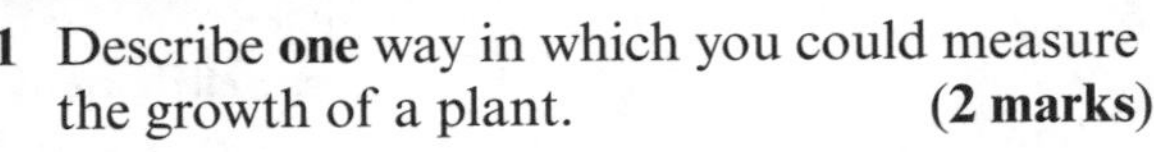

1 Describe **one** way in which you could measure the growth of a plant. **(2 marks)**

2 Describe what a percentile chart is used for. **(2 marks)**

3 Explain why an increase in the size of a balloon is not an example of growth but an increase in the size of a child is. **(2 marks)**

Stem cells

Stem cells have different functions in plants and animals. In humans, they have many potential uses in medicine.

What are stem cells?

Cells in an embryo are unspecialised. They divide to produce all the **specialised cells** in the body, such as neurones and muscle cells. Once the cells have differentiated they cannot divide to produce other kinds of cell.

Stem cells are cells that can divide to produce many types of cell. There are three kinds of stem cell:

- **Embryonic stem cells** are taken from embryos at a very early stage of division (e.g. 8 cells).
- **Adult stem cells** are found in differentiated tissue, such as bone or skin – they divide to replace damaged cells.
- Plants have **meristems** that are found in rapidly growing parts of the plant, e.g. tips of roots and shoots. These cells can divide to produce any kind of plant cell.

Embryonic stem cells

Embryonic stem cells have many uses, including:

- replacing or repairing brain cells to treat people with Parkinson's disease
- replacing damaged cells in the retina of the eye to treat some kinds of blindness
- growing new tissues in the lab to use for transplants or drug testing.

Adult stem cells

Adult stem cells (from bone marrow) can only form a limited number of cell types. They can be used for:

- treatment of leukaemia
- potentially growing new tissues that are genetically matched to the patient.

Using stem cells

embryonic stem cells
easy to extract from embryo
produce any type of cell
embryo destroyed when cells removed – some people think embryos have a right to life
all stem cells
replace faulty cell with healthy cell, so person is well again
! stem cells may not stop dividing, and so cause cancer
adult stem cells
no embryo destroyed so not an ethical issue
if taken from the person to be treated, will not cause rejection by the body
produce only a few types of cell
– advantage
– disadvantage
! – risk

Now try this

1 When gardeners cut off the top of a plant, the plant usually continues to grow sideways but does not grow taller. Explain why. **(2 marks)**

2 Explain **two** disadvantages of research into embryonic stem cells. **(2 marks)**

3 (a) State why embryonic stem cells could be more useful than adult stem cells to replace faulty cells. **(1 mark)**

(b) Describe **one** practical advantage of using adult stem cells from the patient instead of embryonic stem cells to replace faulty cells. **(2 marks)**

If an exam question asks for **two** advantages or disadvantages, make sure you give two **different** examples.

Neurones

Stimuli are detected by sensory **receptors** that send impulses along sensory **neurones** to the central nervous system. Neurones are specialised cells that carry nervous impulses.

Types of neurones

There are three main types of neurones. **Sensory neurones** carry impulses to the central nervous system. **Motor neurones** carry impulses from the central nervous system to effector organs. **Relay neurones** are found only in the central nervous system.

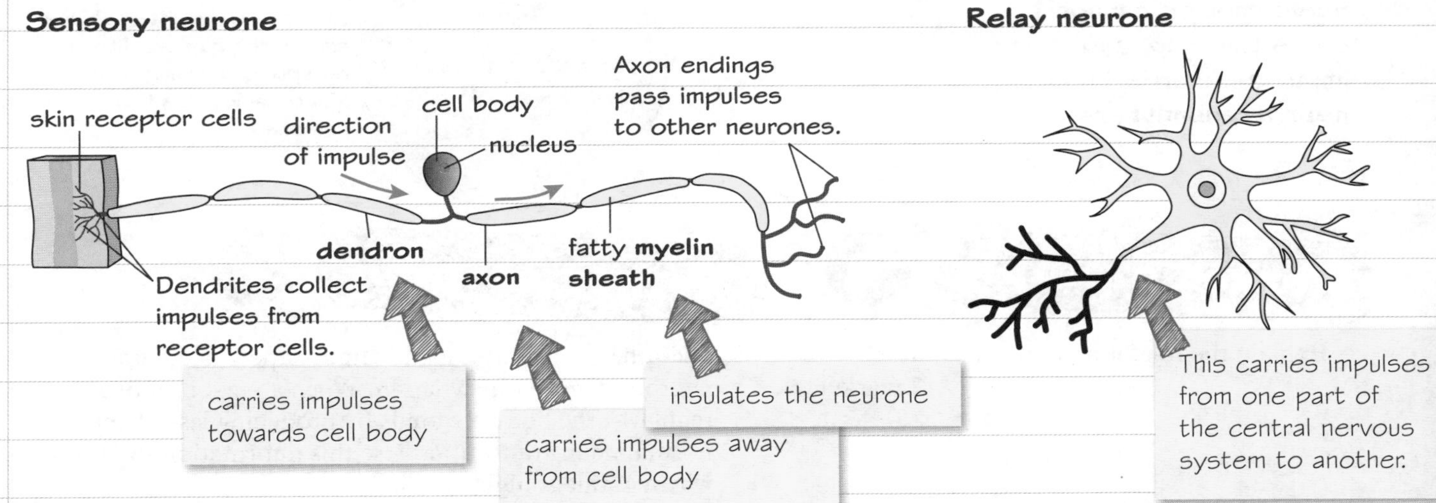

Motor neurone

The electrical impulse jumps from one gap in the myelin sheath to the next, speeding up the rate of transmission.

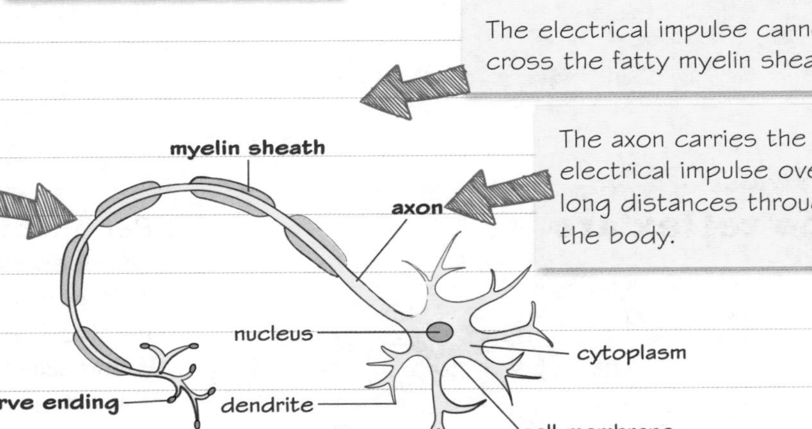

The nerve ending transmits the impulse to an effector, such as a muscle or gland.

Worked example

Complete the table with the name of the structure that corresponds to each function listed. **(3 marks)**

Structure	Function
axon	carries impulses away from the cell body
dendrite	receives impulses from other neurones
myelin sheath	a fatty layer that provides electrical insulation around the neurone

Now try this

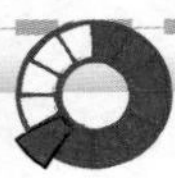

1 Compare the roles of sensory, motor and relay neurones in the nervous system. **(3 marks)**
2 Explain how the structure of a sensory neurone is related to its function. **(2 marks)**

Responding to stimuli

Sensory neurones **carry impulses** from receptors to the central nervous system.

Synapses

The point where two neurones meet is called a **synapse**. There is a small gap between the neurones. The electrical nerve impulse cannot cross this gap, and the impulse is carried by **neurotransmitters**.

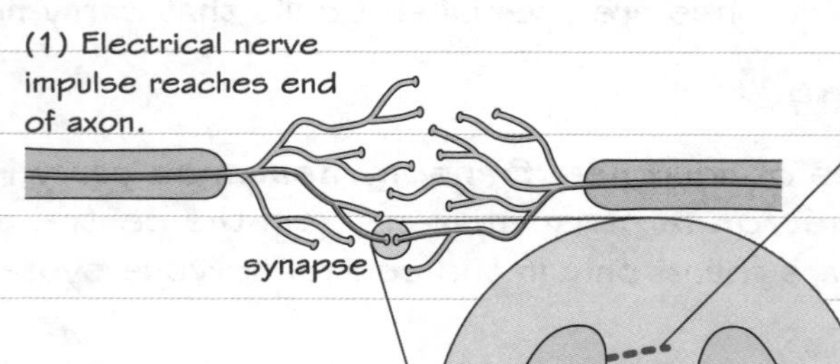

(1) Electrical nerve impulse reaches end of axon.
synapse
(2) Electrical impulse causes chemical neurotransmitter to be released from vesicles in the neurone into gap between neurones.
(3) Neurotransmitter diffuses across the gap and fits into receptors, causing a new electrical impulse in next neurone.

Worked example

Explain the role of a neurotransmitter. **(2 marks)**

It is a chemical released from one neurone that carries the impulse across the synaptic gap to the next neurone. Without it, the electrical impulse cannot cross the synaptic gap.

Worked example

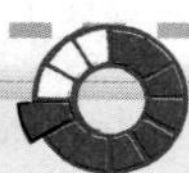

Nicotine in cigarettes is the same shape as a common neurotransmitter in synapses. When a person smokes regularly, the body responds by producing less of the normal neurotransmitter. Use this information to explain why nicotine is highly addictive. **(3 marks)**

When the synapse produces less neurotransmitter, the body does not function as well as usual. When the person smokes a cigarette, the nicotine stimulates the synapse and the person feels better again. Over time, they need to smoke more to get the same feeling.

The reflex arc

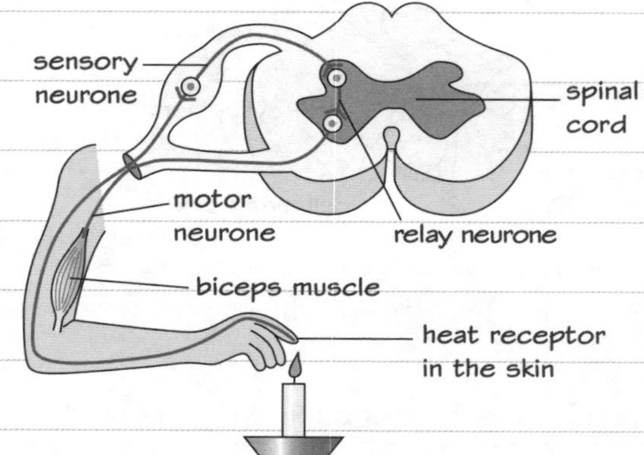

Reflex arcs involve only three neurones, and impulses pass to and from the spinal cord. This provides a fast response that does not involve the brain. Reflex arcs are:

- immediate
- involuntary
- innate
- invariable.

These reflexes help protect us from immediate harm, such as the eye blink reflex which protects the eye if something comes close to it.

If the impulses had to go to the brain to be processed, there would be many more synapses, so the response would take longer.

Now try this

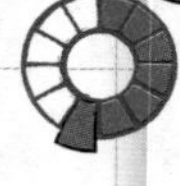

1 Sketch a flow chart to show the neurones in a reflex arc, from the receptor to the effector. **(4 marks)**

2 Synapses make sure that nervous impulses can pass in one direction only. Explain how. **(2 marks)**

Extended response – Cells and control

There will be at least one 6-mark question on your exam paper. For these questions, you will need to think scientifically and structure your answer logically, showing how the points you make are related to each other. You can revise the topic for this question, which is about **stem cells** in medicine, on page 16.

Worked example

Parkinson's disease causes some cells of the nervous system to die and stop releasing neurotransmitter. At the moment, patients with the disease are treated with a drug. This slows the disease, but cannot stop it. Scientists are investigating whether skin cells, turned into stem cells, could be used to treat the disease.

Discuss whether this stem cell treatment could replace drug treatment for Parkinson's disease.

(6 marks)

Stem cells are unspecialised cells that can divide to produce many kinds of specialised cell. This means they could be used to produce new nerve cells to replace those that have died in a person with Parkinson's disease. So, stem cell treatment may be able to cure the disease, while the drug treatment only slows down how quickly the disease develops.

Turning skin cells into stem cells is useful because skin cells are easy to get at. If the skin cells are taken from the person with the disease then, when the stem cells from those skin cells are put back into the body, the person's immune system will recognise the stem cells made from those skin cells as belonging to their body and will not attack them. This should increase the chances of successful treatment.

Using stem cells to treat Parkinson's disease is still being tested. The treatment might not work if the stem cells do not behave properly in the body. They may make the wrong kind of specialised cell, or they may develop into cancer.

Planning an answer

Try planning out your answer before you start writing, e.g. note down a different idea for each paragraph. This can help order the answer, and make sure you do not miss anything important.

A good start is to explain any key words in the question. In this case, explaining what stem cells do will prepare for why this treatment could be better than using drugs.

Command word: Discuss

The answer to a **discuss** question should include all aspects of the issue in the question. You are not expected to draw a conclusion from the discussion.

If stem cell treatment is to replace drug treatment, there must be evidence that it is better. Look for places in your answer where you can compare the two treatments.

When discussing all aspects of using stem cells, remember to include their risks as well as their benefits.

Now try this

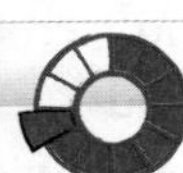

Heart disease can involve heart muscle cells, epithelial (surface) cells and other blood vessel cells. Embryonic stem cells have been shown to replace several types of damaged heart cell in mice with heart disease. Discuss the potential benefits and problems involved with developing this treatment for use in humans with heart damage.

(6 marks)

Meiosis

Meiosis is a type of cell division that produces four **daughter cells**, each with half the number of chromosomes. Meiosis only happens in gamete-producing cells, producing genetically different **haploid** gametes. The other type of cell division, **mitosis**, is covered on page 13.

Stages of meiosis

The parent cell is a diploid cell. So it has two sets of chromosomes.

The parent cell divides in two and then in two again. Four daughter cells are produced.

the other set of chromosomes

one set of chromosomes

pair of chromosomes

Before the parent cell divides, each chromosome is copied.

Each daughter cell gets a copy of one chromosome from each pair.

Each daughter cell has only one set of chromosomes. So these are **haploid** cells. The daughter cells are not all identical – meiosis results in variation.

Remember: haploid cells, produced by meiosis (me-1-osis), have 1 set of chromosomes.

The cells produced by division are always called 'daughter cells', even if they will eventually turn into sperm cells.

Worked example

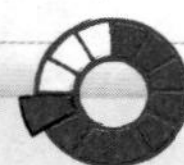

Compare mitosis and meiosis. **(4 marks)**

Meiosis produces four daughter cells, but mitosis only produces two.

Meiosis produces genetically different daughter cells, but in mitosis the daughter cells are identical to each other and to the parent cell.

Meiosis produces haploid cells, but mitosis produces diploid cells.

Mitosis occurs in body cells, but meiosis occurs only in gamete-producing cells.

Be careful to spell mitosis and meiosis correctly. If you write 'meitosis' or 'miosis' it won't be clear which kind of cell division you are referring to.

Now try this

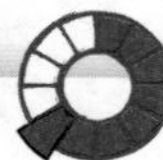

1 Describe the outcome of meiosis of a diploid parent cell. **(3 marks)**

2 Explain the importance of meiosis occurring before fertilisation. **(3 marks)**

You need to say enough for 3 marks here. It may help to think about what would happen if gametes were produced by mitosis and not meiosis.

3 Distinguish between the terms **haploid** and **diploid**. **(2 marks)**

4 The cells produced by meiosis are

☐ **A** diploid
☐ **B** genetically different
☐ **C** embryos
☐ **D** identical **(1 mark)**

DNA

DNA is the **genetic material** found in the chromosomes in the nuclei of cells.

DNA in the cell

The nucleus contains chromosomes.

Most cells have a nucleus.

cell

chromosome

A chromosome consists of a string of genes.

A **gene** is a short piece of **DNA** that codes for a specific **protein**. You have genes for hair structure, eye colour, enzymes and every other protein in your body.

DNA

Each gene is a length of DNA. DNA is a long, coiled molecule formed from two strands. The strands are twisted in a **double helix**.

G C

A T

C G

A T

T A

The two strands of the double helix are joined by pairs of **bases**. There are four different bases in DNA:
A = adenine T = thymine
C = cytosine G = guanine

Bases form **complementary pairs**:
A always pairs with T
C always pairs with G.

Remember: straight A with straight T; curly C with curly G.

Weak hydrogen bonds between the base pairs hold the DNA strands together.

The **genome** is the base sequence of all the DNA in an organism.

DNA structure

DNA is a **polymer** made of many **monomers**, called **nucleotides**, joined together.

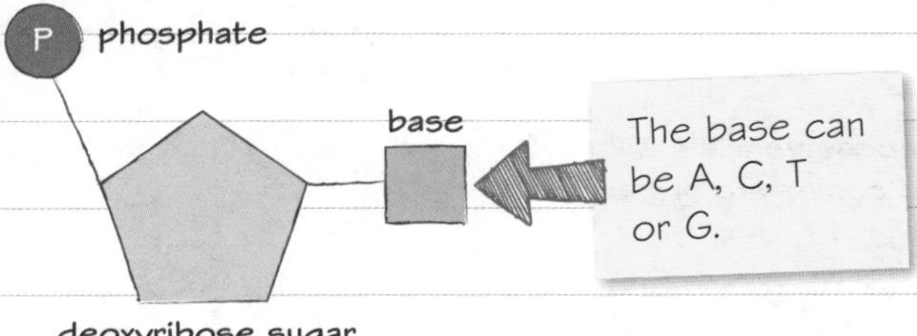

Practical skills DNA from fruit

DNA can be extracted from fruit by:

1. grinding the fruit with sand, using a pestle and mortar, to separate the cells
2. adding a detergent to break open the membranes
3. adding ice-cold alcohol so that the DNA precipitates out.

Now try this

You do not need to know the names of the bases. You just need to know their letters are A, C, T and G.

1 Write a sentence to define each of these words:
(a) gene (b) base. **(2 marks)**

2 The sequence of bases on one strand of DNA is CGAT. Write down the sequence of bases on the complementary strand, and explain how you worked out your answer. **(2 marks)**

3 Describe the structure of DNA. **(4 marks)**

This question has 4 marks, so the answer needs 4 different ideas.

Had a look ☐ Nearly there ☐ Nailed it! ☐

Genetic terms

You need to be able to explain all of the **genetic terms** in bold on this page.

Inside a cell

When **gametes** fuse at fertilisation, they form a diploid **zygote**. Each zygote inherits different alleles (genetic variants) of their genes from their parents. This produces variation in inherited characteristics between different individuals.

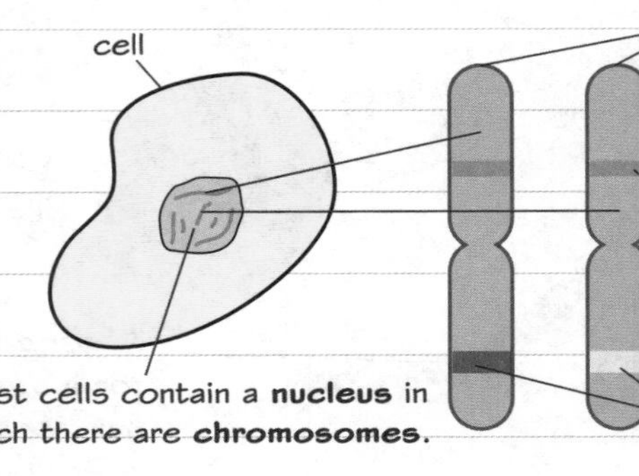

cell
Most cells contain a **nucleus** in which there are **chromosomes**.
There are two copies of each **chromosome** in body cells – each copy has the same genes in the same order along its length (except chromosomes that determine sex).
A **gene** is a short piece of DNA at a particular point on a chromosome – a gene codes for a characteristic, e.g. eye colour.
A gene may come in different forms, called **alleles**, that produce different variations of the characteristic, e.g. different eye colours.

Alleles

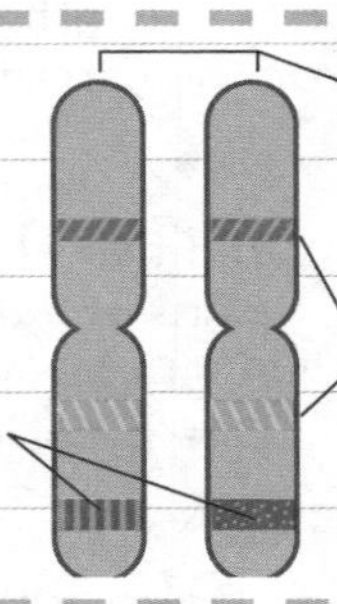

Genetic definitions

The gene for coat colour in rabbits has different alleles. The allele for brown colour (B) is dominant over the allele for black colour (b). The table shows all the possible genotypes and phenotypes for these alleles.

Genotype shows the alleles (forms of the genes) in the individual. Remember that each body cell has two alleles for each characteristic – either two alleles that are the same or two that are different.

Genotype	Phenotype
BB	brown coat
Bb	brown coat
bb	black coat

Phenotype means the characteristics that are produced, including what the individual looks like.

The effect of the **dominant** allele will show when at least one copy is present in the genotype.

The effect of the **recessive** allele will only show when two copies are present in the genotype.

Now try this

Make sure you know the difference between a gene and an allele. This is a common mistake that students make.

1 Define the terms **chromosome**, **gene** and **allele**. **(3 marks)**

2 A pea plant has a recessive allele for white flower colour and a dominant allele for purple flower colour.
(a) Identify if the plant is homozygous or heterozygous for flower colour, and explain your answer. **(2 marks)**
(b) State the phenotype of the plant for flower colour. Explain your answer. **(2 marks)**

Monohybrid inheritance

Monohybrid inheritance can be explained using **genetic diagrams** and **Punnett squares**.

Genetic diagrams

Body cells contain two alleles for each gene. In this example, both parent plants are heterozygous – they have one allele for purple flower colour and one allele for white flower colour.

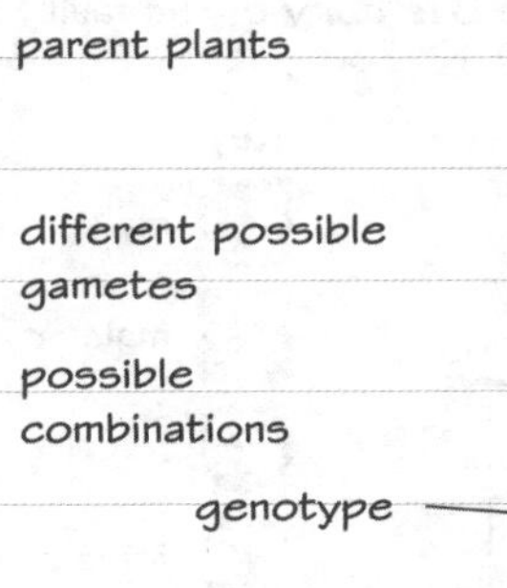
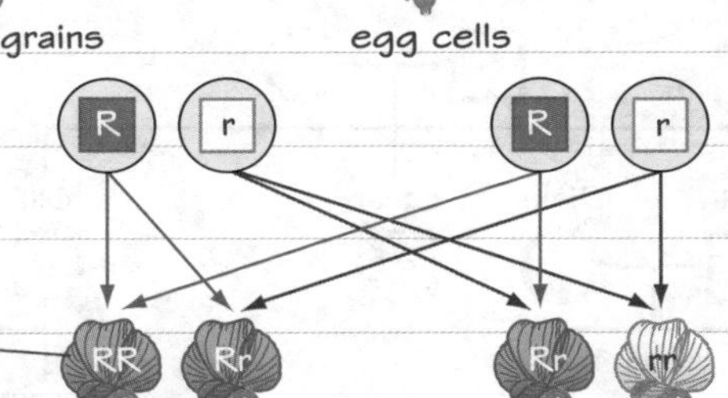

Worked example

Green seed pod (G) is dominant to yellow seed pod (g). Complete the Punnett square to show the possible offspring for plants heterozygous for seed pod colour, and calculate the (a) ratio, (b) probability and (c) percentage of possible offspring genotypes and phenotypes. **(4 marks)**

		Parent genotype Gg	
	Parent gametes	G	g
Parent genotype Gg	G	GG green	Gg green
	g	Gg green	gg yellow

(a) genotype 1 GG : 2 Gg : 1gg
phenotype 3 green : 1 yellow

(b) genotype 1/4 (1 out of 4) GG, 2/4 (1/2) Gg, and 1/4 gg
phenotype 3/4 green and 1/4 yellow

(c) genotype 25% GG, 50% Gg, 25% gg
phenotype 75% green and 25% yellow

A **Punnett square** is a different way of showing the same information about how genotype is inherited and what effect this has on the phenotype.

Genetic diagrams and Punnett squares only show **possible** offspring, not the **actual** offspring from these parents.

Now try this

A heterozygous rabbit with a brown coat was bred with a rabbit with a black coat (homozygous recessive). The four baby rabbits were all black.

(a) Use a diagram to calculate the predicted outcome of this cross. **(3 marks)**

(b) Comment on the difference between this and the actual outcome. **(2 marks)**

Remember to use a Punnett square when answering this question.

Family pedigrees

Cystic fibrosis (CF) is a genetic condition caused by a recessive allele. **Pedigree analysis** can be used to study the inheritance of dominant and recessive alleles.

Family pedigree showing inheritance of cystic fibrosis

You can use a **family pedigree** to show the inheritance of a genetic condition within a family, and to **predict** the chance that someone will inherit the faulty allele.

In the family pedigree below, both Ethan and Mia have CF. Ethan inherited his alleles from Arun and Beth. But they don't have the disease, so they must both be **carriers** (have one copy of the faulty allele).

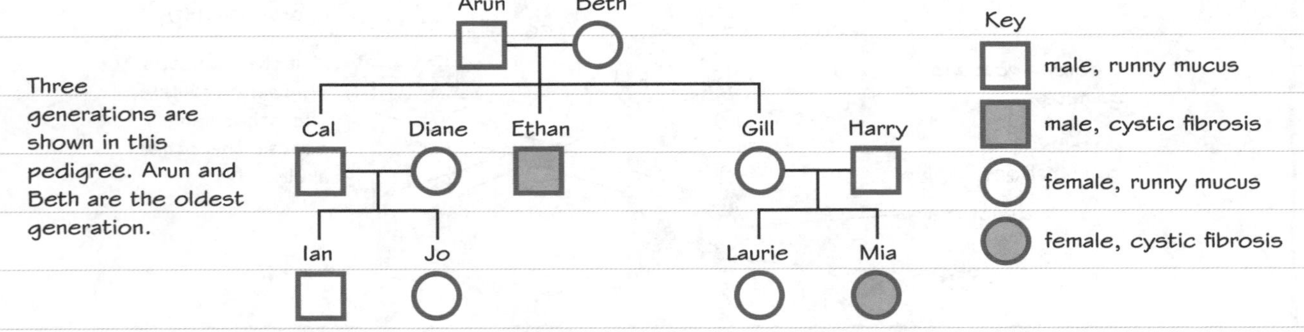

Ethan and Mia must have two copies of the recessive allele, as they have the disease.

Worked example

The ability to taste phenylthiocarbamide (PTC) is a dominant condition. The diagram shows the inheritance of PTC tasting in one family. Describe the evidence that PTC tasting is controlled by a dominant allele. **(2 marks)**

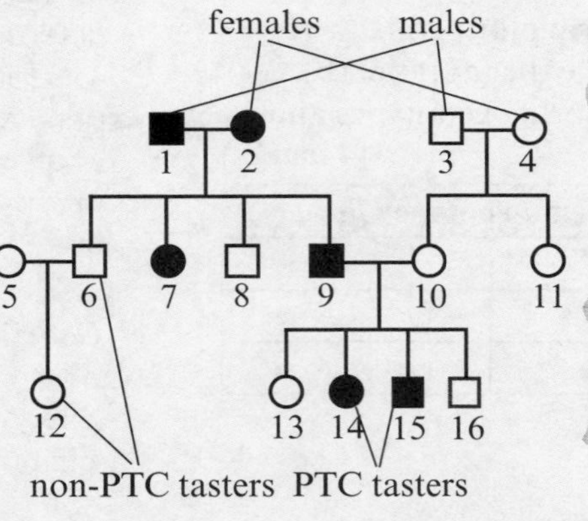

1 and 2 are both PTC tasters but they have two children who are non-tasters. Therefore PTC tasting must be dominant and non-tasting is recessive.

You are asked for **evidence** so make sure you refer to **specific individuals** in the pedigree.

Maths skills: Pedigree analysis

Look at the family pedigree above. You can calculate the **probability** of Gill and Harry having a third child with cystic fibrosis using a Punnett square.

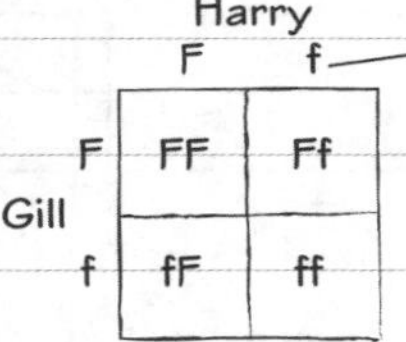

Mia has the disease, so Harry and Gill must both have one copy of the faulty allele.

There is one outcome (ff) that would lead to a child having CF, and three that would not:

P(third child has CF) = $\frac{1}{4}$ or 25%

There is more about Punnett squares on page 25.

Now try this

Look at the family pedigree in the worked example above.

(a) Give the genotype of individual 2. **(1 mark)**

(b) Individuals 9 and 10 have a fifth child. Calculate the probability that the child will be a PTC taster. **(2 marks)**

Remember that the trait is **dominant** so each person only needs **one copy** of the allele in order to be a PTC taster.

Sex determination

The sex of humans is determined at fertilisation and can be expressed using **genetic diagrams** and **Punnett squares**.

Sex chromosomes

The sex of humans is controlled by one pair of **sex chromosomes**. The genotype **XX** produces the **female** phenotype. The genotype **XY** produces the **male** phenotype.

Genotype and phenotype

The genotype is all the genes of the individual. The phenotype is what the individual looks like.

Worked example

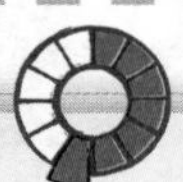

Explain the proportions of the different sex chromosomes in the gametes of men and women. **(3 marks)**

Gametes are haploid because they are produced by meiosis. As the sex chromosomes in a woman's diploid body cell are both X, all the eggs she produces will contain one X chromosome. The sex chromosomes in a man's diploid body cells are XY, so 50% of his sperm will contain one X chromosome and the other 50% will contain one Y chromosome.

Genetic diagrams and Punnett squares

We can use a **genetic diagram** or **Punnett square** to show that the sex of an individual is determined at fertilisation.

Genetic diagram

parent's phenotype	male		female	
parent's genotype	XY		XX	
gametes	X	Y	X	X
possible offspring	XX	XY	XX	XY

Punnett square

		possible female gametes	
		X	X
possible male gametes	X	XX female	XX female
	Y	XY male	XY male

You can usually use either a Punnett square or a genetic diagram to answer a question about inheritance. Use the one that works best for you.

Both diagrams show that, at fertilisation, there is an equal chance of producing a male or a female:

- 50% male XY: 50% female XX
- ratio of 1 : 1 male : female
- 1 out of 2 chance of either male or female.

Use a Punnett square like this.

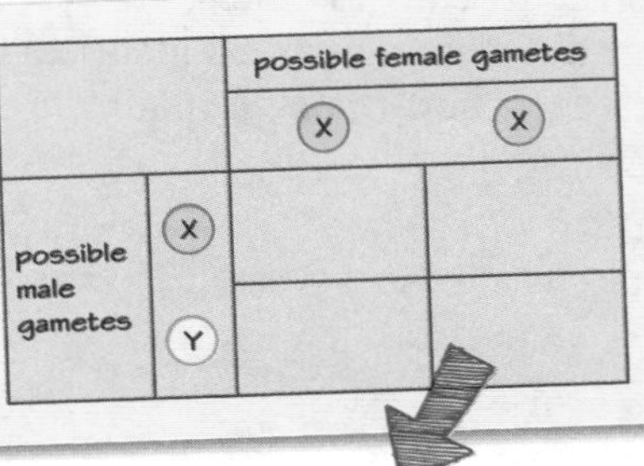

Now try this

1 At which of the following stages is the sex of a baby determined?

☐ **A** when the egg is fertilised
☐ **B** as the foetus develops in the womb
☐ **C** when the baby is born **(1 mark)**

2 A couple have two sons. The woman is pregnant with another child. Draw a genetic diagram to show the % chance of this child being a girl. **(4 marks)**

3 Explain why human eggs all contain one X chromosome. **(2 marks)**

Variation and mutation

Causes of **variation** that influence phenotype include genetic variation and environmental variation. Genetic variation happens through **mutation**.

Causes of variation

Most variations are caused by a combination of genes and environment.

Most phenotypic features in humans, e.g. hair colour, are caused by many genes. Each of these genes may have several alleles, so the phenotype is the result of the **combination** of different alleles for different genes. This combination of alleles that an organism inherits is the result of sexual reproduction.

differences between individuals of the same kind

caused by differences in alleles they have inherited, e.g. eye colour (**genetic variation**)

combination of both causes, e.g. weight, skin colour

caused by differences in conditions in which they developed, e.g. being able to ride a bike, scars (**environmental variation**)

Mutation

A **mutation** or **genetic variant** is created if the sequence of bases in a gene is changed. A mutation in the gene's coding DNA can affect the phenotype of an organism. If the amino acid sequence is altered, the activity of the protein produced may also be altered. However:

- most genetic mutations have no effect on the phenotype
- some mutations have a small effect on the phenotype
- a single mutation can, rarely, significantly affect the phenotype.

A mutation may cause:

- a large change in the protein produced
- a small change in the protein produced
- no change at all in the protein produced.

The bigger the change to the protein, the larger the effect on how the body works.

Non-coding DNA

A mutation in the non-coding DNA can also affect the phenotype. This may increase or decrease the ability of RNA polymerase to bind to DNA. A change like this can increase or decrease the amount of protein produced.

Worked example

Identical twins share the same genes so they are genetically identical. However, there may be small differences between identical twins, for example, in their body mass. Explain why. **(2 marks)**

The small differences between them will be due to environmental variables. One twin may eat a higher-energy diet than the other twin, and so has more body fat. Alternatively, one twin may exercise more than the other twin, so uses up more energy and stores less fat.

Variation due to the environment

These plants are all clones. However, they are not all identical. This could be because they have been exposed to different amounts of light, water, or soil nutrients.

Now try this

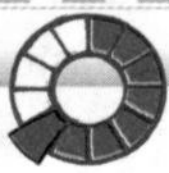

1 Explain what is meant by a mutation. **(2 marks)**

2 Explain why a mutation in a body cell will not affect a person's phenotype. **(2 marks)**

Most mutations occur when DNA replicates before cell division. Most of them cause no change in the phenotype.

The Human Genome Project

The **Human Genome Project** is a collaboration between scientists to decode the **human genome** (the order of bases on all human chromosomes). The project was completed quickly because so many scientists worked on it at the same time. The work was published in 2003 and made freely available to scientists all over the world. The results are being used to develop new medicines and treatments for diseases.

Advantages and disadvantages of the Human Genome Project

Advantages

- Alerting people that they are at particular **risk** of certain diseases, e.g. types of cancer or heart disease. The person may be able to make lifestyle changes to reduce the chances of the disease developing.
- Distinguishing between different forms of diseases such as leukaemia or Alzheimer's disease, as some drugs are beneficial in some forms of these diseases but not in others.
- Allowing doctors to **tailor treatments** for some diseases to the individual, where specific alleles affect how a person will respond to treatment.

Disadvantages

- People who are at risk of certain diseases, e.g. cancer, may have to pay more to obtain life insurance.
- It may not be helpful to tell someone they are at risk of a condition for which there is currently no cure.

Worked example

Describe two possible developments as a result of decoding the human genome, and discuss the implications of these developments. **(4 marks)**

One development is the identification of genes that can cause disease. Knowing if you have a faulty gene could help a person and their family prepare for its effects, but some people would prefer not to know if they have a faulty gene because then they would worry about it.

Another development is gene therapy. This involves replacing faulty alleles in body cells with healthy ones. This would allow the affected person to live a normal life. However, people will have to decide whether the faulty alleles are replaced in gametes, so that the healthy alleles could be passed on to children.

There are many possible answers for this question, because there are many new developments. Other possibilities include: creating personalised medicines, and identifying evolutionary relationships between humans and other organisms. As well as learning about new developments you need to be able to say what the implications are.

Now try this

Humans all share the same genes, but think carefully about alleles and base sequences.

1 The Human Genome Project means that soon we might be able to tell a child that they are at increased risk of developing high blood pressure in early middle age.
Give **one** advantage and **one** disadvantage of this. **(2 marks)**

2 The genome of two different humans is not exactly the same. Explain why. **(2 marks)**

Had a look ☐ Nearly there ☐ Nailed it! ☐

Extended response – Genetics

There will be at least one 6-mark question on your exam paper. For these questions, you will need to think scientifically and structure your answer logically, showing how the points you make are related to each other.

You can revise the topic for this question, which is about **monohybrid inheritance,** on pages 23 and 24.

Worked example

Phenylthiocarbamide (PTC) is a substance that is tasteless to some people but bitter to others. The ability to taste PTC is controlled by a dominant allele. A man and a woman who can both taste PTC have a daughter who can taste PTC, but their son is unable to taste PTC. Explain how, giving the probability of these parents having taster or non-taster children. Use T for the dominant taster allele and t for the recessive allele. **(6 marks)**

The son cannot taste PTC so he must be homozygous for the recessive allele, tt. He must have inherited a t allele from each of his parents, so the genotype of both his parents must be Tt. The Punnett square shows how he inherited this genotype.

You can work out that not tasting PTC is recessive because the son has a different phenotype from both his parents, and you are told that tasting PTC is dominant.

		Father's gametes	
		T	t
Mother's gametes	T	TT taster	Tt taster
	t	Tt taster	Tt non-taster

Even in long-answer questions, a good way to display genetic information clearly is in a Punnett square or genetic diagram.

The Punnett square shows that these parents have a 25% or 1 in 4 chance of having a child who cannot taste PTC. They have a 75% or 3 in 4 chance of having a child who can taste PTC. Their daughter might have the genotype TT or Tt.

When you use a Punnett square or genetic diagram, remember to explain the possible outcomes in terms of probability, ratio or percentage.

Both parents must be heterozygous, Tt, if they have a son who is tt, because he gets one allele from each parent.

Now try this

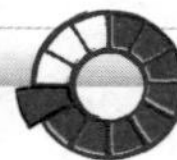

In guinea pigs, rough coat is dominant over smooth coat. Two rough-coated guinea pigs were bred together and they produced seven rough-coated and two smooth-coated offspring. Use a genetic diagram to explain how this occurred. Give the probability of these parents having rough-coated or smooth-coated offspring. Use R for the dominant allele and r for the recessive allele. **(6 marks)**

Evolution

Charles Darwin developed a theory of evolution by means of natural selection. It is still important in modern biology.

Natural selection

Individuals of a species show variation. This can mean that some individuals will be better able to survive in their environment and produce more healthy offspring than others. This is **natural selection**, where the environment (including climate and other organisms) selects which individuals pass on their **alleles** to the next generation.

Theory of evolution in modern biology

The theory of evolution is very important in modern biology:

- It helps us understand the relationships between different species of organisms.
- It explains how new species evolve.
- It explains how different species adapt to changes in their environment.

Darwin's theory

Adults usually produce more young than the environment can support when they are adults (**overproduction**). This produces a **'struggle for existence'** by the young.

→ Some individuals have inherited **advantageous variations** in characteristics that are better adapted to the environment. These individuals will have a better chance of **survival** to adulthood.

→ Individuals with advantageous variations will pass their genes on to their young. The young may inherit the advantageous variations.

→ More individuals will have these advantageous variations in the next generation.

→ Individuals with variations that are not as well adapted to the environment will be less likely to survive.

→ These individuals will not produce young.

One example of Darwin's theory is the development of antibiotic resistance in bacteria.

Worked example

African elephants are hunted illegally by poachers for their tusks. Elephants may be born tuskless due to a mutation. In 1930 in Uganda, only 1% of elephants were born without tusks. In 2010, 15% of females and 9% of male elephants were born without tusks. Explain this change using the theory of evolution. **(3 marks)**

Elephants with tusks are more likely to be poached and killed, so do not survive to pass on their alleles. Tuskless elephants are more likely to survive to pass on the allele for no tusks, so the allele for no tusks increases in frequency in the population.

This is an example of where you may be asked to apply your understanding to a specific example that you may not be familiar with.

Now try this

Remember to use scientific terms like mutation and alleles, as well as explaining the effects on survival and passing on alleles.

In a recent study, scientists tested 321 samples of head lice and found that 82% of the lice were resistant to chemicals used to treat head lice infestations. Use your knowledge of natural selection to explain how this has happened. **(4 marks)**

Had a look ☐ Nearly there ☐ Nailed it! ☐

Human evolution

Some of the evidence we have for evolution leading to modern humans (*Homo sapiens*) comes from **fossils**. These include fossils of bones and teeth.

Species	Ardi (*Ardipithecus ramidus*)	Lucy (*Australopithecus afarensis*)	*Homo habilis* ('handy man')	*Homo erectus* ('upright man')	*Homo sapiens* ('wise'/modern man)
height	120 cm	107 cm	< 130 cm	about 175 cm	wide variety but generally taller than other species
when existed	4.4 million years ago	3.2 million years ago	2.4–1.4 million years ago	1.8–0.5 million years	since c. 200 000 years ago
brain size	350 cm^3	400 cm^3	500–600 cm^3	850–1100 cm^3	approx. 1200 cm^3
other details	tree climber, also walked upright	walked upright, face ape-like	flat face like modern humans, used simple stone tools	long-distance walker, strongly built	user of complex tools

You do not need to remember details such as brain sizes but you do need to remember the names and the general trends.

Homo habilis and *Homo erectus* fossils were found by the archaeologists **Leakey** and his family. They wrote the first description of these early humans.

Stone tools

Stone tools also give us evidence of human evolution. The earliest stone tools are around 2.4 million years old. Over time more complex tools were made, and a greater range of tool types.

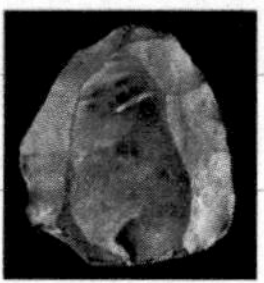

c. 2 million years old. A large stone that has had some chips flaked off it, e.g. simple hand axes

c. 40 000 years old. Made from fine flakes split from larger stones; many types of tool made this way, e.g. arrow head, spear head, scraper, knife

Worked example

Explain how stone tools can be dated from their environment. **(3 marks)**

The amount of radiation in samples of sediment just above and below the layer in which the tools are found can be used to date the sediment and so give a range of dates when the tools were left there.

The stone used to make the tools is much older than the tools, so cannot be used to date when the tool was made.

Now try this

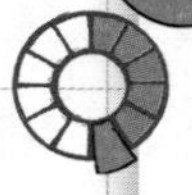

1 Describe **two** ways, that can be seen from fossils, in which human-like species have evolved over the past 4.4 million years. **(2 marks)**

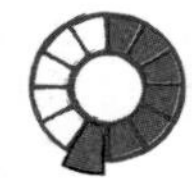

2 Suggest what the development of stone tools implies about human evolution over the last 2.5 million years. **(2 marks)**

Classification

Scientists used to classify living organisms into five big groups, called **kingdoms**. Scientists have now classified organisms into three **domains**.

Five-kingdom system

Carl Linnaeus originally proposed the classification of organisms into just two kingdoms, but this was later developed into five kingdoms. These are:

1. Plants
2. Animals
3. Fungi
4. Protists
5. Prokaryotes.

Three-domain system

Genetic research shows that the organisms which were grouped as prokaryotes in the kingdom system should be separated into two groups, which have been named **Eubacteria** and **Archaea**. This is because the genes of organisms in Archaea work more like those in the eukaryotes, while the genes of organisms in Eubacteria work a little differently.

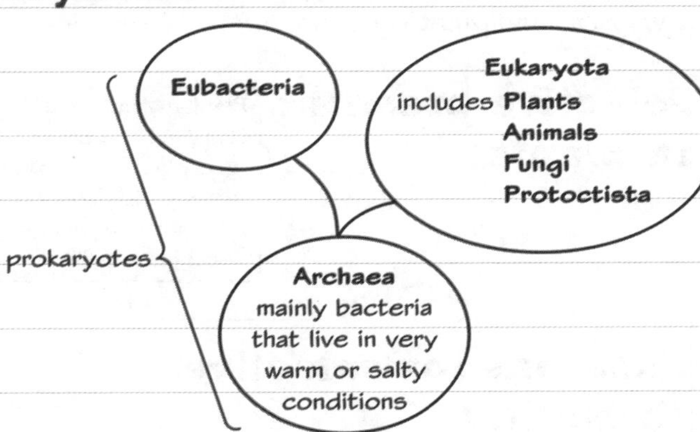

This grouping forms the **three-domain** system of classification.

Worked example

The archaea have cells that look very similar to eubacteria. However, they contain ribosomes that are different in their chemical composition from the ribosomes in eubacteria and eukaryotic cells. The archaea group was only discovered in 1977. Suggest why. **(2 marks)**

Analysing the fine structure of organelles, and the chemicals they contain, requires complex equipment and techniques. These were only developed in the 1970s. Until then, the structure of archaea looked similar to other bacteria because their cell structure is very similar.

Worked example

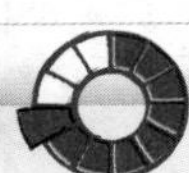

Explain why plants, animals, fungi and protoctista are all classified into the same domain, but prokaryotes are in a different domain. **(3 marks)**

This is because plants, animals, fungi and protoctista all have eukaryotic cells. They have a nucleus with a membrane around it, and membrane-bound organelles such as mitochondria. Prokaryotes do not have mitochondria or chloroplasts, and they do not have a nucleus, so their cell structure is very different.

Note that you need to say why plants, animals, fungi and protoctista are similar, as well as explaining why prokaryotes are different.

Now try this

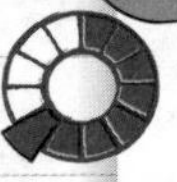

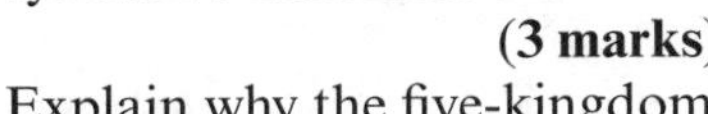

1. Describe the three-domain system of classification. **(3 marks)**
2. Explain why the five-kingdom system of classification has been replaced by the three-domain system. **(2 marks)**

Had a look ☐ Nearly there ☐ Nailed it! ☐

Selective breeding

Selective breeding is when plants or animals with certain desirable characteristics are chosen to breed together, so that offspring will be produced that **inherit** these characteristics. This produces **new breeds** of animals and **new varieties** of plants. Selective breeding has taken place over thousands of years, but more recently **genetic engineering** has been developed as a way of manipulating genetics. You can learn about this on page 33.

Selective breeding in plants

Plants with good features are crossed, e.g. plants that have good yield, are drought tolerant, or need less fertiliser. → Plants grown from seeds of these crosses are selected for their good features and crossed with each other. → Selection and crossing is repeated many times until a high-yielding variety is produced.

Reasons for selective breeding

Plants and animals are selectively bred for:

- ✓ disease resistance
- ✓ increased yield
- ✓ better ability to cope with difficult conditions
- ✓ faster growth
- ✓ better flavour.

Worked example

Describe how a farmer would breed cattle that have a large milk yield. **(4 marks)**

The farmer would choose a cow that produced lots of milk and cross her with a bull whose mother produced a lot of milk. The farmer would then choose the female offspring who produce lots of milk and cross them with a male whose mother had a high milk yield. The farmer would do this for many generations.

Worked example

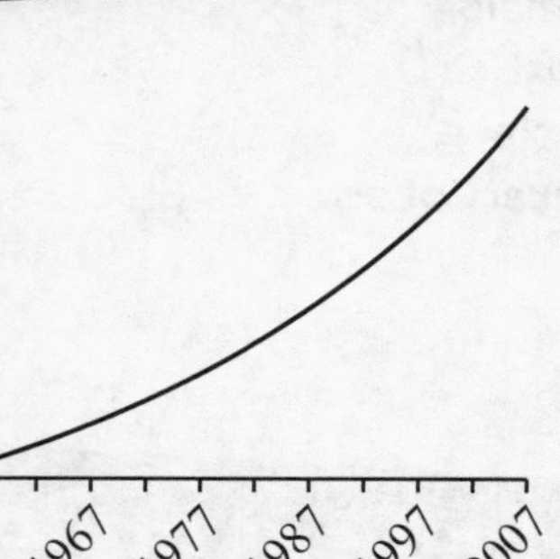

The graph shows the average milk yield per cow from 1957 to 2007. The increase is mainly the result of selective breeding. Over the same period there has been an increase in cows suffering from mastitis (an inflammation of the udders), problems with their legs and reduced fertility. A farming newspaper reported this in an article entitled 'Selective breeding has improved dairy farming'. Evaluate this statement. **(4 marks)**

The graph shows that the milk yield per cow has more than doubled over these 50 years so this is useful for the farmer.

However, the cows are having more problems with their legs and udders which may be because they are producing so much milk, so this is an animal welfare issue.

It also costs the farmer more in veterinary bills.

I think that selective breeding to produce a high-yielding breed of cows is a good thing, but this should not be done at the expense of animal welfare.

advantage

ethical issue

practical issue

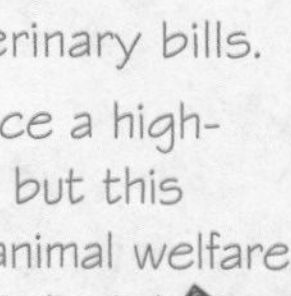

You can draw your own conclusion as long as it clearly relates to the information given and the benefits and risks identified.

Now try this

1 Give **one** advantage and **one** disadvantage of selective breeding. **(2 marks)**

2 Explain the differences between natural selection and selective breeding. **(2 marks)**

3 Describe **one** problem of selective breeding in crop plants. **(2 marks)**

Genetic engineering

Genetic engineering is changing the **genome** (the DNA) of an organism, often by introducing genes from another to create **genetically modified organisms** (GMOs). See page 34 for more details of the processes involved in genetic engineering.

How genetic engineering works

Genes can be transferred from any kind of organism to any other kind of organism, e.g. bacteria, humans, other animals, plants.

The gene for a characteristic is 'cut out' of a chromosome using enzymes. → The gene is inserted into a chromosome inside the nucleus of a cell in a different organism. → The cell of this organism now produces the characteristic from the gene.

Worked example

Human insulin is a hormone used by many patients with diabetes. Bacteria have been genetically modified to carry the human insulin gene. The bacteria are then grown in large quantities to produce the human insulin.

Explain why bacteria had to be modified to make human insulin and describe how the GM bacteria were produced. **(3 marks)**

Bacteria don't normally produce human insulin, so they had to be modified to make it. The gene for making insulin was cut out of a human chromosome. The gene was then inserted into the bacterial chromosome, so that the bacterium made the insulin.

GM crops

GM crop plants have been genetically modified to give them new characteristics, such as:

- resistance to attack by insects
- resistance to **herbicides**, so that fields can be sprayed to kill weeds, but not the crop.

These characteristics can help the crop grow better and produce more food (an increased **yield**).

Evaluating something means identifying advantages and disadvantages, then comparing them to draw a conclusion about whether the example is worth doing. You need to consider issues such as:

- How effective is it and is it worth the effort?
- Is it right/wrong/fair to do this?

Worked example

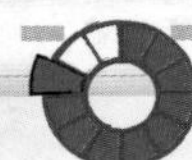

A sheep has been genetically modified to produce a protein in its milk that can be used to treat serious lung diseases in humans. Evaluate the advantages and disadvantages of this. **(4 marks)**

The protein is secreted in milk so the sheep is not harmed to produce the protein, although some people think it is unethical to use animals for the benefit of humans. The protein treats serious lung disease so is very beneficial to human health. However, in developing the genetically modified sheep a lot of embryos would have been made that did not successfully express the gene, so a lot of embryos are wasted to produce a GM sheep. However, I think this is worth doing as it treats serious disease in humans and the sheep is not harmed because the protein is produced in its milk.

Now try this

1 Define **genetic engineering**. **(2 marks)**

2 Describe how a GM crop with herbicide resistance could be developed. **(2 marks)**

3 Some mice have had a gene inserted that gives them a human disease. At the same time, the scientists inserted a gene for a fluorescent protein, so modified mouse cells glow when exposed to ultraviolet light. These mice are then used for testing new treatments for the disease.

(a) Suggest **one** advantage of using a glow gene joined to the disease gene. **(1 mark)**

(b) Describe how lots of mice with the human disease gene could be produced. **(2 marks)**

Stages in genetic engineering

You need to know the processes involved in **genetic** engineering used to make **genetically modified organisms.**

Making human insulin

Human insulin can be made in large quantities from bacteria that have been genetically modified to contain the gene for human insulin.

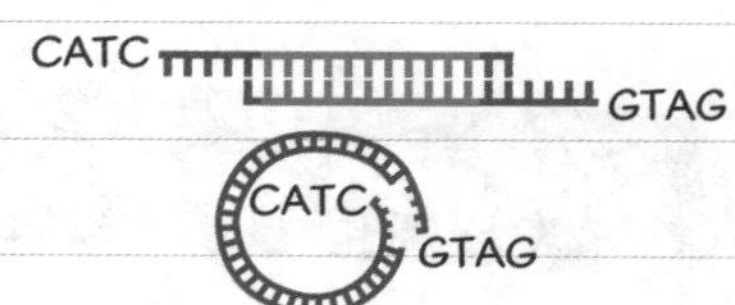

1. DNA from a human cell is cut into pieces using enzymes called **restriction enzymes**. These make staggered cuts across the double-stranded DNA, leaving a few unpaired bases at each end, called **sticky ends**.
2. Bacteria cells contain small circles of DNA called plasmids. The same restriction enzymes are used to cut plasmids open, leaving sticky ends with matching sets of unpaired bases.
3. The pieces of DNA containing the insulin gene are mixed with the plasmids. The bases in the sticky ends pair up. An enzyme called DNA **ligase** is added, linking the DNA back into a continuous circle.
4. The plasmids are inserted into bacteria. The bacteria can now be grown in huge fermenters, where they make human insulin.

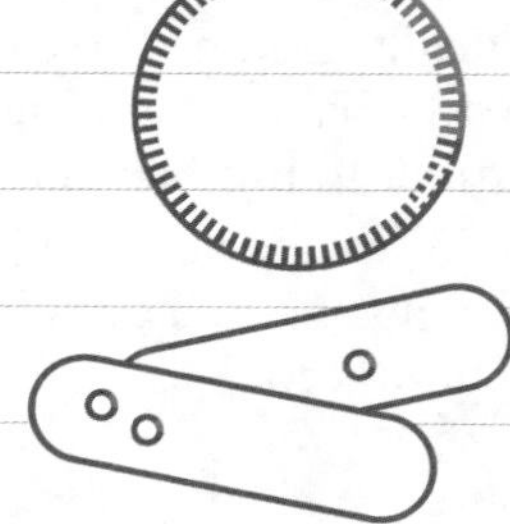

Worked example

Describe the role of enzymes in the formation of genetically modified bacteria. **(3 marks)**

Restriction enzymes cut the required gene out of the human DNA. They leave 'sticky ends' on the gene. The same restriction enzymes cut open the plasmid, creating matching sticky ends. The DNA ligase enzyme joins the matching sticky ends of the human gene and the plasmid, to make one complete modified plasmid.

The plasmid must be removed from a bacterium before the gene is inserted. The plasmid is then inserted into another bacterium. The sticky ends allow the insulin gene and the plasmid to join together properly.

Vectors

A **vector** is the name for anything that carries the new gene into a cell. In the case of insulin above, the vector is a **plasmid**. Other kinds of vectors may be used, such as viruses.

Now try this

There are four stages you should include.

1 Describe all the stages in producing genetically modified bacteria that make insulin. Use a bullet point list or flow chart in your answer. **(4 marks)**

2 Explain why it is useful to make human insulin using genetically modified bacteria. **(2 marks)**

Extended response – Genetic engineering

There will be at least one 6-mark question on your exam paper. For these questions, you will need to think scientifically and structure your answer logically, showing how the points you make are related to each other.

You can revise the topics for this question, which is about the **advantages and disadvantages of genetic modification of crops**, on page 33.

Worked example

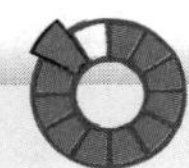

Some crop plants have been modified to produce a toxin when their cells are damaged. This toxin kills caterpillars when they eat it. Evaluate the factors that would need to be considered before growing these crops on a large scale. **(6 marks)**

> Do not be put off if you have not met this specific example before. You are given all the information you need; think about identifying advantages and disadvantages and evaluating them to draw a conclusion.

First you would need to consider whether this method is effective at killing caterpillars and increasing crop yield. Genetically modifying the crop will be expensive, so tests will need to be carried out to find out whether the increased yield brings in enough income to offset the cost of buying genetically modified seeds to plant.

You would also need to know whether the toxin is harmful to people eating the crop. Tests must be carried out to make sure the crop is safe to eat. It is also important that customers are happy to buy this crop. Some people do not like the idea of eating genetically modified food, and choose not to buy it.

Growing this crop might save time and costs for the farmer, because it has a toxin in the crop so the farmer does not need to buy a pesticide to kill the caterpillars. The toxin is always there so caterpillars should be killed as soon as they start to eat the crop. Also, there will not be any harmful pesticide residues left on the crop. If the farmer sprays his crops with a pesticide instead, this could harm useful insects as well as the caterpillars.

> Consider what the alternative to using the GM crop might be, as this will help you to weigh up the different factors that have to be considered.

There is a possible ethical issue if the toxin gene was transferred to wild plants by cross-pollination. This could possibly lead to wild plants containing a toxin, which could kill insects and cause problems in food chains. On balance I think it would be a good idea to grow this crop, but only after the necessary tests have been carried out.

> This can happen if pollen from the GM crop is carried by a pollinating insect and fertilises a closely related wild plant.

> Remember to always state your opinion in questions like this.

Other answers

The answer to this question could have covered other issues, such as how the development of resistance to the toxin by insects will affect the benefits of using GM crops. What matters in this answer is that it covers all that the question asks, and is presented in a clear and logical order.

Now try this

Some children do not produce enough growth hormone, which leads to a slow growth rate in childhood. Scientists can genetically modify bacteria to produce human growth hormone. Explain how they could do this. **(6 marks)**

Health and disease

The World Health Organization defines **health** as 'a state of complete physical, mental and social wellbeing'. There are many different **diseases** with different causes.

What is health?

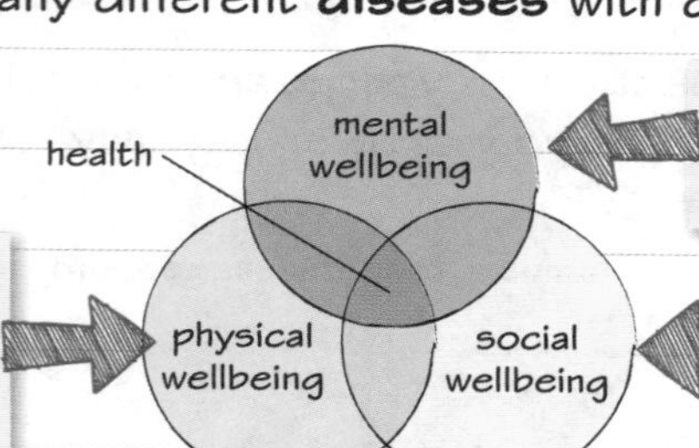

such as how you feel about yourself

such as eating and sleeping well, and being free from disease

such as how well you get on with other people

Types of disease

Diseases may be **communicable** (they can be passed from one person to another) or **non-communicable** (not passed between people). A **pathogen** is an organism that causes an infectious disease.

Communicable	Non-communicable
rapid variation in number of cases over time	Number of cases changes only gradually.
Cases are often localised.	Cases may be more widely spread.
e.g. malaria, typhoid, cholera	e.g. cancer, diabetes, heart disease

Worked example

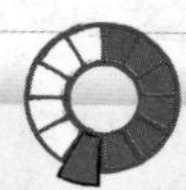

Chalara fungus sometimes kills ash trees directly, but sometimes it makes the tree more likely to be infected by other pathogens which kill the tree. Suggest how this happens. **(2 marks)**

The fungus damages the bark and the cells of the ash tree, making it easier for other pathogens to infect the tree.

HIV is a pathogen that makes it easier for other pathogens to infect a person. In the case of HIV, this happens because of damage to the immune system.

Pathogens

Pathogens include bacteria and viruses, fungi and protists. When a few pathogens **infect** you (get inside your body) they can reproduce very rapidly. Large numbers of pathogens make you ill.

Bacteria are much smaller than human cells.

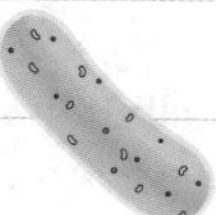

bacterium

Bacteria may release **toxins** (poisons) that make us feel ill. Some types of bacteria invade and destroy body cells.

Viruses are much smaller than bacteria.

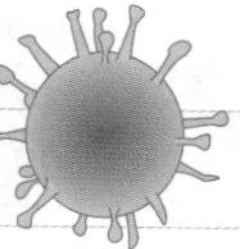

virus

Viruses take over a body cell's DNA, causing the cell to make **toxins** or causing damage when new viruses are released from cells.

fungi

Fungi are eukaryotic organisms.

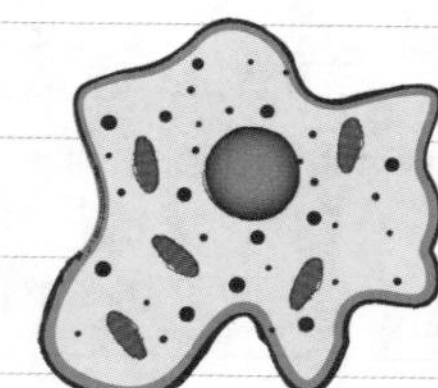

protist

Protists are eukaryotic organisms. Many are free-living but some are pathogens.

Pathogens make you feel ill when they damage cells or change how they work.

Now try this

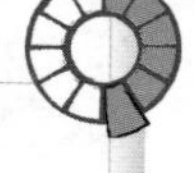

1 A stroke is caused when a blood clot blocks an artery in the brain. This stops oxygen reaching part of the brain and nerve cells die. State whether a stroke is a communicable or non-communicable disease. Give a reason for your answer. **(2 marks)**

2 Explain why an infectious disease may affect a lot of people in a community, and then fall to a very low level again. **(2 marks)**

Common infections

Different kinds of pathogens can cause different diseases with different symptoms. You need to be able to describe the **infections** given on this page.

Infection	Type of pathogen	Main symptoms
cholera	bacterium	• watery, pale-coloured **diarrhoea**, often in large amounts (watery faeces)
malaria	protist	• **fever** (high temperature) • **weakness** • chills • sweating
HIV/AIDS	virus	• mild flu-like symptoms may occur when first infected • often, no symptoms for a long time • eventually, **repeated infections** (e.g. TB) that would not be a problem if the immune system was working properly
tuberculosis (TB)	bacterium	• lung damage seen in **blood-speckled mucus** • **weight loss** • **fever** and chills • night sweats
Ebola	virus	• internal bleeding and fever (**haemorrhagic fever**) • severe headache • muscle pain • vomiting • diarrhoea (frequent watery faeces)
stomach ulcers	bacterium	• **inflammation** in stomach causing pain • bleeding in stomach
ash die-back *Chalara*	fungus	• leaf loss • bark **lesions** (damage) • **dieback** of top of tree (crown)

Worked example

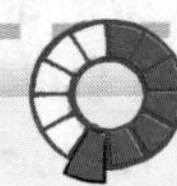

Describe a communicable disease caused by a bacterium and the symptoms of the disease. **(3 marks)**

TB is a communicable disease. Two of its symptoms are blood-speckled mucus and weight loss.

You could choose any of the other communicable diseases in the table. One mark is for naming the disease, then two marks for two symptoms.

Now try this

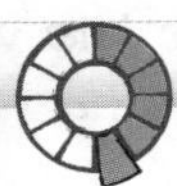

Give **one** example of a pathogen from each of the following groups, describing **two** signs of the infection it causes:

(a) virus **(2 marks)**
(b) protist **(2 marks)**
(c)fungus. **(2 marks)**

Make sure you know which group each pathogen belongs to.

How pathogens spread

Understanding how pathogens are spread can help us to find ways of **reducing** or **preventing** their spread.

Malaria

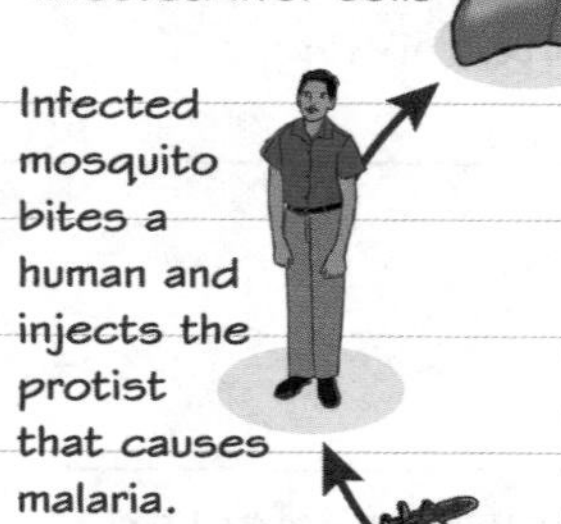

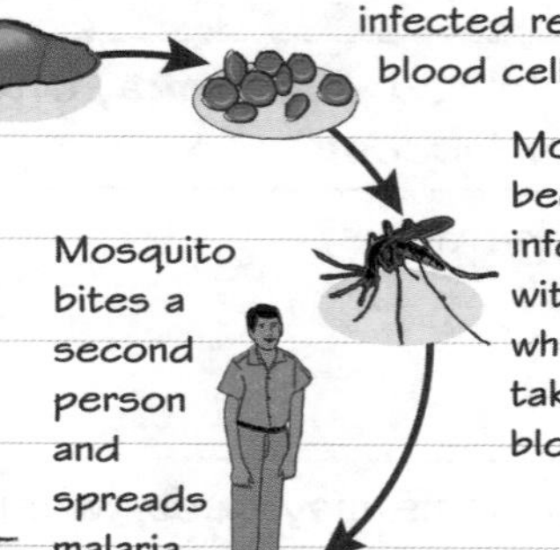

infected liver cells
infected red blood cells
Infected mosquito bites a human and injects the protist that causes malaria.
Mosquito becomes infected with parasite when it takes a blood meal.
Mosquito bites a second person and spreads malaria.

The protist that causes malaria is spread by a **vector**, the mosquito.

Watch out! Vectors are not pathogens. Vectors carry pathogens from one person to another. Vectors are one way infectious diseases are spread.

HIV

The human immunodeficiency virus, **HIV**, enters the blood and reproduces inside white blood cells, causing white blood cell destruction.

Eventually so many white blood cells are destroyed that the immune system cannot work properly. This stage is called **AIDS**. The person becomes ill with a disease, such as tuberculosis, that would not be a problem if their immune system was working properly.

Ebola virus

The **Ebola virus** causes **haemorrhagic fever**. The virus infects humans from other infected people, infected animals or objects that have been in contact with the virus. The virus infects:

- liver cells
- cells from the lining of the blood vessels
- white blood cells.

It multiplies inside these cells and destroys them. This causes fever, severe headache, muscle pain, extreme weakness, vomiting, diarrhoea, and internal and external bleeding.

Reducing the spread of pathogens

Disease (cause)	Ways to reduce or prevent its spread
cholera (bacteria)	• Boil water to kill bacteria before drinking. • Wash hands thoroughly after using toilet to prevent spread by touch.
tuberculosis (bacteria)	• Ventilate buildings adequately to reduce chance of breathing in bacteria in droplets of mucus coughed out by an infected person. • Diagnose infected people promptly and give antibiotics to kill tuberculosis bacteria. • Isolate infected people so they cannot pass the infection to others.
malaria (protists)	• Prevent mosquito vectors biting people, by killing mosquitoes or keeping them off skin (e.g. by using a sleeping net).
stomach ulcers (bacteria)	• Cook food thoroughly to kill bacteria. • Wash hands thoroughly before preparing food to avoid transfer.
Ebola haemorrhagic fever (virus)	• Keep infected people isolated because the virus is easily spread. • Wear fully protective clothing while working with infected people or dead bodies.

Now try this

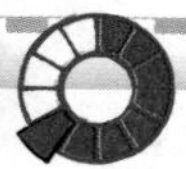

Using malaria as an example, explain the difference between a pathogen and a vector. **(2 marks)**

STIs

Chlamydia and HIV are sexually transmitted infections (**STIs**) spread by sexual activity. *Chlamydia* is a bacterium, and HIV is a virus.

Spread of STIs

Chlamydia bacteria can be spread by contact with sexual fluid from an infected partner. An infected mother can pass *Chlamydia* to her baby during birth.

Common methods of **HIV transmission** are:

- unprotected sex with an infected partner
- sharing needles with an infected person
- transmission from infected mother to foetus
- infection from blood products.

Reducing or preventing STIs

The spread of many STIs can be reduced or prevented by:

- using **condoms** during sexual intercourse
- **screening people**, including pregnant women
- **screening blood** transfusions
- supplying intravenous drug abusers with **sterile needles** (HIV)
- treating infected people using **antibiotics** (for *Chlamydia* and other bacterial infections).

Worked example

All pregnant women in the UK are screened for HIV infection. Explain two advantages of this. **(4 marks)**

If a pregnant woman has HIV, this allows doctors to give her treatment to stop the infection getting worse, and also to try to stop HIV spreading to the baby. Another advantage of this is that it gives doctors a measurement of how many people have HIV. This helps them to target awareness raising and education programmes.

There are other possible answers, for example, if the woman is HIV positive, doctors may discuss with her whether to breast-feed her baby after birth or not, as breast-feeding might pass HIV on to the baby.

Worked example

About 3000 15-year-olds are diagnosed with *Chlamydia* every year. Explain how this rate of infection could be reduced. **(2 marks)**

There could be better health education in schools, so that students learn how Chlamydia is spread and what the symptoms are. This might lead to fewer young people catching Chlamydia and, if they do get it, they will realise that they need to see a doctor.

Health education can include using condoms during sexual intercourse.

Worked example

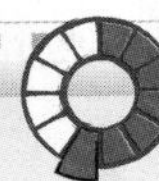

A newborn baby's eyes are infected with *Chlamydia*. Explain how this could happen. **(1 mark)**

During birth, the baby's eyes and lungs may become contaminated with vaginal fluid containing the bacteria.

Pathogens enter the body through natural openings, so the baby's eyes, nose and mouth are places where pathogens can enter the body easily.

Now try this

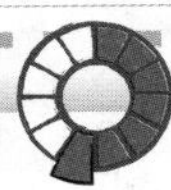

1 State what STIs are and how they are spread. **(2 marks)**

2 During sexual intercourse a man places a condom over his penis. Explain how this reduces his risk of being infected with an STI. **(1 mark)**

Human defences

The human body has both physical barriers and chemical defences to give protection against pathogens. **Physical barriers** make it hard for pathogens to enter the body. **Chemical defences** are chemicals that are produced to kill pathogens or make them inactive.

Chemical and physical defences

Chemical defences

Lysozyme enzyme in tears kills bacteria by digesting their cell walls.

Lysozyme enzyme is also present in saliva and mucus.

Hydrochloric acid in stomach kills pathogens in food and drink.

Physical barriers

Unbroken **skin** forms a protective barrier because it is too thick for most pathogens to get through.

Sticky **mucus** in the breathing passages and lungs traps pathogens. **Cilia** on the cells lining the lungs move in a wave-like motion, moving mucus and trapped pathogens out of lungs towards the back of the throat where it is swallowed.

Mucus traps pathogens.

Cilia move mucus away from lungs.

cilia

epithelial cells

Remember that epithelial cells line the surface of tubes.

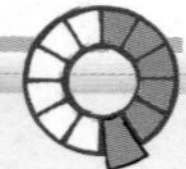

Worked example

Cilia move mucus away from the lungs. Describe how this protects the body from pathogens. **(3 marks)**

It stops pathogens entering the lungs where they can cause disease. When the mucus gets to the throat it is swallowed and the pathogens are destroyed by the acid in the stomach.

Cilia are tiny hairs on the surfaces of cells that move together in a rhythm to move substances along, such as mucus. Their movement is a bit like a crowd of people doing a Mexican wave.

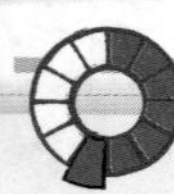

Worked example

Chemicals in cigarette smoke paralyse the cilia in the epithelium of the airways. Smokers are more likely than non-smokers to suffer from lung infections. Explain why. **(2 marks)**

When the cilia are paralysed, they cannot move the mucus containing pathogens back up to the throat. The mucus travels down into the lungs, carrying pathogens with it.

Now try this

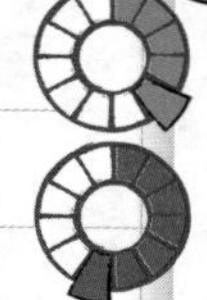

1 Give **one** example of a physical barrier and **one** example of a chemical defence that prevents pathogens entering the body. **(2 marks)**

2 A burn damages the skin, and a bad burn can completely break through the skin. First aid advice for a person who has suffered a burn includes covering the burn with clean cling film or a plastic bag. Explain why. **(2 marks)**

The immune system

The **immune system** helps to protect the body by attacking pathogens if they manage to enter the body. **Lymphocytes** are part of this immune system.

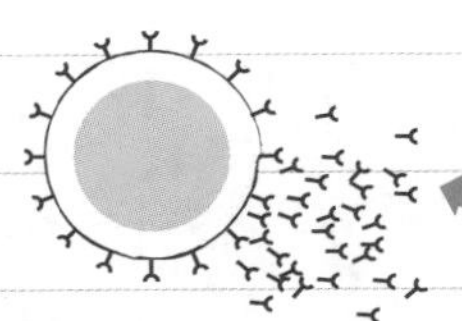

1 Each pathogen has unique **antigens** on its surface.

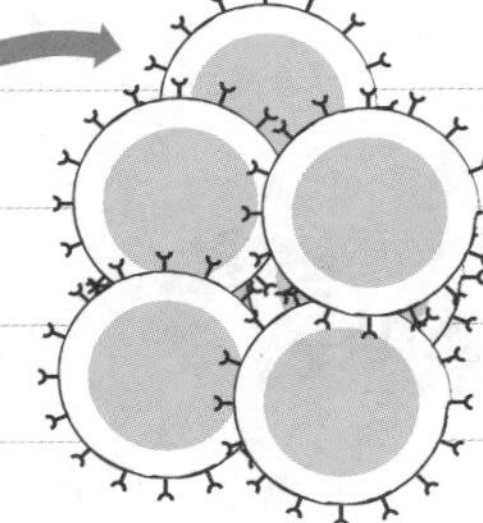

2 A **lymphocyte** with an **antibody** that fits the antigen is activated.

3 The lymphocyte **divides** many times to produce clones of identical lymphocytes.

4 Some of the lymphocytes produce lots of antibodies which stick to the pathogen and destroy it. Other lymphocytes stay in the blood as **memory lymphocytes**, ready to respond immediately if the same antigen returns.

Antibodies

The antibodies produced by a white blood cell are **specific** for one particular kind of pathogen. This means they can only destroy that kind of pathogen. They cannot destroy another kind of bacterium or virus.

Immunity

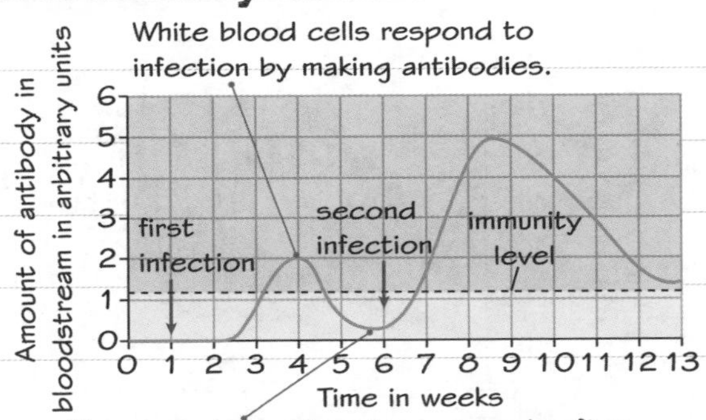

The memory lymphocytes are made after the infection and then respond more quickly to another infection by the pathogen.

This is called the **secondary response**.

Worked example

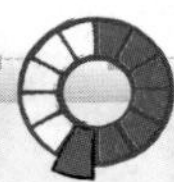

Explain why there is a delay of a few days between a person becoming infected by a pathogen for the first time and antibodies being made. **(2 marks)**

It takes time for lymphocytes to identify the antigens on the pathogen, and to secrete specific antibodies.

The secondary response is much faster because memory cells are already present, and can secrete specific antibodies immediately.

Worked example

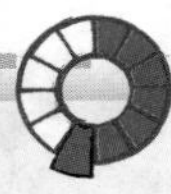

Pathogens have antigens that are foreign to the body. When a pathogen enters the body, the body is exposed to these antigens. Describe two responses triggered by the antigens. **(2 marks)**

The antigens trigger an immune response which causes the production of antibodies.
They also trigger the production of memory lymphocytes.

Now try this

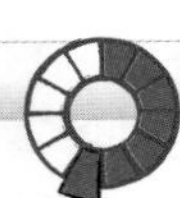

Describe how the immune system helps to fight infection. **(3 marks)**

Had a look ☐ Nearly there ☐ Nailed it! ☐

Immunisation

Immunisation is when you give a person a vaccine to prevent them becoming ill from a disease.

How vaccines work

A **vaccine** contains antigens from the pathogen, often in the form of dead or weakened pathogens. → The person's lymphocytes produce antibodies against the pathogen and also memory lymphocytes. → If the person becomes infected with the real pathogen, the memory lymphocytes will give a very rapid secondary response to the pathogen. This means the person is very unlikely to become ill.

This form of the pathogen is safe. It is **inactive** and cannot cause the disease in the person receiving the vaccine.

Worked example

(a) The graph shows the number of cases of measles in the UK from 1940 to 2014. A vaccine against measles was introduced in 1968. Give evidence from the graph to support this. **(1 mark)**

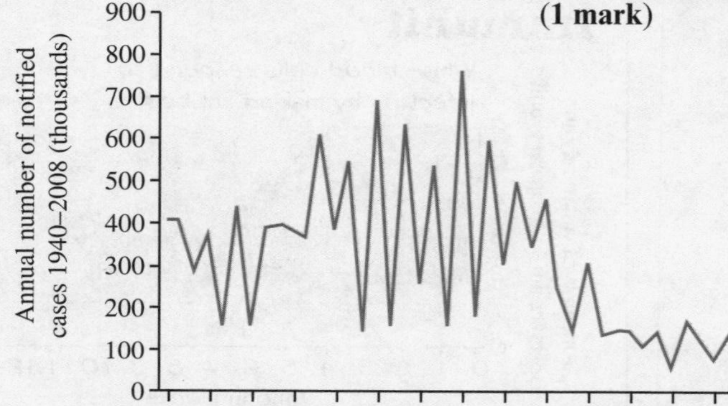

The number of cases of measles drops steadily after 1968.

(b) The number of cases of measles was still quite high for some time after 1968. Suggest a reason why. **(2 marks)**

The vaccine may not have been given to all children in 1968 so there were still children who had not been vaccinated. There would also have been some adults who had not been vaccinated and caught measles.

Worked example

Describe how a vaccine can protect you for life. **(2 marks)**

Once you have received the vaccine, the immune system makes antibodies but also memory lymphocytes which stay in the body for a long time. If you ever catch the disease 'for real', these memory cells divide very quickly and produce huge numbers of antibodies.

Worked example

Explain why immunisation only protects you against one particular disease. **(2 marks)**

Each pathogen has a particular antigen. The immune system produces antibodies that are exactly the right shape to fit on to these antigens. Other pathogens have antigens of a different shape, so one kind of antibody cannot bind to a different kind of antigen.

Now try this

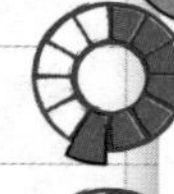

1 Describe the difference between a vaccine and immunisation. **(2 marks)**

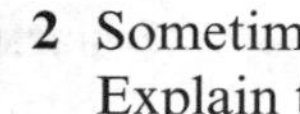

2 Sometimes a person needs to receive a booster vaccination. Explain the advantage of this. **(2 marks)**

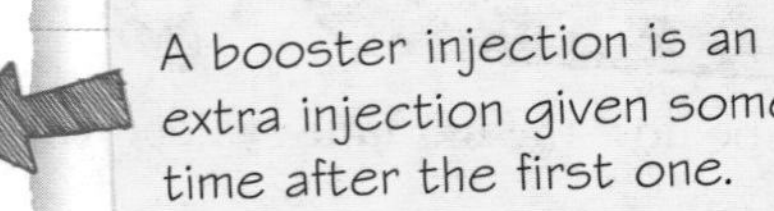

A booster injection is an extra injection given some time after the first one.

Treating infections

Antibiotics

Antibiotics are medicines used to treat bacterial infections. They **kill bacteria** inside the body. Specific bacteria are only killed by a specific antibiotic, so the correct antibiotic must be used. Deaths from bacterial diseases have greatly decreased where antibiotics are used. **Penicillin** is an example of an antibiotic.

Bacterial resistance

Some bacteria are becoming **resistant** to some antibiotics. This means that the antibiotic is no longer effective at killing or inhibiting them. This is partly because people who take an antibiotic to treat an infection often stop taking it too early, because they feel better. This leaves the more resistant bacteria still alive. They reproduce and spread, causing infections that cannot be treated with the antibiotic because most of the bacteria are now resistant.

Bacterial resistance provides evidence for evolution (see page 29).

Worked example

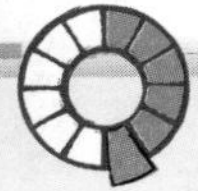

Describe how antibiotics kill bacteria. **(2 marks)**

Antibiotics inhibit cell processes in the bacterium but not the host organism. For example, some antibiotics stop bacterial cell walls forming properly. This does not harm the host animal because animal cells do not have cell walls.

Do not confuse **antibiotics** and **antibodies**. They are very different things!

Effect of antibiotics on bacteria

The effect of antiseptics, antibiotics or plant extracts on the growth of bacteria can be studied on a **bacterial culture** in a **Petri dish**.

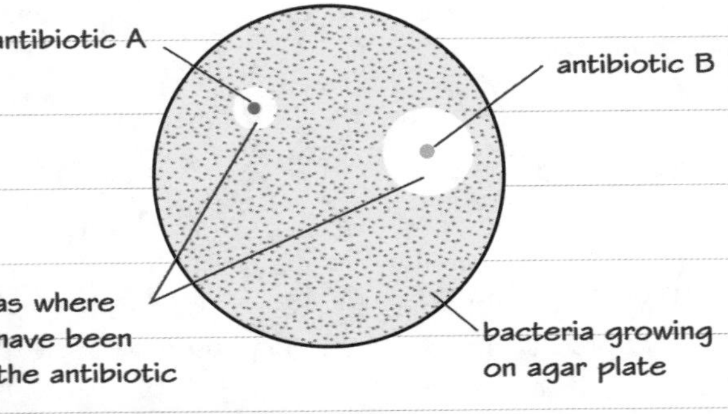

Worked example

Look at the agar plate illustrated to the left. Identify which antibiotic was more effective at killing the bacteria. Explain your answer. **(2 marks)**

Antibiotic B is more effective because the clear area around it is much bigger than the clear area around antibiotic A. The clear area shows where bacteria have been killed.

Now try this

1 Explain why doctors do not prescribe antibiotics if a patient has the flu virus. **(1 mark)**

2 Some antibiotics work by binding to bacterial ribosomes so that proteins cannot be made. Bacterial ribosomes are smaller than human ribosomes and have a different structure. Explain why these antibiotics do not harm human cells. **(2 marks)**

New medicines

Medicines are chemicals that are used to treat the cause or signs of an illness. Scientists are continually developing new medicines. New medicines must be extensively tested before doctors can **prescribe** them to patients. There are several steps of testing.

Development and testing

Discovery

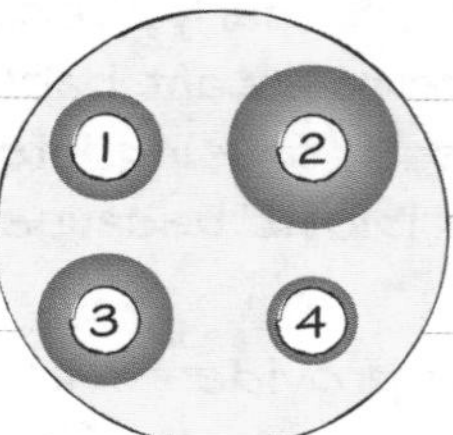

New medicines are **discovered**, e.g. by screening organisms to see if they produce antibiotics that kill bacteria. They are then **developed** through a series of stages.

Preclinical testing (in the lab)

cultures of cells

cultures of tissues

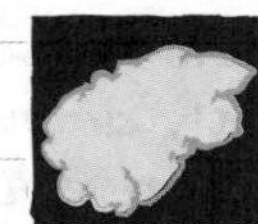

animals

Antibiotics are **tested** in the lab to make sure the medicine gets into cells without harming them, and damages pathogens inside cells.

Clinical trials: stage 1

healthy volunteer

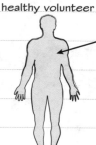

very small **dose** of drug

to check that the drug is not **toxic** (harmful)

Medical drugs are prescribed (given) by doctors to help patients who are ill.

Clinical trials: stage 2

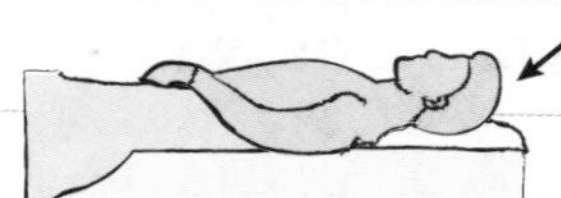

different doses of drug

patient with the disease that the new drug is developed for

to test **efficacy** (whether it works) and to find the **optimum** dose (the dose that works best)

Worked example

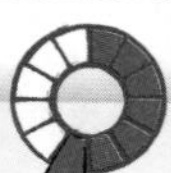

Explain why a new medicine is not tested on lots of people straight away. **(3 marks)**

This is to make sure there are no harmful side effects and also to make sure it works on people. It would be unethical to give the medicine to lots of people before it was tested for safety and to see if it works on people.

Now try this

There are several different possible answers here, but you only need to think of two of them.

1 There are two antibiotics, A and B, that are effective against a particular bacterial infection. Antibiotic A produces a larger clear zone around it than antibiotic B when tested on a Petri dish of bacteria. However, the doctor chooses to prescribe antibiotic B to the patient. Give **two** possible reasons for this. **(2 marks)**

2 Suggest **one** advantage and **one** disadvantage of using animals as models for testing new drugs rather than direct testing on humans. **(2 marks)**

Non-communicable diseases

Many non-communicable diseases are caused by the **interaction of several factors**. Having several factors can increase the risk of developing a disease.

Factors affecting the risk of developing non-communicable diseases

Genes (inherited factors): Different alleles of a gene may be more prone to mutation (e.g. in breast cancer) or how well you absorb nutrients (e.g. in diseases related to diet). These factors may be more common in particular ethnic groups.

Age: The older the body, the more likely that cells may develop mutations which lead to cancer.

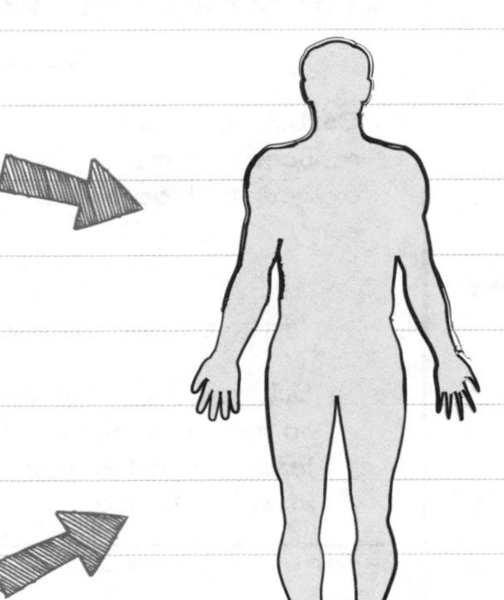

Sex: The female hormone oestrogen has protective effects that men do not get.

Environmental: Air pollution can cause lung diseases; poisons in food and drink can damage the body.

Lifestyle factors: The way we live, including diet, alcohol, smoking and exercise, can affect our risk of developing many diseases. You can find out more about lifestyle factors on pages 46–48.

Some non-communicable diseases caused by these factors include cancer, cardiovascular (heart) disease, lung and liver diseases, and dietary diseases. Rates of these diseases are different between different groups of people in a country, or in different countries, according to individual exposure to these risk factors.

Worked example

The graph shows the number of deaths from heart disease in the USA for one year.

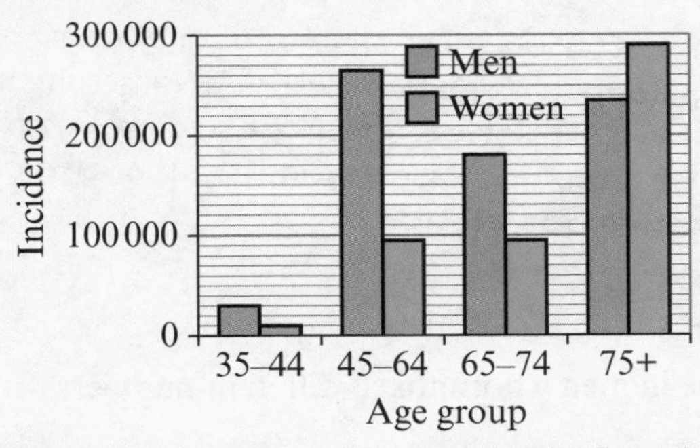

(a) Describe the effect of age on deaths from heart disease shown in the chart. **(2 marks)**

The death rate from heart disease is at least twice as high in men compared to women until the 75+ age group, when more women die from heart disease than men.

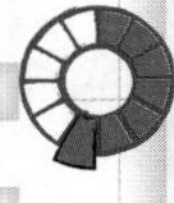

(b) Women produce the hormone oestrogen until the menopause (at about 50). Suggest whether the chart supports the idea that oestrogen can protect against heart disease. **(2 marks)**

The idea is not clearly supported, because the 65–74 category is about the same as the 45–64 group, but it does increase greatly in the 75+ group.

Now try this

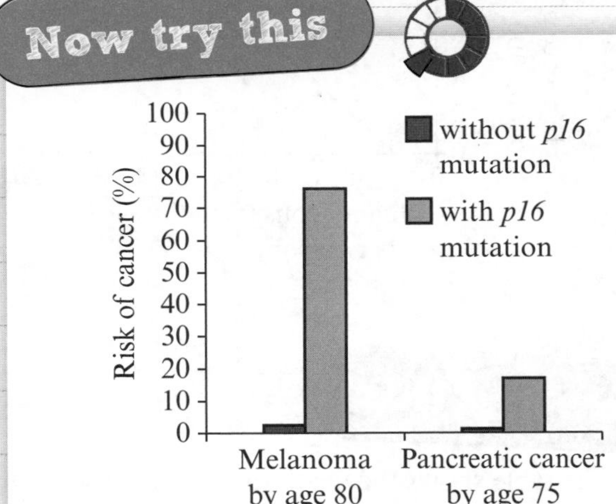

The chart shows the risk of developing some types of cancer in people with a particular mutation in the gene *p16* compared with people without this mutation.

(a) Explain why the risk of developing cancer is given as a percentage. **(2 marks)**

(b) Explain whether this data supports the hypothesis that a mutation in the *p16* gene increases a person's chance of developing cancer. **(2 marks)**

Alcohol and smoking

Lifestyle factors including **drinking alcohol** and **smoking** increase the risk of non-communicable diseases by changing how the body works and increasing the levels of **toxins** (poisons) in your body.

Alcohol and disease

Ethanol (found in alcohol) is poisonous to cells. When absorbed from the gut, it passes first to the **liver** to be broken down. So liver cells are more likely than other cells to be damaged, leading to liver disease, e.g. cirrhosis.

This can lead to low birth weight in babies whose mothers smoke.

The damage caused by smoking

Nicotine is addictive.

Carbon monoxide reduces how much oxygen the blood can carry.

Chemicals in **tar** are **carcinogens** that cause cancers, particularly of mouth and lungs.

Substances in cigarettes can cause blood vessels to narrow, increasing blood pressure. This can lead to **cardiovascular diseases** such as heart attack or strokes.

Worked example

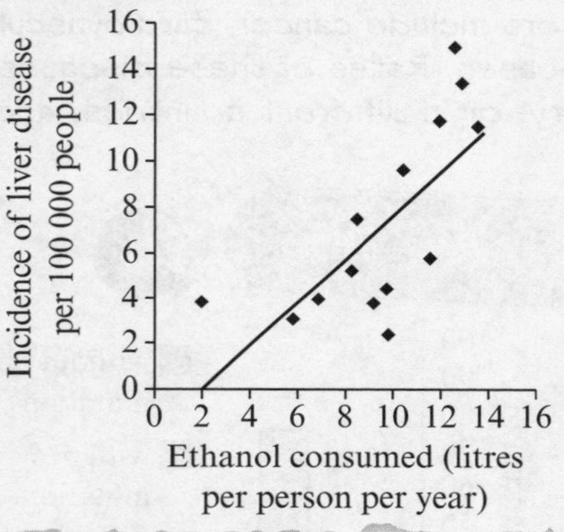

The graph shows the incidence of liver disease and the mean ethanol consumption per person per year for different countries.
Describe and explain the relationship shown in the graph. **(2 marks)**.

As the amount of alcohol consumed increases, the incidence of liver disease also increases. This is because ethanol is poisonous, particularly to liver cells.

This is a scatter diagram. It is good for showing a link (**correlation**) between two factors. This graph shows a **positive correlation**. You might also see a graph where one factor increases while the other decreases – a **negative correlation**.

Now try this

The table shows the death rate from coronary heart disease in men who smoked different numbers of cigarettes.

	Death rate			
Age group	**Non-smokers**	**Light smokers**	**Moderate smokers**	**Heavy smokers**
35–54	0.44	0.95	1.47	1.84
55–64	5.34	6.49	6.75	6.98
65–74	10.13	10.77	10.53	14.85
75 and over	15.19	20.00	22.27	24.93

(a) Describe what this data shows. **(2 marks)**

(b) Explain the relationship between smoking and deaths from coronary heart disease. **(2 marks)**

Malnutrition and obesity

Malnutrition happens when a person eats too much or too little of a nutrient. Too little of some nutrients can lead to **deficiency diseases**. If there is too much energy in the diet and not enough exercise then there will be an increase in body fat. Too much body fat leads to **obesity**.

Measuring obesity

Obesity can be measured by **BMI** or by **waist : hip** ratio.

Waist : hip ratio is calculated using this equation:

$$\text{waist : hip ratio} = \frac{\text{waist measurement}}{\text{hip measurement}}$$

People with a high waist : hip ratio ('apples') are more at risk of some diseases than those with a lower waist : hip ratio ('pears').

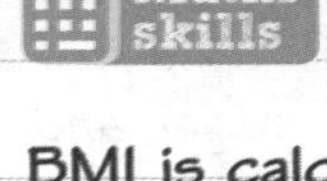

Body mass index (BMI)

BMI is calculated using this equation:

$$\text{BMI} = \frac{\text{weight in kilograms}}{(\text{height in metres})^2}$$

Adults who have a BMI over 30 are said to be obese.

Obesity is linked with many health problems, including type 2 diabetes.

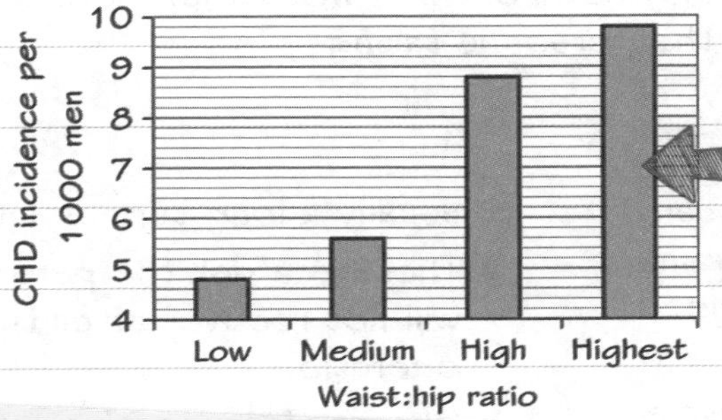

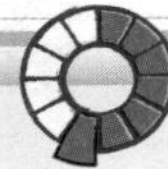

As waist:hip ratio increases, the **risk** of coronary heart disease increases.

CHD = coronary heart disease

Worked example

A scientific study showed that for a 10 g increase in wholegrain fibre that a person ate each day, there was a 10% decrease in their risk of developing bowel cancer. The scientists concluded that people should eat lots of wholegrain foods every day to help them stay healthy.

Do the results of this study support this conclusion? Give a reason for your answer. **(2 marks)**

The results do support this conclusion, because the more fibre the person eats the lower their risk of getting bowel cancer.

You will not be expected to know the effect of fibre in the diet, but you will be expected to answer questions like this about the effect of food on health from data that you are given. You will be given data to work with.

Now try this

Look at page 64 for more information about diabetes.

1 A study of over 43 000 American adult men showed that overweight men who were physically fit had the same risk of diseases such as type 2 diabetes as men who were not overweight, but that unfit overweight men had a much higher risk of these diseases. A simple conclusion from this study is that being fit is important for health as well as not being overweight. Evaluate this conclusion. **(3 marks)**

For an 'evaluate' answer, use your knowledge and the information given to consider evidence for and against, and then draw a suitable conclusion from your arguments.

2 A 45-year-old woman is 160 cm high and weighs 80 kg. Calculate her BMI and state whether or not she is obese. **(1 mark)**

Had a look ☐ Nearly there ☐ Nailed it! ☐

Cardiovascular disease

Cardiovascular disease may be **treated** using medication, surgery, or by making lifestyle changes.

Lifestyle changes

Doctors will advise the patient to:

- give up smoking
- take more exercise
- eat a healthier diet (lower fat, sugar and salt)
- lose weight.

👍 no side effects

👍 may reduce chances of getting other health conditions

👍 the cheapest option

👎 may take time to work or may not work effectively

Medication

If lifestyle changes do not improve the patient's health, the doctor may prescribe **medication**, such as betablockers, to reduce blood pressure.

👍 start working immediately, e.g. to lower blood pressure

👍 easy to do

👍 cheaper and less risky than surgery

👎 need to be taken long term and can have side effects

👎 may not work well with other medication the person is taking

Surgery

If an artery becomes narrowed, the drop in blood flow can cause damage to tissue beyond the blockage. A wire frame, called a **stent**, is inserted into the narrowed part of the artery.

Sometimes **heart bypass** surgery is carried out, when a new blood vessel is inserted to bypass blocked coronary arteries.

If the blood supply to the heart muscle is reduced or stopped, the heart muscle cells cannot get enough oxygen for respiration. This means the cells die and the person has a heart attack.

👍 usually a long-term solution

👎 There is a risk the person will not recover after the operation.

👎 Surgery is expensive.

👎 more difficult to do than giving medication

👎 There is a risk the person will develop an infection after surgery.

Worked example

In a study of patients with narrowed coronary arteries, one group exercised and was given medicine, while a similar group exercised, took medicine and had a stent inserted into the narrowed artery.

In the 'without stent' group, 202 out of 1092 patients died of a heart attack within 5 years. In the 'with stent' group, 211 out of 1111 died of a heart attack within 5 years. Evaluate whether these results suggest that stent surgery is worth doing. **(3 marks)**

The percentage of deaths was 18.5% in the group without stents and 19.0% for the group with stents. This is not a big difference with such large sample sizes. This suggests that stent surgery is not worth doing if other treatments are carried out at the same time.

Maths skills You need to calculate the percentages so you can make an accurate comparison.

Now try this

Surgery is usually only carried out if the person has serious heart disease and other treatments will not work. Explain why. **(2 marks)**

Extended response – Health and disease

There will be at least one 6-mark question on your exam paper. For these questions, you will need to think scientifically and structure your answer logically, showing how the points you make are related to each other.

You may wish to revise the topic for this question, which is about evaluating treatments for heart disease, on page 48.

Worked example

Stanol esters are made from substances found in plants. Some scientists wanted to find out if stanol esters lower blood cholesterol. They carried out an investigation in which they added stanol esters to margarine and gave it to some people. There was also a control group. The scientists measured the blood cholesterol concentration of each person at regular intervals over several months. The graph shows their results.

Explain how the scientists should have carried out this investigation to get reliable results. **(6 marks)**

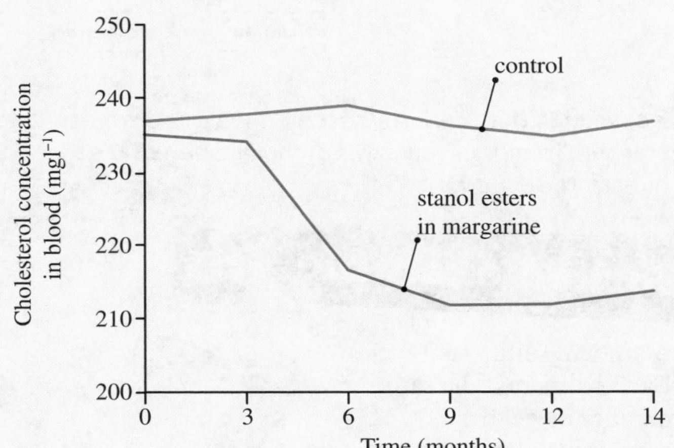

The scientists would need to carry out this investigation with a large number of people who should be volunteers. They should be divided into two groups, matched by age and sex. One group would be given margarine containing stanol esters and the other group should be given the same margarine without stanol esters in it. The people should not know which group they are in. This is so they cannot know whether they have stanol esters in the margarine they are given. This is to avoid bias.

The scientists should measure the blood cholesterol level of each person before the investigation starts and then every month for a year. After this the scientists should compare the results from the two groups to see whether the stanol esters have reduced the blood cholesterol concentration of the people who ate that margarine, compared to the control group.

Other factors that could have been mentioned are body mass, starting level of cholesterol and diet.

The factors the scientists control are all factors that could affect cholesterol levels, but are not the factor being studied.

Command word: Discuss

Plan out your answer to a **discuss** question, e.g. using bullet points that cover all the issues in the question, including any benefits or risks. Arrange your bullet points into a logical order. Then use your plan to produce a well-organised answer.

Now try this

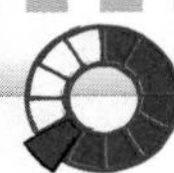

Explain how lifestyle factors might increase the chances of a person developing some kinds of cancer. **(6 marks)**

Photosynthesis

Photosynthesis is the process that plants and algae use to make their own food. Plants are called **producers** because they produce their own food. Photosynthetic organisms are the main producers of food and therefore **biomass**. Biomass is the mass of living material at a particular stage in a food chain.

Photosynthesis equation

Photosynthesis can be summarised by this equation:

carbon dioxide + water —light energy (absorbed by **chlorophyll**)→ glucose + oxygen

- carbon dioxide: from air
- water: from soil
- glucose: a sugar
- oxygen: released to air as **by-product** (not needed)

Chlorophyll is a green substance found in the chloroplasts of most plants.

Remember that during photosynthesis there is an energy transfer from light energy to chemical energy stored in the sugars produced.

Worked example

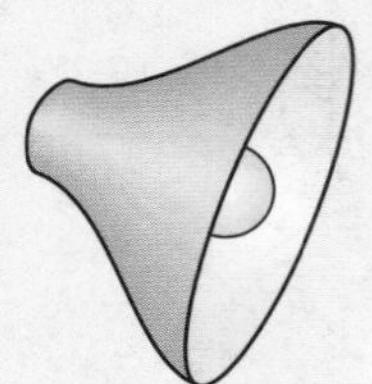

Maths skills Light intensity is proportional to $\frac{1}{d^2}$ where d = distance between the algae and the lamp.

A student set up six bottles. Each contained the same number of gel beads containing algae, and the same volume of hydrogencarbonate indicator solution. The indicator is yellow when carbon dioxide concentration is high, orange-red when carbon dioxide concentration is the same as in normal air, and purple when carbon dioxide concentration is low. The student placed the bottles at different distances from a lamp as shown in the diagram.

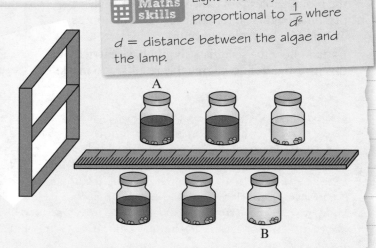

After two hours, the bottles were the colours shown in the diagram. Explain the difference in colour between bottles A and B. **(3 marks)**

Bottle A has high light intensity so photosynthesis is occurring much faster than respiration. Therefore the concentration of carbon dioxide is very low. Bottle B is too far away from the light to photosynthesise very fast, but it is respiring, so the concentration of carbon dioxide is very high.

Worked example

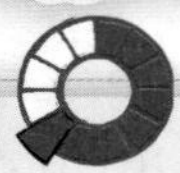

Explain why a leaf covered in black paper makes no food. **(2 marks)**

The black paper stops light from reaching the chlorophyll. Therefore, the leaf cannot use light energy to convert carbon dioxide and water into sugar in photosynthesis.

Plants make food in photosynthesis, but they can only do this in the light. However, plants respire all the time, whether it is dark or light.

Now try this

Remember that 'endothermic' reactions take in energy from the surroundings.

1 Explain why photosynthesis is an endothermic reaction. **(2 marks)**

2 Explain the role of plants as producers of food. **(4 marks)**

Limiting factors

Low temperature, dim light and low carbon dioxide concentration all limit the rate of photosynthesis. They are all **limiting factors** for photosynthesis.

Effect of light on rate of photosynthesis

You can measure the effect of light intensity on the rate of photosynthesis by measuring the rate at which oxygen is given off by a piece of pondweed. You can use the apparatus shown to investigate the effect of light intensity using bright and dim lights.

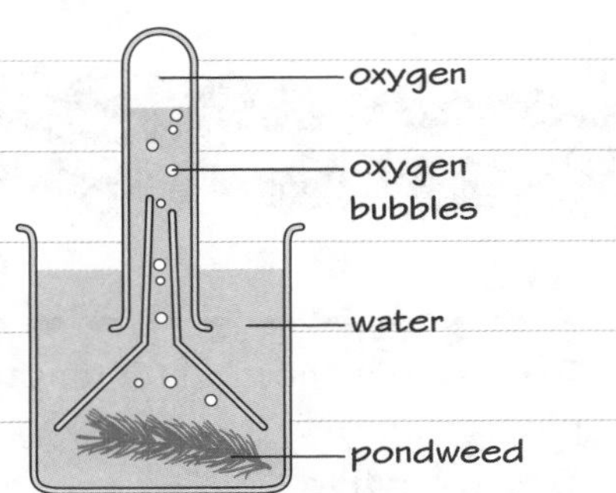

Worked example

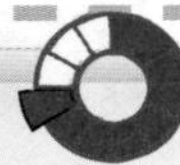

The graph shows the rate of photosynthesis of a plant at different light intensities. Explain the shape of the graph at points A and B. **(4 marks)**

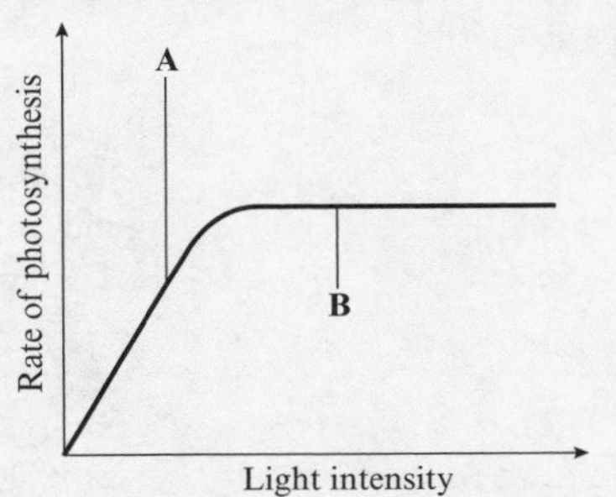

At A, light intensity is limiting photosynthesis, because increasing the light intensity increases the rate of photosynthesis. At B, light intensity is no longer the limiting factor because increasing the light intensity does not change the rate of photosynthesis. Another factor, such as carbon dioxide concentration or temperature, is limiting the rate of photosynthesis.

Increasing only carbon dioxide concentration or temperature, while keeping other factors constant, will produce a similar graph to this.

Carbon dioxide concentration

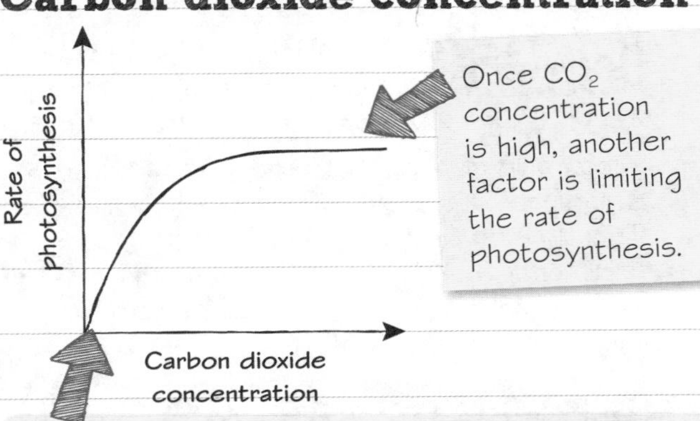

CO_2 is a limiting factor because it increases the rate of photosynthesis. There is more CO_2 to use to make sugars.

Temperature

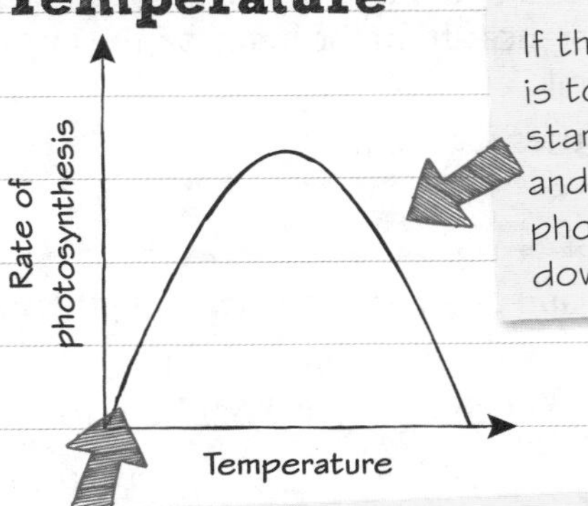

If the temperature is too high, enzymes start to denature and the rate of photosynthesis slows down.

Temperature is a limiting factor because it increases the kinetic energy of molecules and increases the rate of enzyme activity making photosynthesis faster.

Now try this

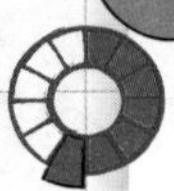

1 Define the term **limiting factor**. **(2 marks)**

2 Explain why measuring the oxygen given off by a plant is a way of measuring the rate of photosynthesis. **(2 marks)**

Had a look ☐ Nearly there ☐ Nailed it! ☐

Light intensity

You can investigate the **effect of light intensity on the rate of photosynthesis** by measuring the change in pH of solution around algal balls. The pH changes because carbon dioxide forms an acidic solution, and photosynthesis changes the concentration of carbon dioxide in the solution.

Core practical

Aim

To investigate how light intensity affects the rate of photosynthesis.

Apparatus

- screw-top bottles (one for each distance)
- algal balls (20 per bottle)
- universal indicator solution
- measuring cylinder
- bright lamp
- small tank of water (if lamp is hot)
- metre ruler
- pH colour chart for indicator

Change in the colour of the solution can also be measured as change in absorbance using a colorimeter.

Method

1 Place 20 algal balls and the same volume of indicator solution in each bottle, and replace the screw tops.
2 Check the colour of the indicator against the colour chart and record the starting pH.
3 If using a hot lamp, place the water tank next to the lamp.
4 Use the metre ruler to place the uncovered bottles at specific distances from the lamp, on the opposite side of the tank.
5 Leave the bottles for 1–2 hours.

Results

Record distance and pH of each bottle, using the colour chart for identifying the pH.

Calculate the rate of change in pH per hour for each bottle.

Draw a graph of your results to show distance against rate of change in pH.

Conclusion

As distance increases, light intensity falls and so the rate of photosynthesis decreases. This is because energy transferred by light is needed for photosynthesis to take place.

Instead of distance, light intensity could be measured directly using a light meter.

Safety: Take care if using a lamp with a hot bulb.

Using a water tank if the lamp is hot helps control for temperature, which will also vary with distance from the lamp.

Maths skills A graph using light intensity should show that rate of change in pH is directly proportional to light intensity.

You could also calculate the inverse square distances and draw a graph to show the relationship of that with rate of pH change. This should show that the rate of change in pH is inversely proportional to distance from the lamp.

Now try this

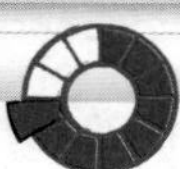

1 Explain why rate of change in pH is a measure of the rate of photosynthesis in this experiment. **(3 marks)**
2 Explain why using a water tank is important if the lamp has a hot bulb. **(2 marks)**
3 A student covered one bottle of algal balls in kitchen foil and placed it close to the tank as an addition to the method above. Explain the purpose of this bottle. **(2 marks)**

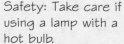

Specialised plant cells

Some plant cells are **specialised** to carry out specific functions.

Phloem

Phloem contains **sieve tube elements** which have very little cytoplasm so that there is a lot of space to transport sucrose and other nutrients. It also contains **companion cells** which have lots of mitochondria. These supply energy from respiration for **active transport** of sucrose into and out of the sieve tubes.

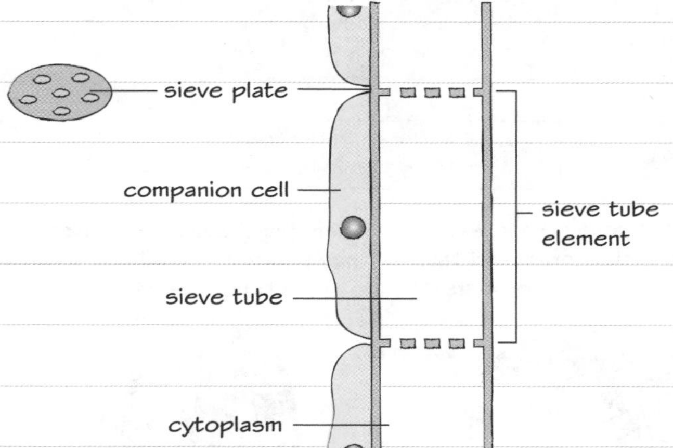

Sucrose is **translocated** around the plant in the phloem sieve tubes (see page 55 for more on translocation).

Xylem

Xylem vessels are **dead** cells which have no cytoplasm or cell contents. This means there is more space for water containing mineral ions to move through.

They have holes called **pits** in their walls to allow water and mineral ions to move out.

The walls are strengthened with **lignin** rings, which makes them very strong and prevents them from collapsing.

They have no end walls so they form a long tube that water can flow through easily.

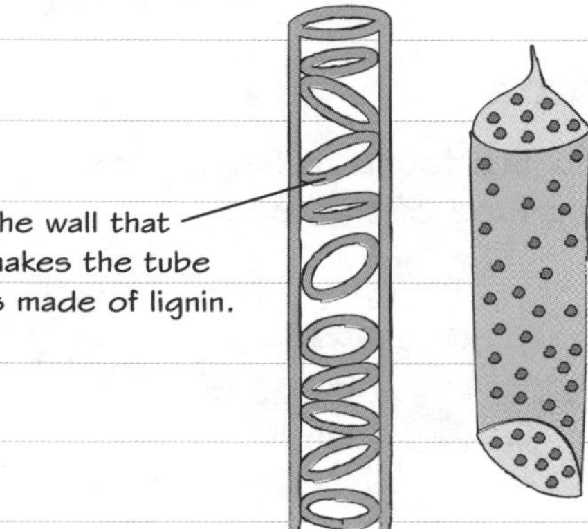

The wall that makes the tube is made of lignin.

Worked example

The diagram shows two root hairs on the outside of a root. Describe how these cells are adapted to take up water and mineral ions. **(2 marks)**

Root hair cells have long extensions that stretch out into the soil. This gives them a large surface area where osmosis can take place, which means that more water molecules can cross the cell membrane into the cell at the same time. This also gives a large surface area for mineral ions to enter the root hair cell by diffusion and active transport.

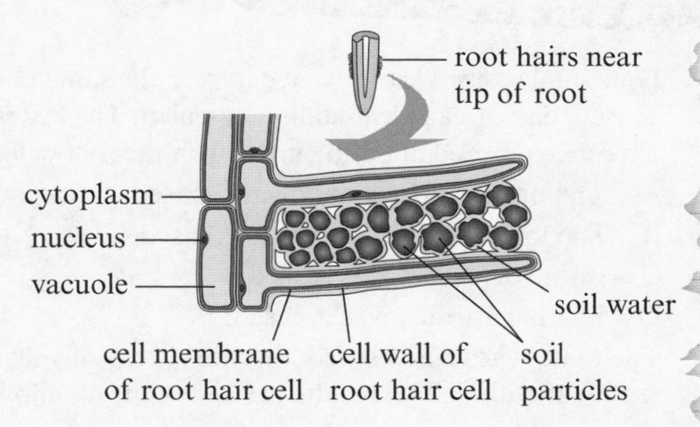

Worked example

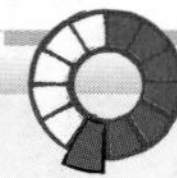

Explain the meaning of the phrase **active transport**. Use minerals entering roots as your example. **(3 marks)**

Mineral salts cannot enter the root cells from soil water by diffusion because there is a higher concentration of mineral salts in the cells than in the soil. So the root cells have to use energy to transport mineral salts into the cells against their concentration gradient.

Now try this

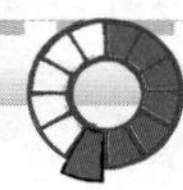

1. Explain how the structure of root hair cells is adapted to support active transport of mineral ions into roots. **(2 marks)**
2. Describe **three** adaptations of xylem vessels and explain how these adapt the cells for transporting water through the plant. **(6 marks)**

Transpiration

Transpiration is the loss of water by evaporation from leaves. The movement of water from the roots to the leaves is called the **transpiration stream**.

water vapour evaporates from leaves mainly through the stomata

↑

draws water out of the leaf cells and xylem

↑

draws water up the stem through the xylem from the roots

↑

causes water to enter the roots by osmosis

Stomata

Stomata are found mainly on the lower surface of the leaf.

When guard cells take in water by osmosis, they swell and this causes the stoma to open.

When guard cells lose water, they become flaccid and the stoma closes.

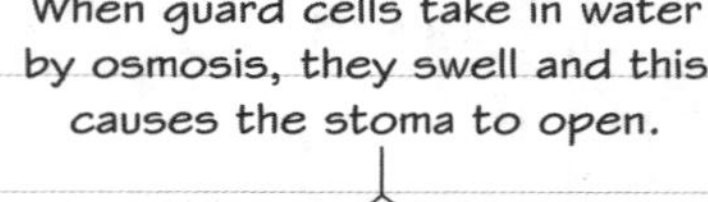

The cell wall is thicker on one side of the cell than on the other.

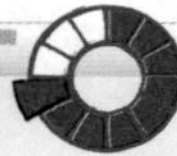

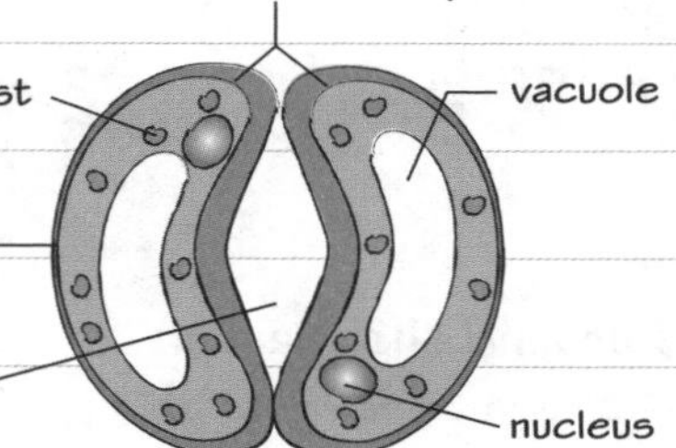

stoma open

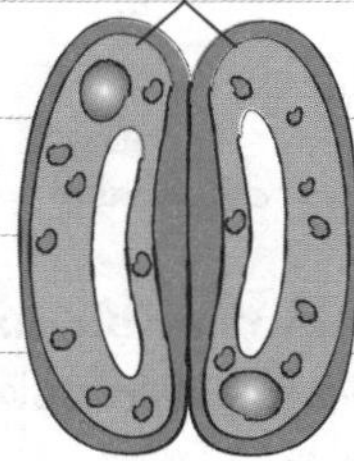

stoma closed

Worked example

Four similar-sized leaves were cut from the same plant. The cut end of each leaf-stalk was sealed. The leaves were then covered in petroleum jelly in the following way:

A – The upper surface only was greased.
B – The lower surface only was greased.
C – Both surfaces were greased.
D – Neither surface was greased.

The mass was re-measured and the % loss of mass of each leaf found. The results are shown in the table.

Leaf	% loss in mass
A	40
B	4
C	2
D	43

Describe and explain these results. **(4 marks)**

Leaves A and D lost the most mass, while B and C lost very little mass. This shows that the leaves that lost the most mass were those that did not have grease on the lower surface. Leaves lose most water through the stomata which causes a loss in mass. Leaves B and C had petroleum jelly covering the stomata so they could not lose water that way. Therefore they lost almost no water.

Do not confuse transpiration with osmosis. Water enters the roots by osmosis, but transpiration is the evaporation of water that 'pulls' water up the plant.

Now try this

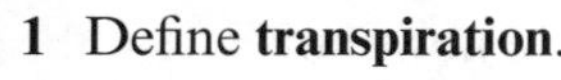

1 Define **transpiration**. **(2 marks)**

2 Describe how water moves from the soil into a plant and evaporates from the leaf. **(4 marks)**

3 Suggest the advantage to the plant of stomata being present mainly on the underside of the leaf. **(2 marks)**

Translocation

Translocation is the transport of **sucrose** around a plant.

Phloem

Dissolved sugars are transported around a plant in phloem.

dissolved sucrose needed for growth in growing regions, e.g. bud

Sucrose is produced in leaves from glucose formed during photosynthesis.

phloem in plant veins

Dissolved sucrose is carried around the plant in phloem.

storage organ (e.g. potato)

Dissolved sugars are converted to starch and stored in storage organs so they can be used later.

Plants have two separate transport systems: **xylem** and **phloem**.

Worked example

Carbon dioxide containing radioactive carbon was supplied to a leaf of a plant as shown in the diagram.

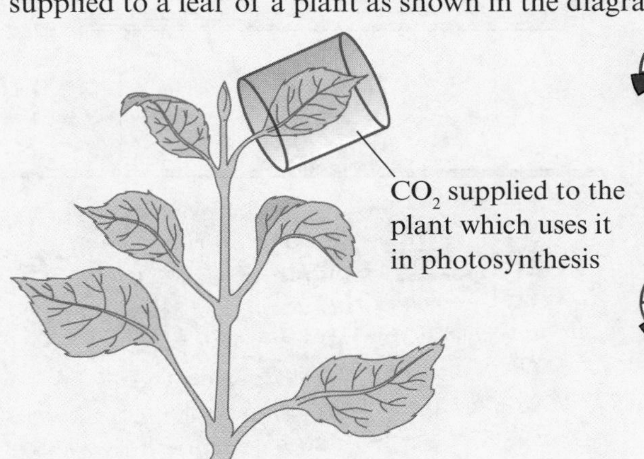
CO_2 supplied to the plant which uses it in photosynthesis

Remember that the phloem transports sugars and organic substances in the plant.
See page 53 for more information.

(a) Radioactive carbon atoms will be detected in the phloem. Explain why. **(2 marks)**

The carbon dioxide will be used in photosynthesis to make glucose. Some of the glucose will be converted to sucrose which is transported round the plant in the phloem.

(b) Describe how you could use this investigation to show that sucrose can move both up and down the plant. **(2 marks)**

This will be shown if radioactivity can be detected both above and below the leaf being supplied with $^{14}CO_2$.

Now try this

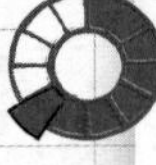

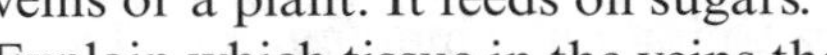

1 An aphid is an insect pest that inserts needle-like mouthparts into the veins of a plant. It feeds on sugars.
Explain which tissue in the veins the aphid feeds from. **(2 marks)**

2 A young potato plant grows rapidly during the spring and summer, but in the autumn develops new potatoes, which are storage organs. Compare the direction of movement of sucrose in the plant at different times of the year. **(3 marks)**

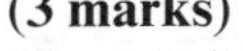

3 Name the **two** transport systems found in plants. **(2 marks)**

Remember that sugars can be transported both upwards and downwards in a plant.

Had a look ☐ Nearly there ☐ Nailed it! ☐

Water uptake in plants

Environmental factors can change the rate of water uptake in plants.

Factors that affect transpiration

Factor	Effect on transpiration
light intensity	High light intensity causes the stomata to open. This increases the rate of evaporation of water from the leaf so more water is taken up to replace this.
air movement	Wind blows moist air away from the stomata, keeping the diffusion gradient high. So the more air movement there is, the higher the transpiration rate.
temperature	The higher the temperature, the more energy water molecules have, so they move faster which means a faster rate of transpiration.

Using a potometer

You can measure the **rate** of transpiration using a **potometer**.

1. Note the position of the air bubble on the ruler at the start of the investigation.
2. Note the position of the bubble on the ruler after a known number of minutes.
3. Divide the distance moved by the bubble by the time taken.

Rate of transpiration can be measured as

$$\frac{\text{distance moved}}{\text{time taken}}$$

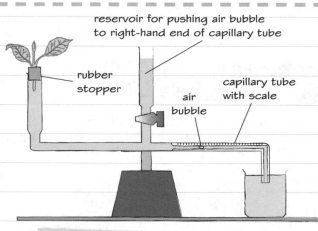

reservoir for pushing air bubble to right-hand end of capillary tube
rubber stopper
air bubble
capillary tube with scale

Maths skills You can find the **volume** of water taken up by finding the volume of the capillary tube between the bubble's start and finish points using the formula $\pi r^2 d$ where r = radius of tube, d = distance moved by bubble and $\pi = 3.14$.

Worked example

Three identical potted plants were watered and their mass recorded. Their pots were wrapped and sealed in plastic bags so that only the plant was in the open air. Each plant was then placed in different conditions. After 6 hours, the bags were removed and the mass of the plants was recorded again. The table shows the results.

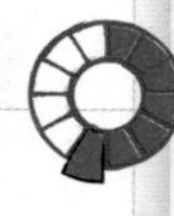

(a) Calculate the percentage change in mass for plants B and C. **(2 marks)**

Mass in g	Plant A (cool still air)	Plant B (warm still air)	Plant C (warm windy)
at start	436	452	448
at end	412	398	332
% change	5.5	11.9	25.9

(b) Explain the differences in results. **(2 marks)**

The results show that evaporation of water from the plant was faster in warm air than cool air and even faster in windy air than in still air. This is because evaporation from the stomata is faster in hot and windy conditions.

Now try this

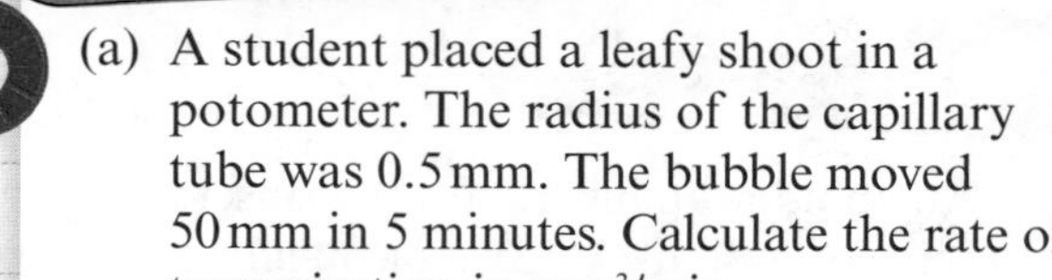

(a) A student placed a leafy shoot in a potometer. The radius of the capillary tube was 0.5 mm. The bubble moved 50 mm in 5 minutes. Calculate the rate of transpiration in mm^3/min. **(2 marks)**

(b) Explain how this measurement would be different if you repeated this with the same shoot in warmer conditions. **(2 marks)**

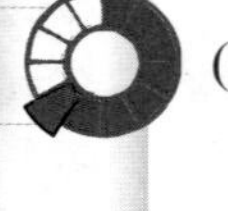

(c) Describe how you would carry out this investigation to make sure the results in the two different conditions could be compared fairly. **(2 marks)**

Extended response – Plant structures and functions

There will be at least one 6-mark question on your exam paper. For these questions, you will need to think scientifically and structure your answer logically, showing how the points you make are related to each other. You can revise the topics for this question, which is about the **effect of environmental factors on transpiration rate**, on page 56.

Worked example

A student watered a pot plant and placed it on a sunny windowsill. By the early afternoon the plant had wilted. A few hours later, when the air was cooler, the plant had recovered and was fully upright again.
Explain these observations. **(6 marks)**

wilted plant

recovered plant

Plants absorb water from the soil through their roots and lose it through stomata in their leaves by evaporation in a process called transpiration.

A good start to this answer is to identify the processes that are involved in the question.

The rate of evaporation of water from the leaf surface increases as temperature increases, because water molecules move around faster. When the temperature was high, the rate of evaporation from the leaves was faster than the rate of absorption of water from the soil. This meant there was not enough water in the plant cells to keep them strong and support the upper part of the plant, so the plant wilted.

To answer this question well, you need to explain clearly how changes in temperature affect the rate of evaporation of water from leaves. Remember to link ideas to explanations with words such as **because**, or **this means that**.

When the plant was cooler, the rate of evaporation of water from the leaves decreased. The rate of absorption of water from the soil was then fast enough to fill all cells with water so the stems were strong enough to stand upright and support the leaves.

There are two situations that need to be explained in this answer: when the plant was hot, and when it was cooler. Arranging the explanations in time sequence helps to make the answer clearer.

Now try this

A student was studying a pond on a sunny day. During the morning he noticed bubbles coming from the pondweed. The rate of bubbling increased until early afternoon, and then decreased again towards evening.
Explain these observations. **(6 marks)**

Had a look ☐ Nearly there ☐ Nailed it! ☐

Hormones

Hormones are 'chemical messengers' that target organs in the body.

Production and transport

Hormones are produced by **endocrine glands** and released into the blood. They travel around the body in the blood until they reach their **target organs**. Each hormone causes its target organ(s) to respond, e.g. by releasing another chemical substance.

Different hormones have different target organs and cause different responses.

Hormones and nerves

Nerves and hormones both help you to respond to changes in the environment and in your body. Hormones usually have a long-lived effect while nerves have a short-term effect. Nerve impulses work quickly while hormones take longer to work.

Release of hormones

Different endocrine glands produce different hormones.

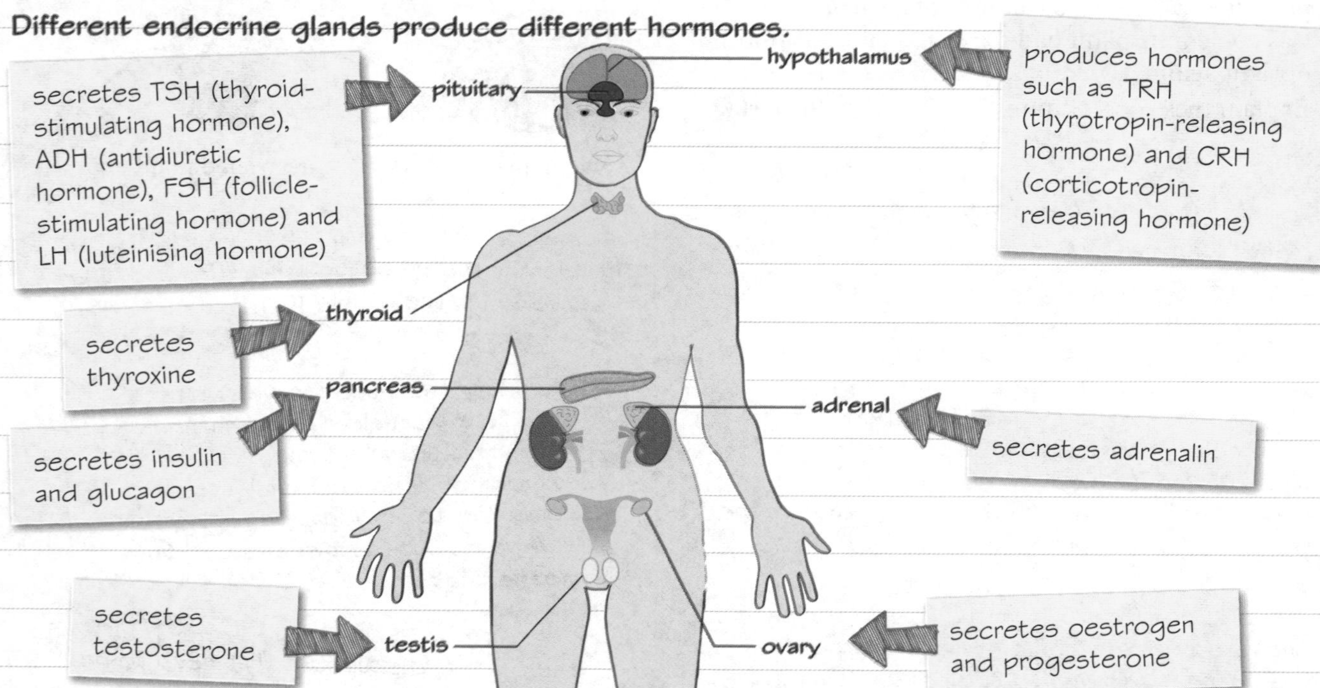

Worked example

Complete the table to show the target organs for each hormone.

(9 marks)

Hormone	Target organ(s)
TRH and CRH	pituitary gland
TSH	thyroid gland
ADH	kidney
FSH and LH	ovaries
insulin and glucagon	liver, muscle and adipose tissue
adrenalin	various organs, e.g. heart, liver, skin
progesterone	uterus
oestrogen	ovaries, uterus, pituitary gland
testosterone	male reproductive organs

Remember: a target organ is where the hormone **acts**, not where it is produced.

Now try this

1 State what a **hormone** is. **(2 marks)**

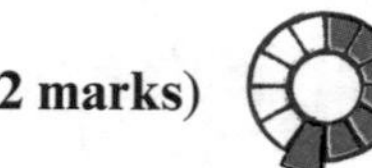

2 Give **two** differences between nervous and hormonal communication. **(2 marks)**

Adrenalin and thyroxine

Adrenalin and thyroxine are hormones produced in the human body. The production of thyroxine is an example of **negative feedback**.

Negative feedback

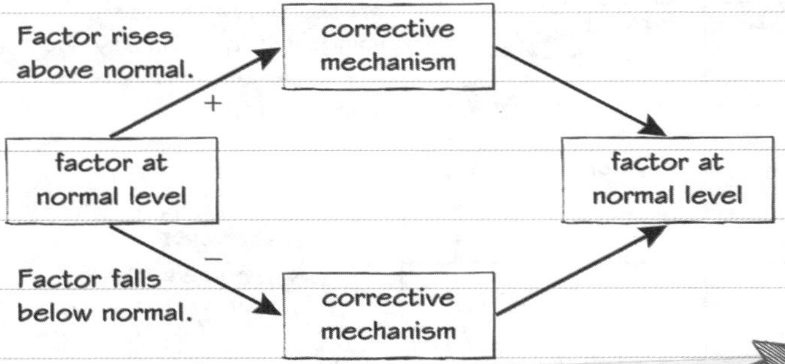

Make sure you can explain negative feedback by describing how a change from the normal level brings about changes to restore the normal level.

Worked example

Explain how the production of thyroxine is controlled by **negative feedback**. **(4 marks)**

When the concentration of thyroxine in the blood is too low, this stimulates a corrective mechanism. The hypothalamus secretes more TRH which causes the pituitary to produce more TSH. As a result, the thyroid produces more thyroxine. If the thyroxine concentration is too high, this inhibits TRH production by the hypothalamus, so less thyroxine is produced.

Adrenalin

Adrenalin is a hormone that is released from the adrenal glands in response to sudden stress. It brings about the **'fight or flight' response.**

Thyroxine

Thyroxine is a hormone that controls **metabolic rate.** This is the rate at which the cells respire. It is measured as the rate of energy transfers in the body.

Effects of adrenalin

Adrenalin has many target organs including the liver, the heart and blood vessels. It:

- increases heart rate
- constricts some blood vessels to make blood pressure higher
- dilates other blood vessels to increase blood flow to muscles
- causes the liver to convert glycogen to glucose, which is released into the blood.

Remember that explain means 'give a reason' – try to use 'because' in your answer.

Remember to name the target and the effect that adrenalin has for both cells or organs.

Worked example

Explain two ways in which adrenalin is useful for preparing skeletal muscles for fight or flight. **(4 marks)**

The heart beating faster means that oxygen is carried round the body faster; because this allows faster respiration in muscle cells, energy is released for cell contraction faster.

The blood glucose concentration also increases, making more glucose available for respiration in muscle cells.

This question could have been answered in other ways, such as increased blood flow to muscles to supply oxygen and nutrients for faster respiration.

Now try this

1 (a) Name **two** target cells or organs that adrenalin affects. **(1 mark)**

(b) Explain how adrenalin affects them. **(1 mark)**

2 Beta-blockers are drugs that reduce the effect of adrenalin on the heart. Explain why they are prescribed for people with heart disease. **(1 mark)**

This is testing whether you can apply what you know to a specific example.

3 People who produce too much thyroxine tend to eat a lot but generally do not get fat. They also feel very warm most of the time. Explain why. **(3 marks)**

The menstrual cycle

Between puberty and about the age of 50, women have a **menstrual cycle** that occurs about every 28 days. During the cycle, changes take place in the ovaries and the uterus.

If fertilisation occurs, the uterus lining remains thick so that the embryo can embed in the lining and obtain the nutrients it needs.

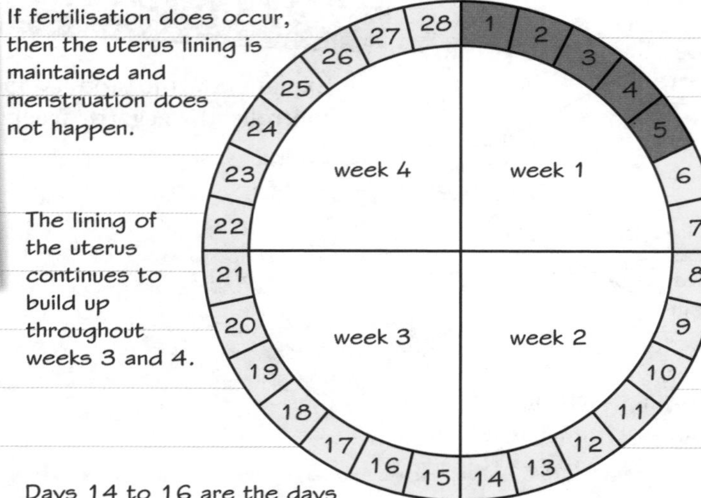

If fertilisation does occur, then the uterus lining is maintained and menstruation does not happen.

Menstruation is the breakdown of the uterus lining. It begins on day 1 of the cycle and usually lasts about 5 days.

The lining of the uterus continues to build up throughout weeks 3 and 4.

Oestrogen and **progesterone** are two of the hormones that control the menstrual cycle.

Days 14 to 16 are the days when **fertilisation** is most likely to take place.

Ovulation is the release of an egg from an ovary. This usually takes place around day 14.

During the second week, the lining of the **uterus** is gradually built up.

Contraception

Contraception is the prevention of fertilisation. **Hormonal contraception** includes hormone pills, implants or injections, and works by releasing hormones to prevent ovulation and thicken mucus at the cervix, preventing sperm from passing. **Barrier methods** include male and female condoms, the diaphragm, caps and sponges. These work by stopping the sperm from reaching the egg.

Worked example

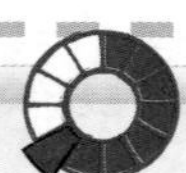

Give one advantage and one disadvantage of hormonal contraceptives compared with condoms. **(2 marks)**

Hormonal contraceptives are more effective than condoms at preventing pregnancy. However, condoms protect against sexually transmitted diseases (STDs) but hormonal contraceptives do not.

Different methods have different success rates at preventing pregnancy when used correctly:
condoms = 98%
diaphragm/cap = 92–96%
hormonal methods = >99%.

Now try this

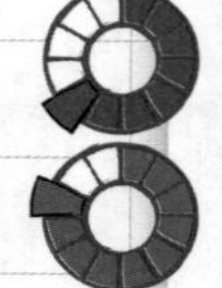

1 Explain why failure to menstruate may be the first sign of pregnancy. **(2 marks)**

2 The box opposite shows some information on the use of the contraceptive pill.
Use the information to evaluate the benefits and problems that may arise from using contraceptive pills to control fertility. **(3 marks)**

Around one third of UK women of reproductive age take a contraceptive pill. If used properly, the pill is 100% effective against pregnancy. Studies over 40 years show that the pill reduces the risk of many cancers. If the woman smokes heavily or is obese, taking the pill greatly increases the risk of thrombosis (blood clot). The increase in risk of thrombosis for women who do not smoke or are not overweight is very small, and is far less than the risk of thrombosis during pregnancy.

Control of the menstrual cycle

Four hormones control the menstrual cycle: **oestrogen**, **progesterone**, **FSH** and **LH**.

Changes during the menstrual cycle

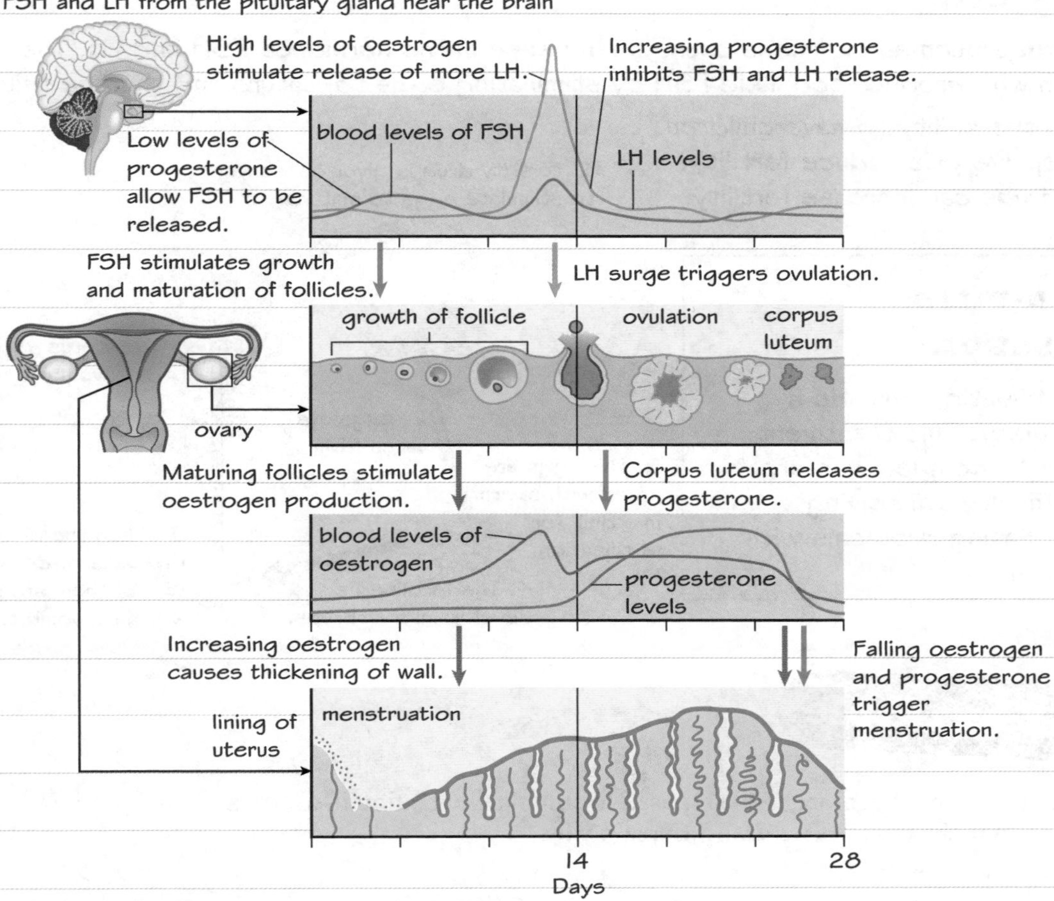
FSH and LH from the pituitary gland near the brain
High levels of oestrogen stimulate release of more LH.
Increasing progesterone inhibits FSH and LH release.
blood levels of FSH
LH levels
Low levels of progesterone allow FSH to be released.
FSH stimulates growth and maturation of follicles.
LH surge triggers ovulation.
growth of follicle
ovulation
corpus luteum
ovary
Maturing follicles stimulate oestrogen production.
Corpus luteum releases progesterone.
blood levels of oestrogen
progesterone levels
Increasing oestrogen causes thickening of wall.
Falling oestrogen and progesterone trigger menstruation.
lining of uterus
menstruation
14
28
Days

Worked example

The question asks you to explain how the hormones interact, so it is important to say how they affect each other.

Explain how FSH, LH, oestrogen and progesterone interact to control the menstrual cycle. **(6 marks)**

FSH is secreted from the pituitary gland. This causes a follicle in the ovary to mature. As it matures it secretes oestrogen which inhibits FSH and starts to thicken the lining of the uterus. A high concentration of oestrogen causes a surge in LH from the pituitary gland. This causes ovulation when the egg is released from the follicle. The ruptured follicle becomes a corpus luteum which secretes progesterone and some oestrogen. These cause the uterus lining to thicken even more. Progesterone inhibits FSH and LH. If the egg is not fertilised, the corpus luteum breaks down, and the progesterone concentration falls. This triggers menstruation. FSH is no longer inhibited, so it can be secreted from the pituitary gland again.

Now try this

1 State where the following hormones are made in the human body:
(a) oestrogen and progesterone
(b) FSH and LH. **(2 marks)**

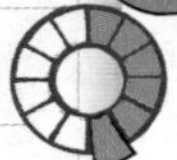

2 Describe the hormonal changes that cause menstruation. **(1 mark)**

3 Give the role of the following hormones in the menstrual cycle:
(a) oestrogen
(b) progesterone
(c) FSH
(d) LH. **(4 marks)**

Assisted Reproductive Therapy

Hormones are used in Assisted Reproductive Therapy (**ART**) including **IVF** treatment and **clomifene** therapy.

Fertility drugs

Fertility drugs such as clomifene cause an increase in the hormones FSH and LH. The drugs can help women who produce too little FSH by stimulating eggs to mature and then be released.

Fertility is the ability to have children.

- Contraceptive pills reduce fertility.
- Fertility drugs can increase fertility.

IVF (*in-vitro* fertilisation)

IVF is fertilisation outside a woman's body. This treatment is offered to couples who are having difficulty conceiving a child (i.e. having problems with fertilisation).

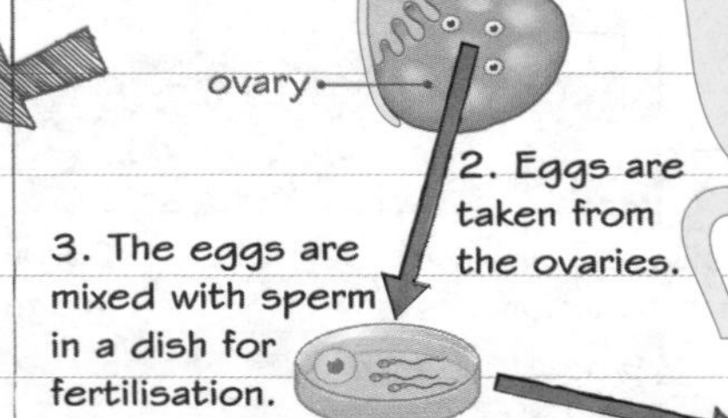

1. Fertility drug is given to woman to stimulate eggs to mature.
ovary
uterus (womb)
2. Eggs are taken from the ovaries.
3. The eggs are mixed with sperm in a dish for fertilisation.
4. The fertilised eggs develop into embryos.
5. When the embryos are tiny balls of cells, one or two of them are placed in the woman's womb to develop.

Worked example

(a) Follicle-stimulating hormone (FSH) is used to stimulate ovulation in a woman undergoing IVF treatment, even if she ovulates naturally. Explain why. **(2 marks)**

This is to stimulate the maturation of many eggs. Normally only one egg would mature in a normal cycle.

(b) The table shows the proportion of IVF treatments carried out in the UK in 2008 that resulted in a baby for mothers of different ages.

Age of mother	<35	36–37	38–39	>40
Proportion of successful treatments	33.1%	27.2%	19.3%	10.7%

Use this data to suggest the advantage of using FSH in IVF treatment. **(3 marks)**

The data shows that IVF treatment is not always successful, especially as women get older. Stimulating more eggs to mature means more embryos can be produced. This means that there are enough embryos for the woman to undergo several cycles of IVF if necessary.

Now try this

1 Describe how IVF can make it possible for a couple to have a baby when the woman does not normally release matured eggs from her ovaries. **(4 marks)**

2 Explain how clomifene can be used to treat fertility problems. **(3 marks)**

You need to outline all the steps in IVF treatment.

Blood glucose regulation

Homeostasis maintains some conditions inside the body at a more or less constant level, in response to internal and external change. Blood glucose regulation is an example of homeostasis. It is controlled by two hormones: **insulin** and **glucagon**.

Pancreatic control

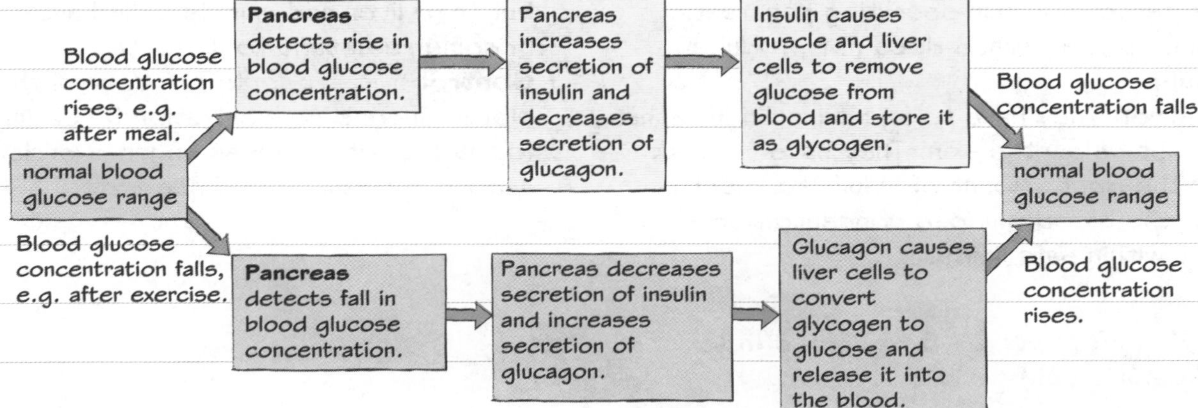

Worked example

The graph shows the changes in blood glucose concentration of a person who has just had a meal.

Blood glucose conc. (mg/dl): 0, 80, 100, 120, 140

Hours: 1, 2, 3, 4, 5

A, B

Describe what is happening. **(3 marks)**

Remember to refer to the graph when you answer a question like this.

At A the person has just had a meal, so the carbohydrates are being digested and absorbed into the blood as glucose. This causes the blood glucose concentration to increase.

At B insulin has been secreted from the pancreas which converts glucose to glycogen. Therefore, the blood glucose concentration is falling.

From three to five hours the blood glucose level slowly increases as glucagon converts glycogen back into glucose.

Now try this

1 Name the endocrine gland that secretes glucagon, and the target organ that the hormone affects. **(2 marks)**

It may help to look back at page 58.

2 The control of blood glucose concentration is an example of negative feedback. Explain why. **(2 marks)**

3 Insulin converts glucose to

☐ **A** glycogen
☐ **B** glucagon
☐ **C** galactose
☐ **D** starch **(1 mark)**

Diabetes

A person who cannot control their blood glucose concentration properly has a condition known as **diabetes**. There are two main types of diabetes.

Type 1 diabetes

- **Cause**: The immune system has damaged the person's insulin-secreting pancreatic cells, so the person does not produce **insulin**.
- **Control**: They have to inject insulin into the fat just below the skin. They have to work out the right amount of insulin to inject so that the blood glucose concentration is kept within safe limits.

Type 2 diabetes

- **Cause**: The person does produce insulin but their liver and muscle cells have become **resistant** to it.
- **Control**: Most people can control their blood glucose concentration by eating foods that contain less sugar, exercising and using medication if needed.

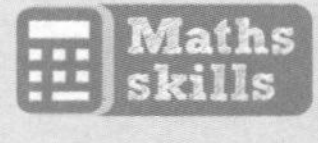

Maths skills A person's **Body Mass Index (BMI)** is calculated using the equation:

$$\text{BMI} = \frac{\text{weight (kg)}}{(\text{height (m)})^2}$$

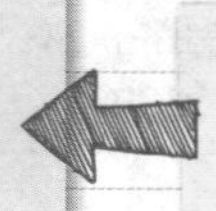

You may be asked to evaluate data on the correlation between type 2 diabetes and BMI or waist:hip ratio.

There are other factors that are linked to diabetes, apart from obesity. These include ethnic group, type of diet eaten and activity levels. The effect of exercise and diet is covered on page 63.

Worked example

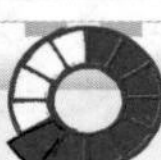

Use the graph to evaluate the correlation between obesity and type 2 diabetes. **(3 marks)**

'Evaluate' here means that you have to say how strong the link is between these factors, giving reasons.

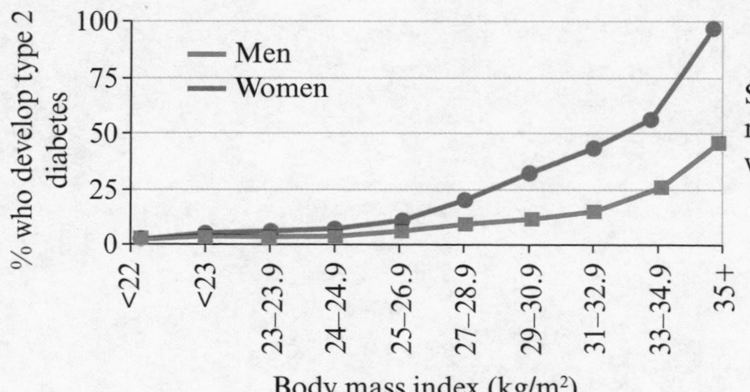

Sample sizes
men: 51 529
women: 114 281

The risk of developing type 2 diabetes increases as BMI increases. It increases faster for women than for men. The large sample sizes and the smooth curves suggest that this is a strong correlation.

Now try this

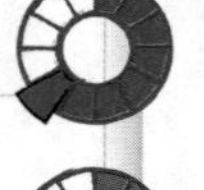

1 Explain why the dose of insulin needed by a person with type 1 diabetes will vary, depending on the food eaten and exercise. **(3 marks)**

2 Many health professionals advise that weight control is needed to prevent a huge increase in cases of diabetes over the next decade or two. Evaluate this advice. **(4 marks)**

You need to give the pros and cons and then come up with your own conclusion. It may help to remember there are 2 types of diabetes with different causes.

Extended response – Control and coordination

There will be at least one 6-mark question on your exam paper. For these questions, you will need to think scientifically and structure your answer logically, showing how the points you make are related to each other. You can revise the topics for this question, which is about the **link between BMI and type 2 diabetes**, on page 64.

Worked example

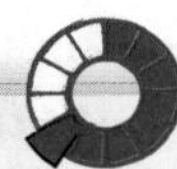

The chart shows the results of a survey of adults in the UK who have type 2 diabetes. The results are grouped by BMI category for men and for women.

Use the chart to explain how controlling weight could affect the occurrence of type 2 diabetes.

(6 marks)

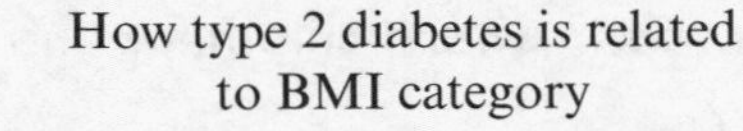

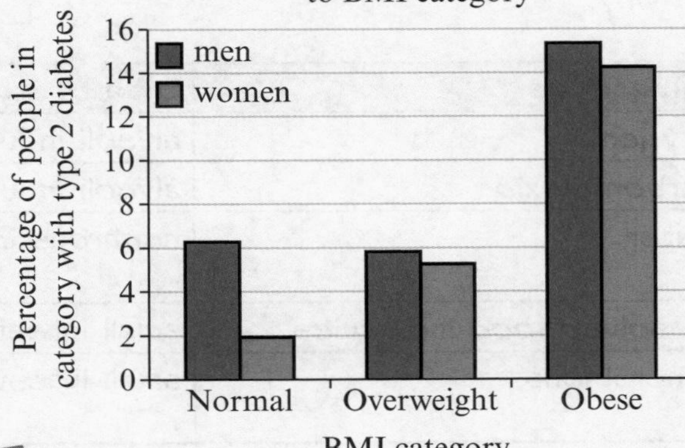

BMI is a measure of body mass (weight) and is calculated by dividing mass by height squared. As the amount of body fat increases, BMI increases. People with a BMI over 30 are said to be obese.

> A good start to answering this question is to show that you understand what terms in the question (BMI and type 2 diabetes) mean.

People with type 2 diabetes are not able to control their blood glucose concentration properly because their pancreatic cells do not produce enough insulin, or because their muscle and liver cells do not respond properly to insulin.

The chart shows that the risk of developing type 2 diabetes increases as BMI increases, both for men and for women. The risk for women increases more rapidly than for men when their BMI reaches the overweight category.

> Before directly answering the question, you should explain what the graph shows. As well as describing trends shown by the bars on the graph, make sure you identify differences. Using values from the graph can show how well you understand the graph.

If people control their weight so they do not develop a high BMI, then they should keep their risk of developing type 2 diabetes low. This beneficial effect will be greater for women than for men.

> As there are two categories on the graph (men and women), you need to explain how they respond differently to controlling weight.

The chart shows men and women in the UK only. So this conclusion may not be the same for people in other countries, or different ethnic groups, or for other variables such as exercise or diet.

> This point evaluates the link, showing why the the conclusion may not apply in different studies.

Now try this

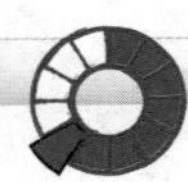

Explain how blood glucose concentration is regulated in a healthy person. **(6 marks)**

Had a look ☐ Nearly there ☐ Nailed it! ☐

Exchanging materials

Substances such as oxygen, carbon dioxide, water, dissolved food molecules, mineral ions and urea need to be **transported** into and out of organisms.

How substances are exchanged

The table below shows different substances that need to be transported, where this happens and why.

More complex organisms

As organisms get **bigger**, their surface area:volume ratio gets **smaller**. This means they cannot rely on diffusion. They need to have specialised **exchange surfaces** and **transport systems**.

Substance	Site of exchange	Reason for exchange
oxygen	alveoli in lungs	needed for respiration
carbon dioxide	alveoli in lungs	waste product of metabolism
water	nephrons in kidney	needed for cells to function properly
dissolved food molecules	small intestine	needed for respiration
mineral ions	small intestine	needed for cells to function properly
urea	nephrons in kidney	waste product of metabolism

Worked example

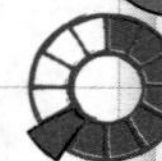

(a) Complete the table to calculate some measurements relating to cubes of different side lengths. **(3 marks)**

Length of one side of cube (cm)	Surface area of cube (cm^2)	Volume of cube (cm^3)	Surface area : volume ratio
1	6	1	6
2	24	8	3
3	54	27	2
4	96	64	1.5

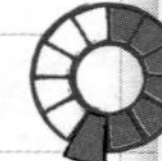

(b) Describe the pattern shown by this data. **(1 mark)**

As the length of one side of the cube increases, the surface area:volume ratio falls.

Maths skills Do not just talk about the surface area – it is the surface area:volume ratio that is important.

Adaptations for exchange

Special organs are adapted to make exchange efficient. For example, the **lungs** are adapted to exchange gases, and the **small intestine** is adapted to exchange solutes.

Now try this

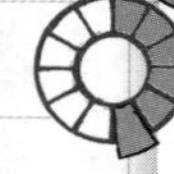

1 Name **one** human organ that is specially adapted for exchanging substances with the environment. **(1 mark)**

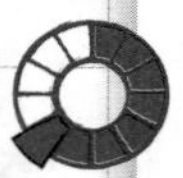

2 The flatworm shown in this diagram is multicellular but it does not have an exchange system or a transport system. Explain why. **(3 marks)**

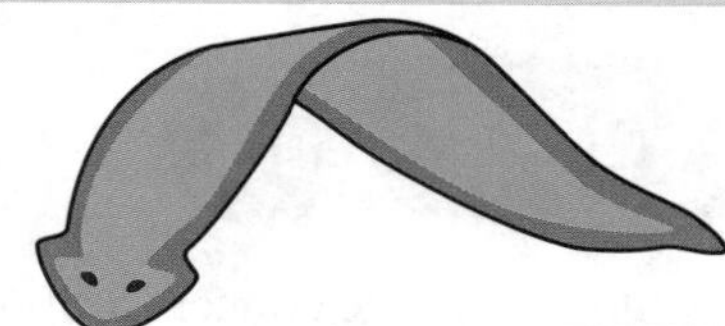

Look at the shape of the flatworm and think about its surface area : volume ratio.

Alveoli

Alveoli are adapted for **gas exchange** by diffusion between air in the lungs and blood in capillaries.

The lungs

The **lungs** are part of the breathing system. The breathing system takes air into and out of the body. In the lungs:

- oxygen diffuses from the air into the blood
- carbon dioxide diffuses from the blood into the air.

(There is more about diffusion on page 10.)

The lungs are **adapted** for efficient gas exchange.

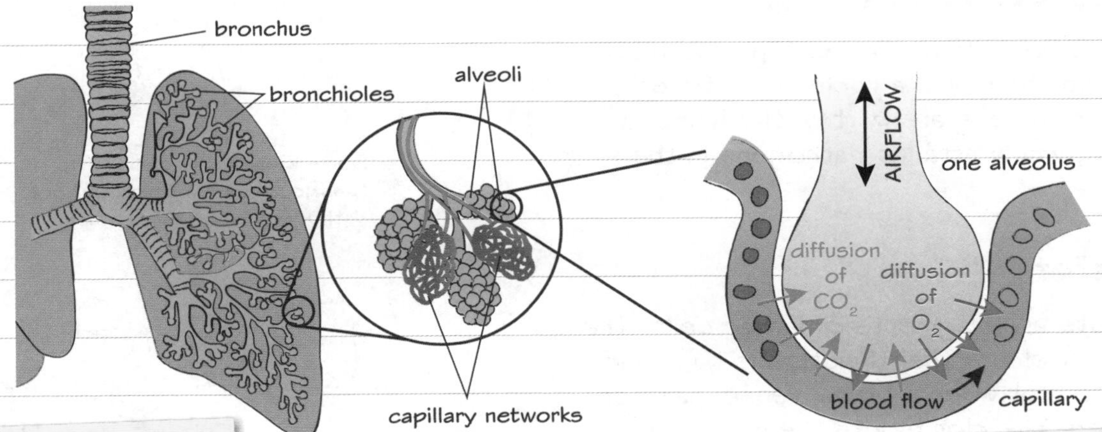

Millions of tiny **alveoli** (air sacs) create a large surface area for **diffusion** of gases.

Each alveolus is closely associated with a **capillary**. Their walls are one cell thick, minimising diffusion distance.

Ventilation of alveolus (breathing) and continual blood flow through capillaries maintains high concentration gradients, to maximise rate of diffusion.

Worked example

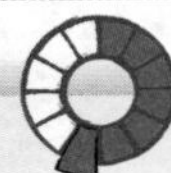

In some lung infections, fluid builds up inside the alveoli. Explain why this makes it harder for the person to breathe properly. **(3 marks)**

This reduces the surface area of the alveoli over which diffusion can take place and so decreases the rate of gas exchange.

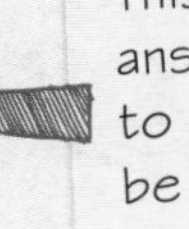

This question says 'Explain...', so your answer needs to specifically refer to how the function of the lung will be disrupted. You should try to use correct scientific language throughout.

Now try this

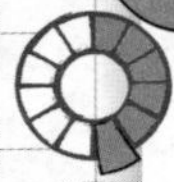

1 State **one** way in which the air breathed out is different from the air breathed in. **(1 mark)**

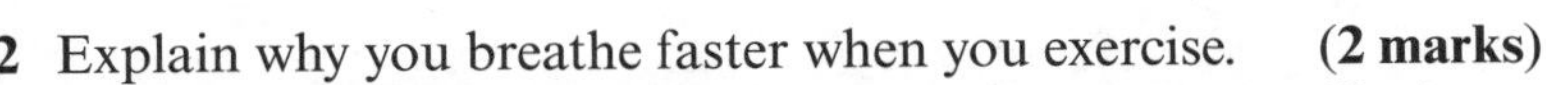

2 Explain why you breathe faster when you exercise. **(2 marks)**

3 Explain the advantage of the alveoli having many capillaries around them. **(2 marks)**

You need to consider both what the body needs to take in and what it needs to lose.

Blood

Blood is made of four main parts: **plasma**, **red blood cells**, **white blood cells** and **platelets**. Each part of the blood has a particular **function** (job).

Blood plasma

Plasma is the liquid part of blood:

- It carries the blood cells through the blood vessels.
- It contains many dissolved substances, such as carbon dioxide and glucose.

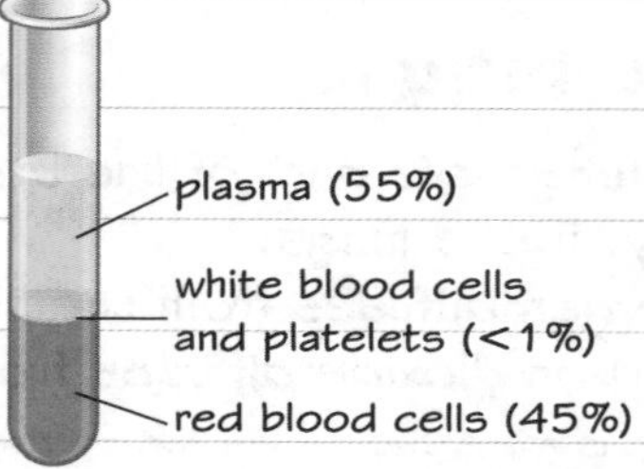

White blood cells

White blood cells are larger than red blood cells, and they have a nucleus. All types of white blood cells are part of the **immune system**, which attacks pathogens in the body.

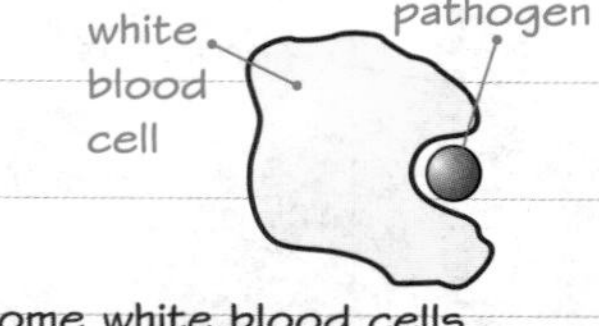

Some white blood cells (**phagocytes**) surround (**ingest**) pathogens and destroy them.

Platelets

Platelets are fragments of larger cells. They have no nucleus. Their function is to cause blood to clot when a blood vessel has been damaged. The clot blocks the wound and prevents pathogens getting into the blood.

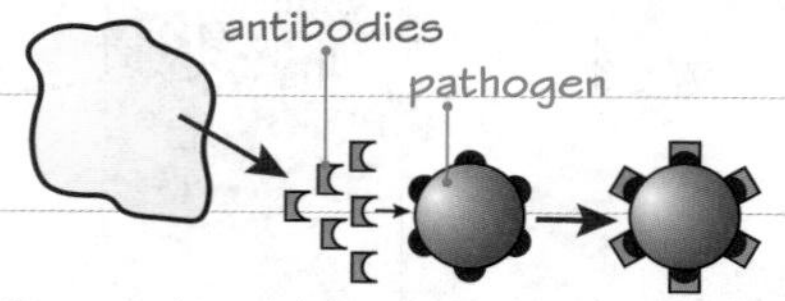

Some white blood cells (**lymphocytes**) produce chemical **antibodies** that attach to pathogens and destroy them.

Worked example

Explain how the structure of a red blood cell (erythrocyte) is related to its function. **(4 marks)**

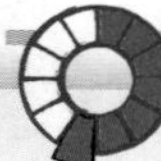

Red blood cells contain haemoglobin which carries oxygen. The biconcave shape of a red blood cell means it has a large surface area. This means it is easier for oxygen to diffuse into and out of the cell. The cell has no nucleus. This means the cell has room for more haemoglobin to carry more oxygen.

Biconcave means the cell is dimpled on both sides so that it is thinner in the middle than at the edges.

Now try this

1 Respiring cells need oxygen and glucose. State which parts of the blood carry each of these substances. **(2 marks)**

2 Describe **two** ways in which white blood cells help to protect the body against disease. **(2 marks)**

3 Explain how platelets help to protect the body against infection. **(3 marks)**

Blood vessels

There are three types of **blood vessels** – **arteries**, **veins** and **capillaries**. Their structure is related to their function.

Veins

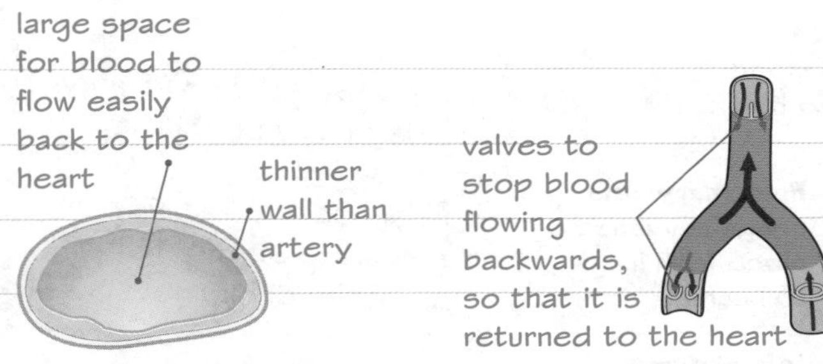

vein (cross-section)

vein (long section)

Arteries

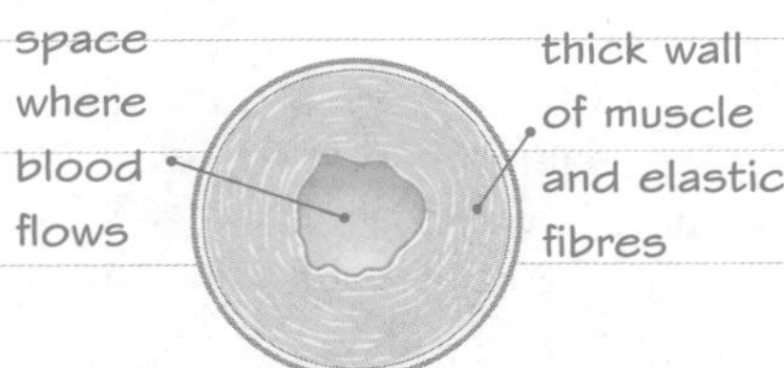

artery (cross-section)

Exchange in capillaries

Substances are exchanged between the body cells and blood in **capillaries**. Capillaries are **adapted** for their function of exchanging substances between the blood and body cells.

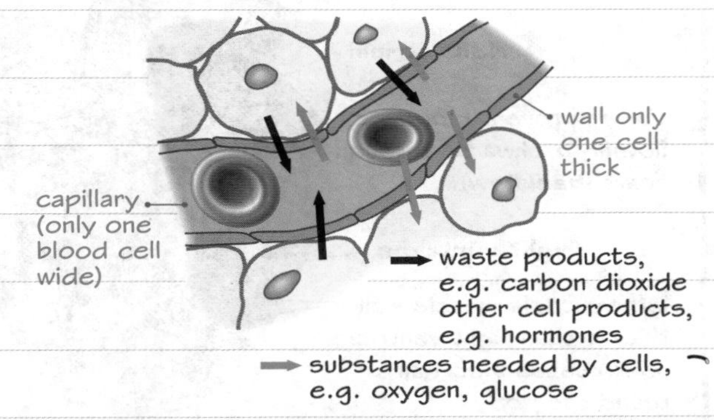

Worked example

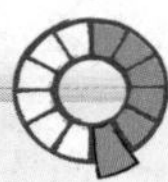

Describe the role of arteries, veins and capillaries in the human circulatory system. **(3 marks)**

Arteries carry blood away from the heart – all arteries except the pulmonary arteries carry oxygenated blood to the body. Veins carry blood towards the heart – all veins except the pulmonary veins carry deoxygenated blood from the body back towards the heart. Capillaries exchange materials, such as oxygen, glucose and carbon dioxide, with body tissues.

Remember: a key function of blood is to deliver oxygen and glucose to cells for respiration and to remove the carbon dioxide produced by respiration from cells.

Now try this

Think carefully about the key function of blood, and what body cells need to survive.

1 (a) Which is the narrowest blood vessel?

☐ **A** artery ☐ **B** capillary ☐ **C** heart ☐ **D** vein **(1 mark)**

(b) Which type of blood vessel has the thickest wall?

☐ **A** artery ☐ **B** capillary ☐ **C** vein ☐ **D** all the same **(1 mark)**

2 Explain why almost every body cell is very close to a capillary. **(4 marks)**

Had a look ☐ Nearly there ☐ Nailed it! ☐

The heart

The **heart** pumps blood round the body. The heart and the blood vessels together make up the **circulatory system**, which delivers oxygen and nutrients such as glucose to all parts of the body.

Structure and function of the heart

Pulmonary artery carries deoxygenated blood from heart to lungs.

Aorta carries oxygenated blood from heart to body.

Vena cava brings **deoxygenated blood** from body to heart.

Pulmonary vein brings oxygenated blood from lungs to heart.

right atrium

left atrium

Valves prevent blood flowing backwards through heart (**backflow**).

left ventricle

right ventricle

left ventricle muscle wall thicker than right ventricle as it pushes blood all round the body

deoxygenated blood

oxygenated blood

The sides of the heart are labelled left and right as if you were looking at the person. So the left side of the heart is on the right side of the diagram.

Remember - arteries take blood away from the heart, veins bring it back in to the heart.

Blood circulation

Valves in the heart make the blood flow in the right direction.

Blood enters the atria. → The atria contract, forcing blood into the ventricles. → The ventricles contract, forcing blood into the arteries. → Blood flows through arteries to the organs and returns to the heart through veins.

There are two circulation systems: one through the lungs and one through all the other organs.

Worked example

Explain how the structure of the heart adapts it to circulate blood effectively to the lungs and body. **(4 marks)**

Blood returning to the heart from the body is low in oxygen. This enters the right side of the heart and is pumped to the lungs by the right ventricle. It becomes saturated with oxygen in the lungs. The blood then enters the left side of the heart so it does not mix with the blood that is low in oxygen on the other side of the heart.

The left ventricle has a very thick muscular wall to pump the blood all the way round the body. There is a double circulation because blood passes through the heart twice to make a complete circulation round the body, making sure the pressure of the blood never falls too low.

Now try this

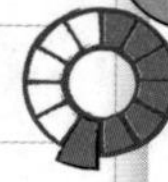

1 Give **one** reason why there are valves in the heart. **(1 mark)**

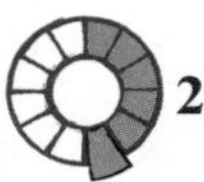

2 The muscular wall of the left ventricle of the heart is much thicker than that of the right ventricle. Give a reason for this difference. **(3 marks)**

Aerobic respiration

Cellular respiration is a process that releases energy from glucose for use in cellular activities. The main type of cellular respiration is **aerobic respiration**, which uses oxygen.

Aerobic respiration

Aerobic respiration is a series of chemical reactions that take place mostly inside mitochondria in the cell. This is an **exothermic** process because it releases energy. This happens continuously in living cells. It is the main source of energy for cells.

Make sure you use the phrase 'releases energy' when describing the function of respiration.

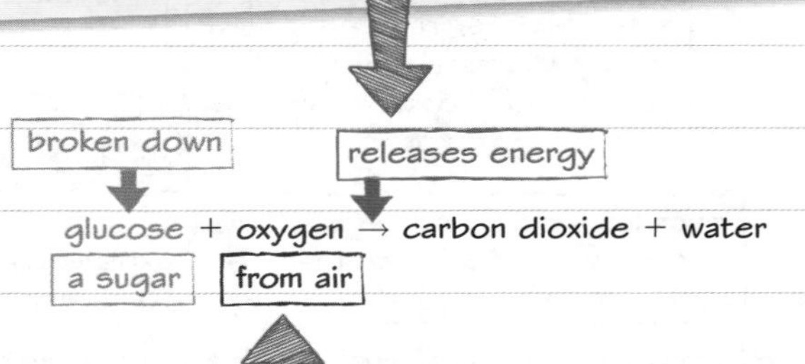

The need for air makes this respiration 'aerobic'.

Use of energy from respiration

In animals:

- for metabolic processes to build larger molecules from smaller ones, e.g. proteins from amino acids, large carbohydrates (e.g. starch, glycogen) from small sugars (e.g. glucose), fats from fatty acids and glycerol
- to enable muscle contraction
- in birds and mammals, also to maintain steady body temperature in colder surroundings.

In plants:

- to build larger molecules from smaller ones, e.g. sugars, nitrates and other nutrients into amino acids, which are then used to make proteins.

Worked example

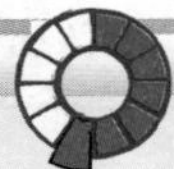

The graph shows the concentration of oxygen and carbon dioxide dissolved in the water of an aquarium that contains only pond weed. Explain the shape of the carbon dioxide curve. **(2 marks)**

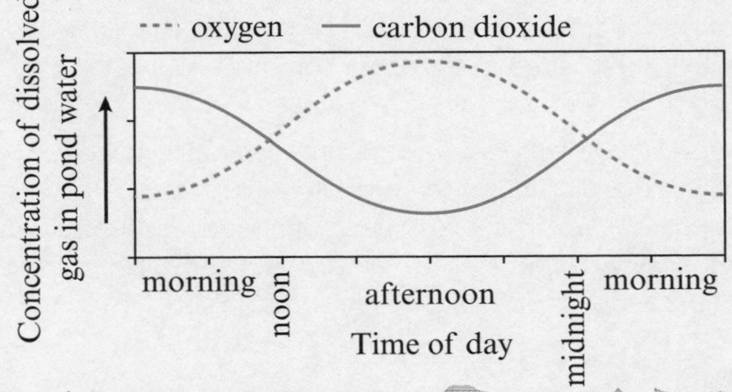

Carbon dioxide concentration increases during the evening and night because it is produced by respiration. Plants need to respire all the time. Carbon dioxide concentration falls during the hours of daylight when carbon dioxide is used in photosynthesis.

This question requires you to use knowledge about photosynthesis and respiration.

Now try this

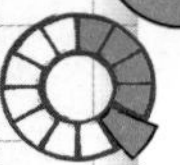

1 Define the term **aerobic respiration**. **(2 marks)**

2 The reactions of respiration are controlled by enzymes. Explain why body temperature and cell pH are carefully controlled in humans. **(2 marks)**

Anaerobic respiration

During exercise, energy needed for muscle contraction comes from **aerobic respiration**. If there is not enough oxygen, or aerobic respiration is not possible, **anaerobic respiration** takes place.

How anaerobic respiration works

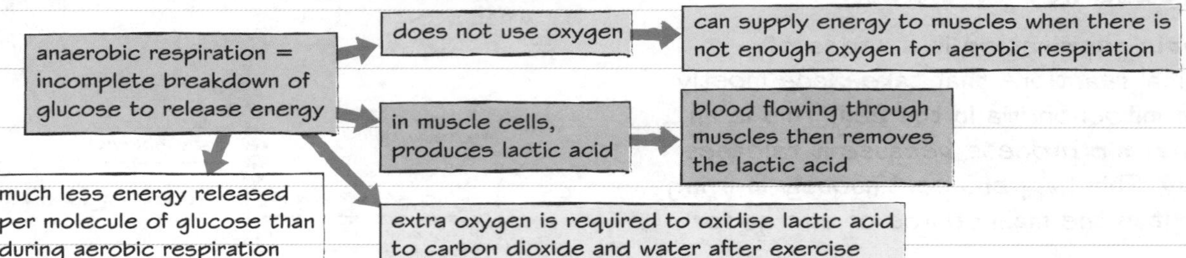
anaerobic respiration = incomplete breakdown of glucose to release energy
does not use oxygen
can supply energy to muscles when there is not enough oxygen for aerobic respiration
in muscle cells, produces lactic acid
blood flowing through muscles then removes the lactic acid
much less energy released per molecule of glucose than during aerobic respiration
extra oxygen is required to oxidise lactic acid to carbon dioxide and water after exercise

In **plants and fungal cells**, there is a different form of anaerobic respiration that produces ethanol. However, like anaerobic respiration in animals:

- it involves breakdown of glucose
- no oxygen is used
- less energy is released per glucose molecule.

Advantages

- Anaerobic respiration is useful for muscle cells because it can release energy to allow muscles to contract when the heart and lungs cannot deliver oxygen and glucose fast enough for aerobic respiration.
- Respiration can continue in organisms that have no, or very limited, oxygen supply.

Disadvantages

- Anaerobic respiration releases much less energy from each molecule of glucose than aerobic respiration.
- Lactic acid is not removed from the body. It builds up in muscle and blood, and must be broken down after exercise.

Worked example

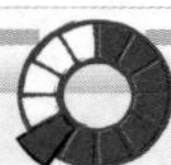

The graph shows how the concentration of lactic acid in the blood varies with level of exercise.

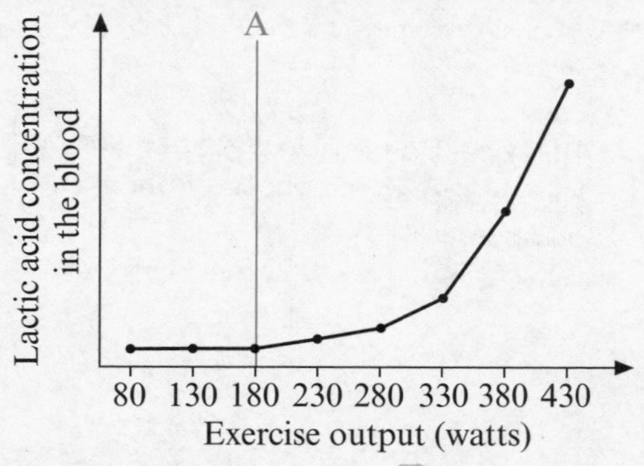
A
Lactic acid concentration in the blood
80 130 180 230 280 330 380 430
Exercise output (watts)

Explain what the graph shows. **(2 marks)**

As the exercise level increases up to point A, blood lactic acid concentration stays the same. This is because the energy for the exercise is coming from aerobic respiration.

As the exercise level increases beyond point A, blood lactic acid concentration increases. This is because aerobic respiration cannot supply enough energy and the rate of anaerobic respiration increases, producing more lactic acid.

Now try this

1 Explain why human muscle cells cannot function on anaerobic respiration alone. **(2 marks)**

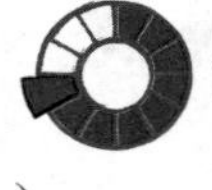

2 After vigorous exercise has ended, an athlete will continue to breathe deeply for several minutes longer. Explain why. **(2 marks)**

Rate of respiration

You need to understand how **the rate of respiration in living organisms can be investigated.** This experiment can use small invertebrates (e.g. maggots, woodlice) or germinating peas.

See pages 71 and 72 for more information on respiration.

Worked example

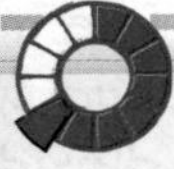

The diagram below shows a **respirometer**, which is used for investigating rates of respiration.

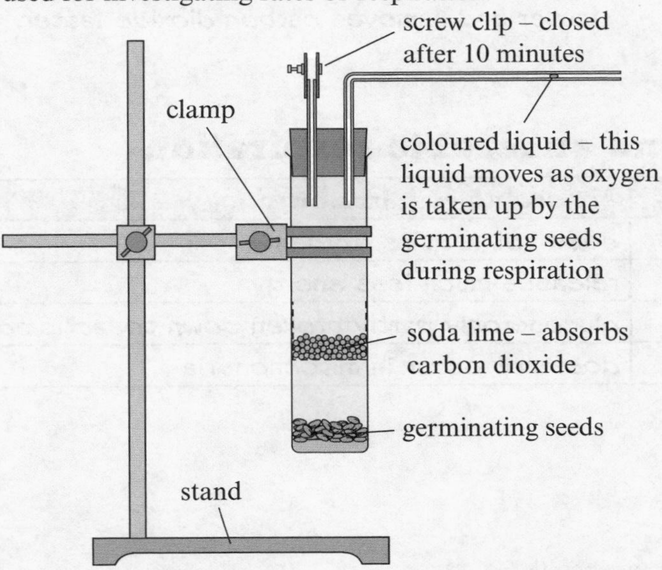

(a) Explain the function of the soda lime in the respirometer. **(2 marks)**

Soda lime absorbs carbon dioxide produced by the seeds during respiration, so that it does not affect the movement of the liquid blob.

(b) Explain why the liquid blob moves during the experiment. **(2 marks)**

Oxygen is absorbed from the surrounding air by the germinating seeds. This reduces the volume of air in the container, so the blob will move towards the container.

(c) Explain how this respirometer could be used to measure rate of respiration of the seeds. **(2 marks)**

The distance that the liquid blob moves over a particular time could be measured, and a rate calculated by dividing distance by time.

Safety and living organisms

You must consider the safe and ethical use of working with living organisms. You should minimise harm to small invertebrates, and return them to their natural environment as soon as possible. You should also wash your hands thoroughly after working with living organisms.

Factors affecting respiration

Remember that respiration is affected by environmental factors, particularly temperature, and that microorganisms also respire. So the investigation will need controls.

Control for temperature – e.g. by keeping respirometers in a water bath during the experiment.

Control for something other than the study organisms affecting the gas volume – e.g. by using the same volume of an inert substance as the organisms, such as glass beads instead of seeds.

$$\text{rate} = \frac{\text{change}}{\text{time}}$$

In this case the change measured could be the amount of oxygen absorbed. In other investigations, you might measure the amount of carbon dioxide produced.

Now try this

The table shows the results of an investigation in which respirometers containing woodlice were kept at different temperatures. The uptake of oxygen in each respirometer was measured at the end of 20 min.

(a) Calculate the rate of oxygen uptake per min for each respirometer. **(2 marks)**

(b) Draw a graph to show rate of oxygen uptake against temperature. **(4 marks)**

(c) Describe the relationship shown by this experiment. **(2 marks)**

Temperature (°C)	Volume of gas absorbed in 20 min (cm^3)
5	0.6
10	1.0
15	1.7
20	2.4

Changes in heart rate

On this page you will revise the formula **cardiac output = stroke volume × heart rate.**

Effects of exercise

When you exercise, your **heart rate** increases. The harder you exercise, the more your heart rate increases. Heart rate can be measured by taking your **pulse** at the wrist. It is usually measured as number of **beats per minute**.

During exercise, muscle cells are respiring faster. This means that they need more oxygen and glucose, and release more carbon dioxide. A faster heart rate means that blood is pumped faster around the body. The blood takes oxygen and glucose to cells faster and removes carbon dioxide faster.

Differences between aerobic and anaerobic respiration

Aerobic respiration	Anaerobic respiration
requires oxygen	does not require oxygen
releases a lot of energy	releases much less energy
glucose broken down completely	glucose only partly broken down to lactic acid
occurs mostly in mitochondria	does not occur in mitochondria

Worked example

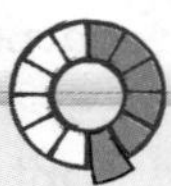

The table shows the stroke volume and heart rate for two people at rest. **(2 marks)**

	Stroke volume (cm^3)	Heart rate (beats per minute)
trained athlete	90	55
untrained person	60	70

Calculate the cardiac output for these two people.

athlete: $90 \times 55 = 4950\,cm^3$ per minute

untrained person: $60 \times 70 = 4200\,cm^3$ per minute

Maths skills cardiac output = stroke volume × heart rate

Stroke volume is the volume of blood pumped by the heart in one beat.

Now try this

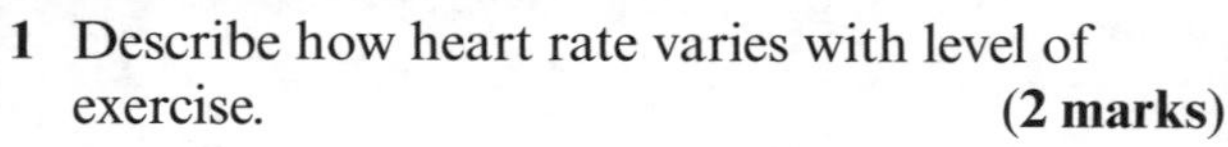

1 Describe how heart rate varies with level of exercise. **(2 marks)**

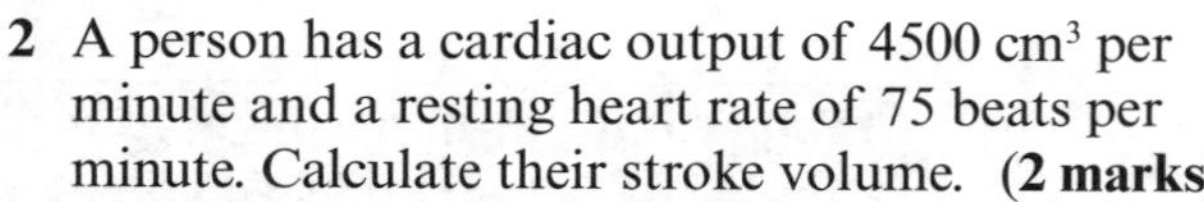

2 A person has a cardiac output of 4500 cm^3 per minute and a resting heart rate of 75 beats per minute. Calculate their stroke volume. **(2 marks)**

3 Suggest why a trained athlete may have a similar cardiac output to an untrained person but a lower resting heart rate. **(2 marks)**

Maths skills Rates are always how often something happens in a particular time period. So heart rate is how many times the heart beats in a minute.

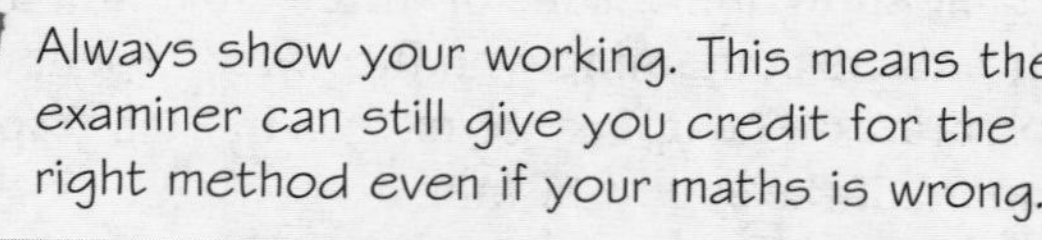

Always show your working. This means the examiner can still give you credit for the right method even if your maths is wrong.

Extended response – Exchange

There will be at least one 6-mark question on your exam paper. For these questions, you will need to think scientifically and structure your answer logically, showing how the points you make are related to each other. You can revise the topics for this question, which is about **aerobic respiration** and **anaerobic respiration**, on pages 71 and 72.

Worked example

Athletes who run a marathon race usually keep their rate of running within their 'aerobic zone' for most of the race, but sprint anaerobically as they get close to the end of the race. Compare and contrast the use of aerobic and anaerobic respiration in marathon racing. **(6 marks)**

Aerobic respiration and anaerobic respiration both release energy from the breakdown of glucose. This energy can be transferred to muscle cells so that they can contract. Aerobic respiration requires oxygen from the air and produces carbon dioxide and water. Anaerobic respiration in muscle cells needs no oxygen and produces lactic acid.

In a long race, staying aerobic is important because more energy is released from each glucose molecule than in anaerobic respiration. This means the athlete has more energy for contracting muscle cells.

At the end of the race, aerobic respiration will not be able to supply extra energy for the sprint because there is a limit to how fast the heart and lungs can deliver oxygen to the muscle cells. So anaerobic respiration provides the extra energy needed for the sprint.

Command words: Compare and contrast

The answer to a **compare and contrast** question should identify similarities and differences between all the examples in the question. No conclusion needs to be drawn from the comparison.

This answer is well organised because it starts by identifying how aerobic and anaerobic respiration are similar, and then describes how they are different.

The question describes two stages of running, so the answer needs to describe what happens in both stages.

Remember that it is not a case of **either** aerobic **or** anaerobic respiration in human muscle cells. Aerobic respiration continues as fast as it can with the oxygen available, while anaerobic respiration provides the extra energy needed.

Now try this

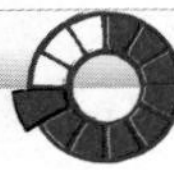

Explain how the lungs are adapted for their role in gas exchange. **(6 marks)**

Ecosystems and abiotic factors

Ecosystems are organised into different levels. At the level of **communities**, both **abiotic** (non-living) and **biotic** (living) factors can have an effect. You will look at abiotic factors on this page, and biotic factors on page 77.

Levels of organisation in an ecosystem

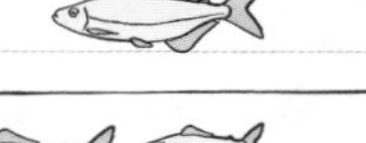
organism

Organism is a single living individual.

population

Population is all the organisms of the same species in an area.

community

Community is all the populations in an area.

ecosystem

Ecosystem is all the living organisms (the community) and the non-living components in an area.

All organisms in an ecosystem are **dependent** on other organisms for food, shelter and so on. Remember to think about the impact of this in questions about ecosystems.

Factors that affect distribution

Factors in the environment affect living organisms and their **distribution** (how widely spread they are). Changes in these factors may change their distribution.

environmental factors

living factors, e.g.
- prey
- competitor
- predator

non-living factors, e.g.
- light
- average temperature
- average rainfall
- oxygen levels in water
- pollution

A change in average temperature or rainfall may change the distribution of organisms in an area.

Oxygen levels are high in unpolluted water and low in polluted water.

A **pollutant** is energy or a chemical substance that has a harmful effect on living organisms.

Changes in non-living factors can be measured using equipment, e.g. oxygen meter, thermometer, rainfall gauge. For example, temperature affects the rates of reactions and light intensity affects the rate of photosynthesis.

Worked example

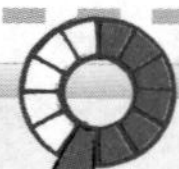

Nettle plants grow taller when they are under a large tree compared with those growing in open ground. Suggest a reason for this difference. **(2 marks)**

Under the tree, the light intensity is lower. Therefore, the plant grows taller so it has more leaves to trap available light.

There are other abiotic factors that you could suggest instead.

Now try this

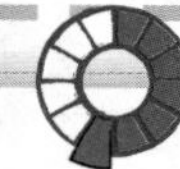

The pygmy shrew is Britain's smallest mammal. It weighs only 2.3–5 g and its length (head and body) is only 40–55 mm. It can only live in southern Britain because the rest of the country is too cold. Explain why the pygmy shrew cannot survive in cold places. **(2 marks)**

This question is asking you to apply principles you have learned (for example, about surface area and volume ratios) to an unfamiliar example.

Biotic factors

You have seen how abiotic factors can affect communities. So can **biotic factors**. Biotic factors are factors that involve living organisms, e.g. **competition** and **predation**.

Deterring predators

Many organisms are the food of other organisms. Some animals and plants have special features that deter **predators**.

Some animals advertise that they are poisonous with very bright colours.

Some animals use colours to make them look more frightening, like these big 'eyes'.

Some plants have big thorns.

Other plants are poisonous.

Competition

Organisms need a supply of materials from their surroundings, and sometimes from other living organisms, so that they can survive and reproduce. This means there is **competition** between organisms for materials that are in limited supply.

... between plants

... between animals

Animals may compete with each other for:

- food
- mates for reproduction
- **territory** (space for feeding, reproduction and rearing young).

Worked example

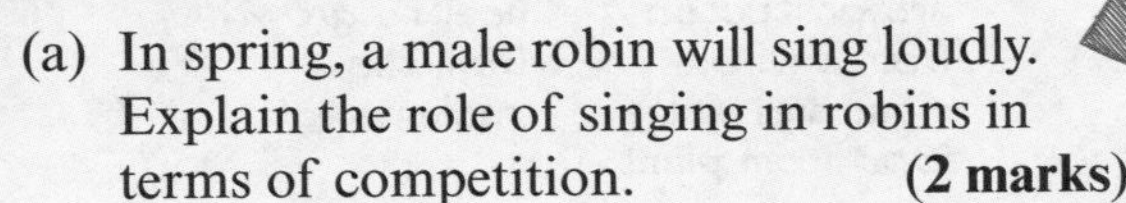

You will be expected to know the factors that organisms are competing for in a real-life situation

(a) In spring, a male robin will sing loudly. Explain the role of singing in robins in terms of competition. **(2 marks)**

A male robin competes with other male robins for mates (females) and for territory. Singing loudly warns other males to keep out of the robin's territory and attracts females who choose to mate with him.

(b) The ground underneath the trees in a wood is mostly bare. Explain why very few plants can grow here. **(2 marks)**

There is very little light underneath the trees in a wood. Plants are unable to compete with the trees for the sunlight.

Now try this

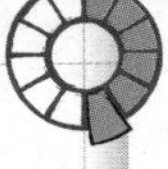

1 A farmer plants the seeds of his crop plants so that they are well separated from each other. Explain why. **(2 marks)**

2 Milk snakes are non-poisonous snakes. Coral snakes are highly poisonous. Milk snakes that live in the same area as coral snakes have a bold red and black pattern that is similar to that of the coral snakes. Suggest the advantage to the milk snake of this patterning. **(2 marks)**

Remember to use technical vocabulary in your answer, for example, 'competition'.

Although you may be given an unfamiliar context like this you should be able to apply principles you already know to answer the question.

Parasitism and mutualism

Parasitism and **mutualism** are examples of **interdependence** where the survival of one species is closely linked with another species.

Parasitism

A **parasite** feeds on another organism (the **host**) while they are living together. This harms the host but benefits the parasite.

Parasite	Host	Description
flea (animal)	other animals, including humans	Fleas feed by sucking the animal's blood after piercing its skin.
head louse (animal)	humans	Head lice feed by sucking blood after piercing the skin on the head.
tapeworm (animal)	other animals, including humans	Tapeworms live in the animal's intestine and absorb nutrients from the digested food in the intestine.
mistletoe (plant)	trees, e.g. apple	Mistletoe grows roots into the tree to absorb water and nutrients from the host.

Mutualism

When two organisms live closely together in a way that benefits them both, they are called **mutualists**.

Remember to explain how **each** organism benefits in a mutualistic relationship.

Oxpecker bird benefits by getting food.

Herbivore benefits from loss of skin parasites.

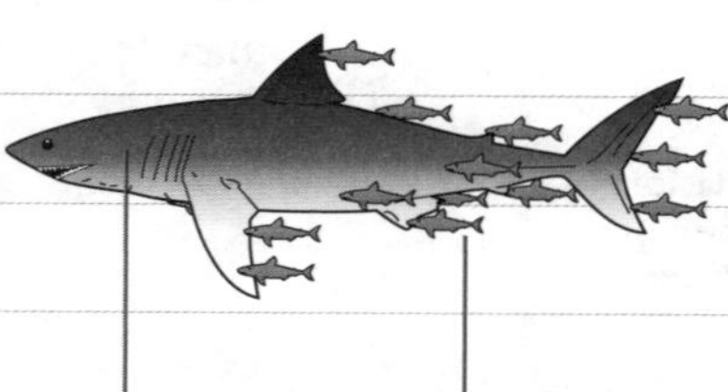

The larger fish benefits from loss of dead skin and parasites.

Cleaner fish benefit by getting food.

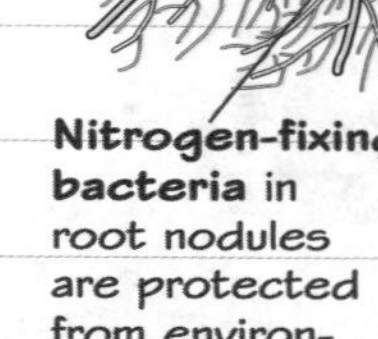

Nitrogen-fixing bacteria in root nodules are protected from environment and get food from plant.

Legume plant gets nitrogen compounds for healthy growth from the bacteria.

Worked example

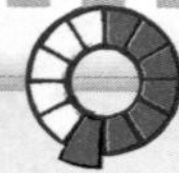

A tapeworm is a parasite that lives in the intestines of mammals. Explain why it is classed as a parasite. **(3 marks)**

The tapeworm depends on the host for its food supply. It absorbs digested food from the host's intestines, so it benefits from the relationship. The host is harmed because it is losing some of the nutrients in the food it eats.

Remember mutualism is a '+ +' relationship as both organisms benefit. Parasitism is a '+ −' relationship because one organism benefits but the other one is harmed.

Now try this

1 Define the term **parasite**. **(1 mark)**

2 Explain what is meant by mutualism. **(1 mark)**

3 Explain how the relationship between nitrogen-fixing bacteria and legumes is mutualistic. **(2 marks)**

Fieldwork techniques

The number of organisms in an area can be found using **fieldwork** techniques.

Sampling with quadrats

When studying organisms in the field, most areas are too large to count every individual organism. So we take **samples** and use them to draw conclusions about the whole area. Samples are often taken with square frames called **quadrats**.

- Quadrats are placed randomly in the area.
- The number of study organisms is counted in each quadrat.
- The number of organisms in the whole area is estimated using the equation:
  number in whole area =

$$\text{mean number of organisms in one quadrat} \times \frac{\text{total area (m}^2\text{)}}{\text{area of one quadrat (m}^2\text{)}}$$

Belt transects

The effect of abiotic and biotic factors on where organisms live (their **distribution**) can be studied using **quadrats** placed along a **belt transect**.

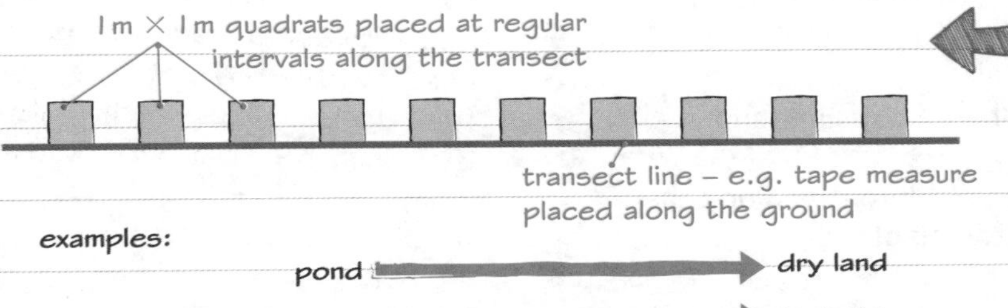

Choosing the number of quadrat samples to take is usually a trade-off – a trade-off between taking more samples (more accurate) and taking fewer samples (quicker).

Changes in **factors** (such as temperature, light intensity, trampling) are also recorded at each quadrat position. This makes it easier to link a change in distribution with a change in physical factor.

Worked example

Ten 1 m^2 quadrats were placed randomly on a school field that was 100 m by 200 m. The number of daisies recorded in each quadrat was: 20, 6, 33, 0, 26, 21, 18, 7, 2, and 9. Estimate the total number of daisies on the field. **(3 marks)**

mean number of plants per quadrat

$$= \frac{\text{total daises sampled}}{\text{number of quadrats}} = 14.2$$

$$\text{estimated number of daisies} = \text{mean} \times \frac{\text{area of field}}{\text{area of quadrat}}$$

$= 14.2 \times (20000/1) = 284000$ daisies

This is an estimate because if the quadrats were placed in different positions, the number sampled might be different.

Quadrats are only useful for sampling plants, or animals that do not move about much.

Maths skills To calculate the mean number of plants for one quadrat, divide the total number of plants by the number of quadrats.

Now try this

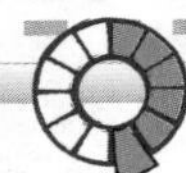

1 State when you would sample randomly with a quadrat. **(1 mark)**

2 Describe the situation in which you would choose to sample along a transect rather than sample randomly. **(1 mark)**

Organisms and their environment

You can use **quadrats** and **belt transects** to investigate the relationship between organisms and their environment in the field. See page 79 for more information on these techniques.

Core practical

Aim

To investigate how the change in light intensity from inside a woodland to a nearby open meadow affects where cowslips grow.

Apparatus

- quadrat
- light meter
- long measuring tape

Method

1. Set out a measuring tape from inside the woodland out to the open meadow.
2. Take a quadrat sample at regular intervals.
3. At each quadrat sample, count the number of cowslips, and measure the light intensity.
4. Compare the number of plants with the light intensity along the transect to see how light affects where the plants live.

Random sampling will not help here because you need to sample organisms in relation to how the environmental factor changes.

An alternative to counting plants is estimating percentage coverage of the quadrat (e.g. 10%, 50%) by the species. This is useful if it is difficult to count individual plants, or if some plants are much bigger than others.

Results

The results were recorded in a table.

quadrat number	Q1	Q2	Q3	Q4	Q5
distance from woodland centre (m)	0	10	20	30	40
light intensity (lux)	2.3	2.1	3.5	9.5	13.5
number of cowslips	0	0	1	5	15

The results are displayed in this graph.

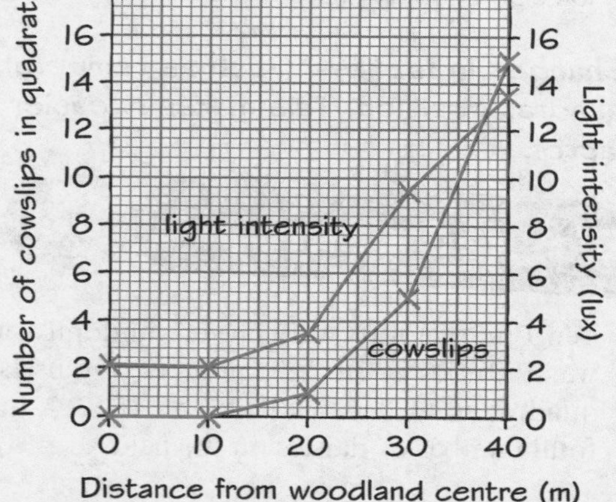

Conclusion

Cowslips seem to grow better in open meadow because the greater the light intensity the greater the number of cowslips.

Repeating the transect several times and comparing the results for each transect would help to average out any random variation.

Now try this

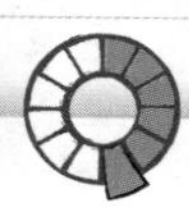

In the woodland/meadow transect above, the number of primroses was also counted. The results were: Q1 2; Q2 4; Q3 7; Q4 1; Q5 0.

(a) Display these values, and the light intensity values, on one graph to show how they change over distance along the transect. **(2 marks)**

(b) Describe what your graph shows about the relationship between light intensity and number of primroses. **(2 marks)**

Human effects on ecosystems

Humans have positive and negative effects on **biodiversity** in ecosystems. Biodiversity is the variety of living organisms in an area.

Fish farming

Fish farming involves growing one kind of fish in an area. The fish are fed and the waste they produce is removed from their tanks.

- 👎 The waste can pollute the local area, changing conditions so that some local species die out.
- 👎 Diseases from the farmed fish (such as lice) can spread to wild fish and kill them.
- 👍 Farming fish reduces fishing of wild fish.

Non-indigenous species

Non-indigenous species (organisms that are not found naturally in the area) may be introduced on purpose or accidentally to an area.

- 👎 They may reproduce rapidly as they have no natural predators in the new area.
- 👎 They may **out-compete** native species for food or other resources.
- 👍 They may provide food for native species.

Eutrophication

Fertilisers added to fields for crops may get into streams and rivers. This adds phosphates and nitrates to the water, which is called **eutrophication**. Eutrophication can lead to loss of biodiversity in nearby water.

Eutrophication causes water plants and algae to grow more quickly. → Plants and algae cover the water surface, and block light to deeper water. → Deeper plants cannot get light, so they die. → Bacteria decompose dying plants and take oxygen from the water. → There is not enough oxygen left in the water for fish, so they die.

Worked example

Unionid mussels are found in many waters in the USA. Zebra mussels are small freshwater mussels that are non-indigenous species introduced to this area in 1991. The graph shows the changes in population sizes of both kinds of mussel over a 20-year period in the Hudson River.

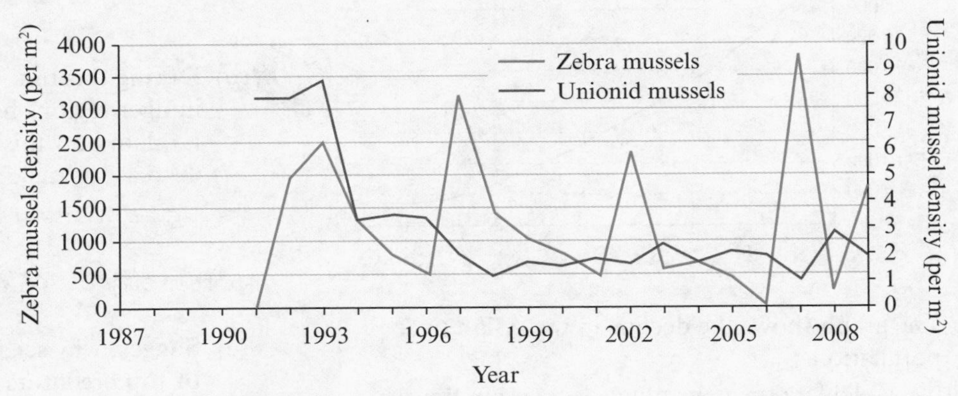

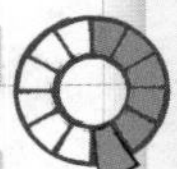

(a) Describe the changes in the unionid mussel population between 1991 and 2009. **(2 marks)**

The population falls rapidly until 1994 then more slowly until 1998. After this the population stays at approximately the same low level.

(b) The zebra mussels do not eat the unionid mussels. Suggest another reason why the zebra mussels have caused this reduction in the unionid mussel population. **(2 marks)**

They may outcompete the unionid mussels for a resource, such as food supply or breeding places.

Now try this

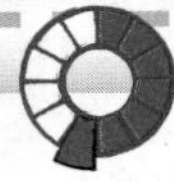

Focus on giving the stages that cause a decrease.

Explain how eutrophication can lead to a decrease in biodiversity. **(3 marks)**

Biodiversity

Maintaining **biodiversity** is very important, at both the **local** and **global** level, because many organisms have important roles in ecosystems and can be useful to humans.

Reasons for maintaining biodiversity

- Moral reasons: Humans should respect other living organisms.
- Aesthetic reasons: People enjoy seeing the variety of living organisms that live in different habitats.
- Ecosystem structure: Some organisms have an important role in ecosystems, such as microorganisms in decay processes and nutrient recycling. If this planet loses species, food chains become more unstable.
- Usefulness: Some species are particularly useful to humans, for example plants that produce life-saving drugs, or wild varieties of plants grown for crops (as a source of genes if the environment changes).

Reforestation

Reforestation is replanting forests where they have been destroyed, for example to create farmland.

Advantages include:

- ✓ restores habitat for species that are endangered. Restoring rain forest, for example, helps to conserve many species.
- ✓ reduces the concentration of carbon dioxide in the air as the trees photosynthesise
- ✓ Tree roots bind the soil together and reduce the effects of soil erosion.
- ✓ affects local climate, for example reducing the range of temperature variation.

Conservation is important so that we do not lose valuable species. We can conserve organisms by protecting their habitat, preventing poaching, or keeping insurance populations in zoos and seed banks.

Worked example

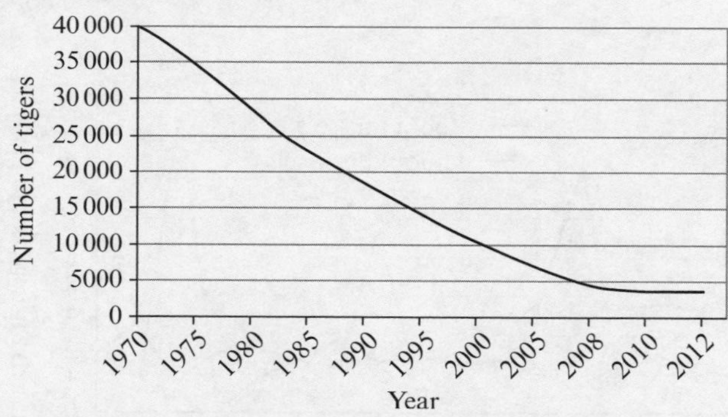

The graph shows the decline in the Asian tiger population.

(a) Calculate the percentage decrease in the population between 1970 and 2008. Show your working. **(2 marks)**

decrease = 40000 − 3500 = 36500

% decrease = (36500 ÷ 40000) × 100 = 91.25%

(b) During this time the human population in tiger regions has increased significantly. Explain how this could be linked to the decrease in tiger population. **(1 mark)**

The tigers' habitat might be being destroyed.

Another suggestion would be hunting by humans.

(c) Suggest why scientists monitor the populations of top predators such as the tiger very closely when they are assessing the biodiversity in an area. **(2 marks)**

If there are plenty of top predators, then there is plenty of energy being transferred through the food web in that ecosystem, which suggests high biodiversity.

Now try this

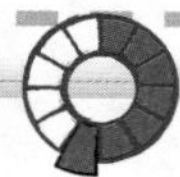

Farmers in the UK are encouraged to keep hedges around their fields.
Explain how keeping hedges around fields can help maintain biodiversity. **(2 marks)**

The carbon cycle

Living organisms need substances from the environment. As the amount of these on Earth is limited, they are recycled through both living (**biotic**) and non-living (**abiotic**) parts of the ecosystem. The **carbon cycle** shows how the element carbon passes between the environment and living organisms.

Controls of the carbon cycle

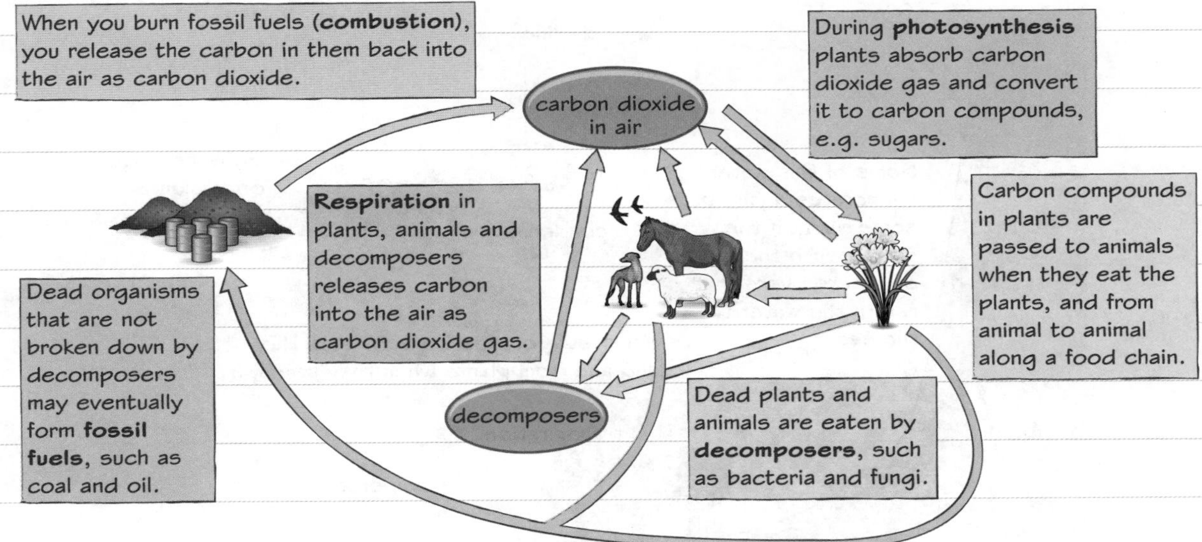

When you burn fossil fuels (combustion), you release the carbon in them back into the air as carbon dioxide.
carbon dioxide in air
During photosynthesis plants absorb carbon dioxide gas and convert it to carbon compounds, e.g. sugars.
Respiration in plants, animals and decomposers releases carbon into the air as carbon dioxide gas.
Carbon compounds in plants are passed to animals when they eat the plants, and from animal to animal along a food chain.
Dead organisms that are not broken down by decomposers may eventually form fossil fuels, such as coal and oil.
decomposers
Dead plants and animals are eaten by decomposers, such as bacteria and fungi.

In the air, carbon is part of carbon dioxide gas. In organisms, it is part of complex carbon compounds. The carbon cycle is important because it recycles carbon dioxide released in respiration to be taken in by plants in photosynthesis, to make organic molecules in living organisms.

Worked example

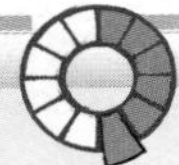

A large forest is cleared by burning. What effects will this have on the amount of carbon dioxide in the air (a) immediately, and (b) over a longer period? **(4 marks)**

(a) Large amounts of carbon dioxide will be released into the air by the burning (combustion) of the trees.

(b) Less carbon dioxide will be removed from the air than before because the trees would have used some for photosynthesis. So the amount of carbon dioxide in the air is likely to remain high.

Two key processes in the carbon cycle are **respiration** and **photosynthesis**. These processes are important in maintaining oxygen and carbon dioxide concentrations in the air. Combustion can change this balance.

Remember that plants photosynthesise in the light but, like all other living organisms, they respire all the time.

Now try this

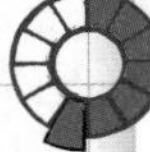

1 Describe the importance of decomposers in the carbon cycle. **(1 mark)**

Decomposers respire using dead plant and animal matter, releasing carbon dioxide into the atmosphere.

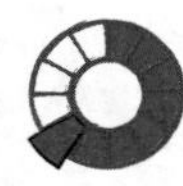

2 Explain the effect of respiration, photosynthesis and combustion in the carbon cycle in transferring carbon dioxide to and from the atmosphere. **(3 marks)**

In each case, explain whether these release carbon dioxide into the atmosphere, or remove it.

Had a look ☐ Nearly there ☐ Nailed it! ☐

The water cycle

The **water cycle** describes how water moves between different parts of our planet.

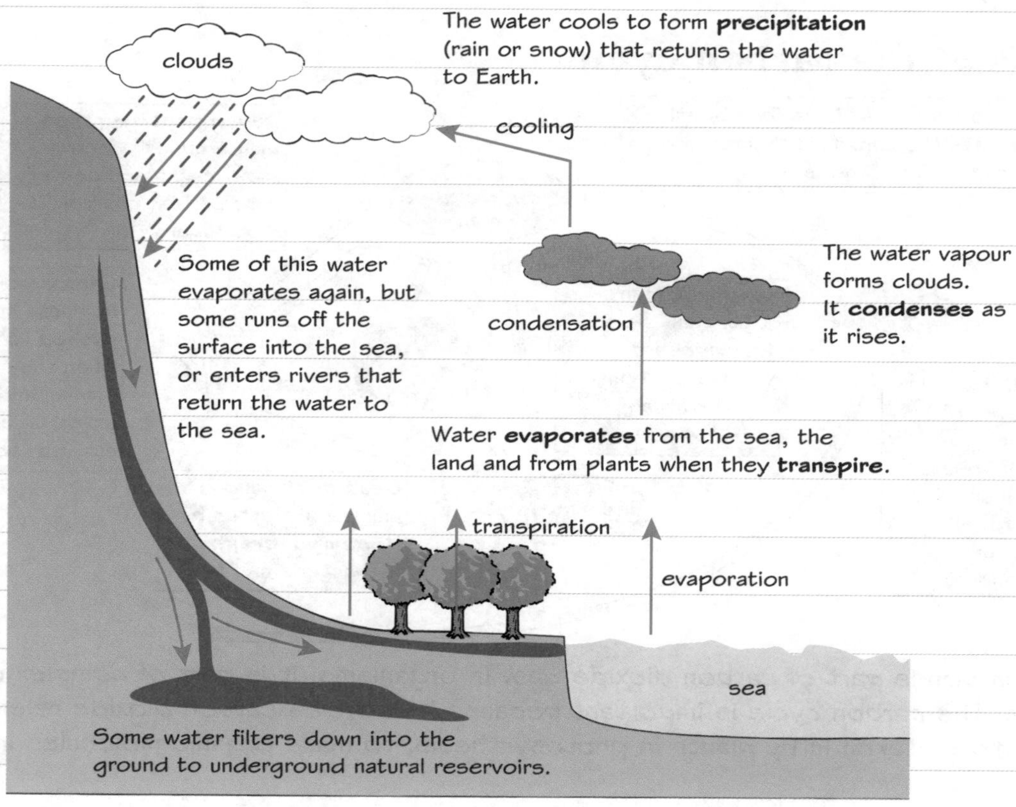

clouds
The water cools to form **precipitation** (rain or snow) that returns the water to Earth.
cooling
Some of this water evaporates again, but some runs off the surface into the sea, or enters rivers that return the water to the sea.
condensation
The water vapour forms clouds. It **condenses** as it rises.
Water **evaporates** from the sea, the land and from plants when they **transpire**.
transpiration
evaporation
sea
Some water filters down into the ground to underground natural reservoirs.

Desalination

Some areas of the world where it is hot and dry suffer **drought**. They do not get enough precipitation to use as their source of **potable** water (that is, fit to drink). If they are close to the sea, they can produce drinking water from salty water by **desalination**. One way to do this is **distillation**.

Worked example

Explain how distillation can be used to produce potable water in a village that is close to the sea. **(4 marks)**

- Salty water is heated until the water evaporates, forming steam.
- The steam is condensed in another container to give pure water.
- The salt and other impurities are left behind in the salty water.
- If renewable energy, such as sunlight, is used this saves electricity costs and avoids pollution.

Now try this

You need to describe each stage in turn.

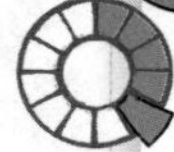

1 Describe the processes that take place in the water cycle. **(5 marks)**

2 Suggest the effect on the water cycle of planting a large number of trees. **(3 marks)**

The nitrogen cycle

Plants need **nitrates** to grow well. They can get nitrates in several different ways, including through the **nitrogen cycle**.

Controls of the nitrogen cycle

A diagram of the **nitrogen cycle** shows how the element nitrogen moves between living organisms and the environment.

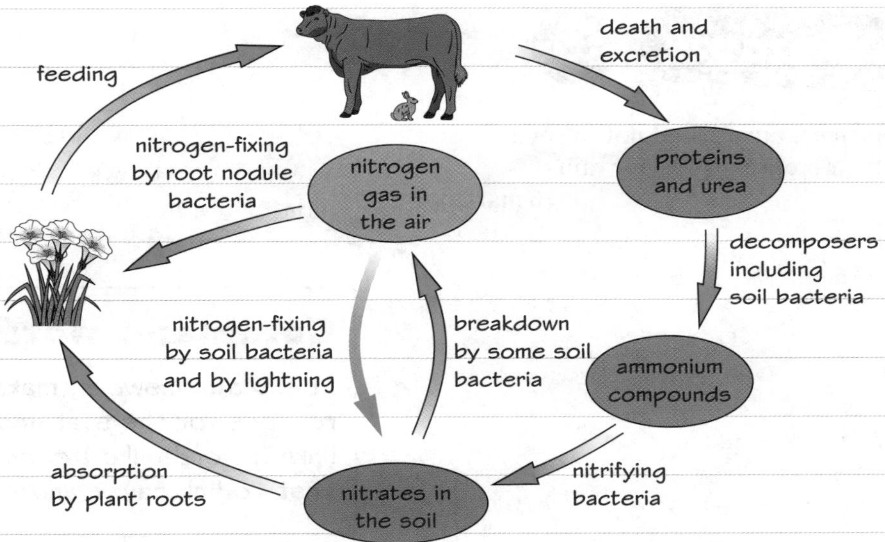

Worked example

Plants grow well in fertile soil. Explain how bacteria help to keep soil fertile. **(4 marks)**

Plants need nitrogen for making proteins but they can only take in nitrogen in the form of nitrogen compounds such as nitrates. Soil bacteria act as decomposers, releasing ammonia from proteins in dead bodies and from urea. Bacteria in the soil make nitrates. Plants can then use the nitrates to make proteins. Nitrogen-fixing bacteria in the soil and roots of some plants convert nitrogen gas from the air into nitrogen compounds that the plant can use.

In the nitrogen cycle it is important to focus on the role of bacteria.

Fertilisers and crop rotation

Farmers add **nitrates** to crop plants in **fertiliser** to help the plants grow better.

Crop rotation means growing different crops each year, on a rotation basis. This improves soil fertility (the amount of nutrients in the soil) because:

- different crops remove different nutrients from the soil
- plants such as clover have nitrogen-fixing bacteria in their roots, and can be ploughed back into the soil, to add nitrates.

Explain why this could be an advantage to a farmer.

Now try this

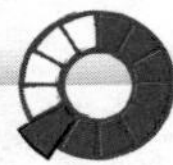

1 Explain why a farmer might decide one year to grow clover plants in a field instead of a crop. **(2 marks)**

2 Farmers plough crop stubble back into the soil after harvesting. Explain how this helps to improve soil fertility. **(4 marks)**

Extended response – Ecosystems and material cycles

There will be at least one 6-mark question on your exam paper. For these questions, you will need to think scientifically and structure your answer logically, showing how the points you make are related to each other. You can revise the topics for this question, which is about **the impact of human interactions on ecosystems** and **plant uptake of nitrates**, on pages 81 and 85.

Worked example

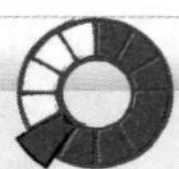

Explain why farmers are advised not to spread fertilisers on their crops when heavy rain is due. **(6 marks)**

Fertilisers contain nitrates and other mineral ions that plants need for healthy growth. Mineral ions in fertilisers dissolve in water, and are absorbed from the soil through plant roots.

If it rains heavily, then the mineral ions could be washed away from the crops and drain into nearby water, such as streams or rivers. This means that there will be fewer mineral ions for the crop plants so they will not grow so well. This will have been a waste of money for the farmer.

Extra mineral ions added to the streams and rivers will cause eutrophication. This will cause rapid growth of algae and water plants. The extra growth blocks light to organisms deeper in the water, meaning these organisms die, and takes oxygen from the water for respiration.

Bacteria that decompose dying plants and animals will also take oxygen from the water. If not enough oxygen is left in the water, fish and other animals may die and biodiversity may be reduced.

Remember the importance of mineral ions in plant growth when discussing fertilisers. This is a good way to start this answer.

Command word: Explain

In **explain** answers, make sure you give reasons for the statements you make. Use linking words like **because** or **this means that** to link cause and effect.

Use appropriate science words, such as eutrophication, in your answers, and make sure it is clear what you mean when you use them.

In questions about the environment, remember to consider how the interdependency of organisms, including microorganisms, can result in changes to biodiversity in the ecosystem.

Remember to consider the **advantages** and the **disadvantages** to ecosystems and biodiversity of fish farming.

Now try this

Wild salmon take up to five years to reach adult size. Farmed salmon are kept in conditions so they reach this size in less than two years. Explain the impact of fish farming on ecosystems. **(6 marks)**

Formulae

You should be able to recall the formulae of elements, simple compounds and ions.

Elements

An **element** is a substance made from atoms with the same number of protons in the nucleus. Each element has its own **chemical symbol**, which:

- consists of one, two or three letters
- starts with a capital letter
- has any other letters in lower case.

For example, mercury is Hg. You can find the symbols for the elements in the periodic table.

Atoms and molecules

An **atom** is the smallest particle of an element that still has its chemical properties.

A **molecule** consists of two or more atoms chemically joined together.

Most elements that are gases at room temperature exist as molecules with two atoms. They are shown by **chemical formulae**, e.g. H_2, N_2, O_2, F_2, Cl_2.

Compounds

A **compound** consists of two or more different elements chemically joined together.

carbon hydrogen

CH_4

one four

The chemical formula for methane is CH_4. Each methane molecule has:

- one carbon atom (no number in the formula)
- four hydrogen atoms (4 in the formula)
- atoms of two elements joined together
- (1 + 4) = 5 atoms in total.

Remember: you will not find the formulae for any compounds in the periodic table.

Worked example

(a) Give the meaning of the term **ion**. **(2 marks)**

An ion is a charged particle formed when an atom, or group of atoms, loses or gains electrons.

(b) An ion is represented as Na^+.
Explain what this shows. **(2 marks)**

The ion has one positive charge, and it is formed from a sodium atom.

(c) An ion is represented as SO_4^{2-}.
Explain what this shows. **(2 marks)**

The ion has two negative charges, and it is formed from one sulfur atom and four oxygen atoms chemically joined together.

(d) Write the formula for sodium sulfate. **(1 mark)**

Na_2SO_4

In general:

- metal atoms lose electrons to form positive ions
- non-metal atoms gain electrons to form negative ions.

Look at page 91 for help with atomic structure. For a reminder about how ions form, look at page 96.

Write the charge as a superscript at the top right of the symbol.

You should show a single positive charge as + and a single negative charge as –.

A sodium ion has a single positive charge and a sulfate ion has two negative charges. Two sodium ions balance the charge on one sulfate ion.

Now try this

Use the periodic table on page 94 to help you answer these questions.

1 The formula for bromine is Br_2.
Explain what this shows. **(2 marks)**

2 In copper carbonate, there is one carbon atom and three oxygen atoms for every copper atom.
State the formula for copper carbonate. **(1 mark)**

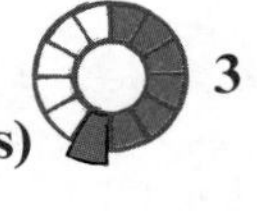

3 The formula for magnesium hydroxide is $Mg(OH)_2$.
Explain what this tells you about the number and type of each atom or ion present. **(2 marks)**

Multiply the number of atoms inside the brackets by 2.

Equations

Chemical **equations** are used to model the changes that happen in chemical reactions.

Word equations

In a chemical reaction:

- **reactants** are the substances that undergo a chemical change in a reaction
- **products** are the new substances formed.

In general: reactants → products

Two or more reactants or products are separated by a + sign. In a **word equation**, you write the name of each substance, not its formula.

An example word equation

Iron(III) oxide reacts with carbon to form iron and carbon monoxide:

iron(III) oxide + carbon → iron + carbon monoxide

reactants on the left — products on the right

Write it all on one line if possible. If you are going to run out of space, write words below others as shown here.

Balanced equations

- All substances are shown by their formulae.
- The numbers of atoms of each element in the reactants and products are the same.
- You may need to write a number in front of a reactant or product to balance an equation. For example, hydrogen reacts with chlorine to form hydrogen chloride.

1. Write all the symbols and formulae: $H_2 + Cl_2 \rightarrow HCl$
2. Count the atoms of each element on each side, and write numbers if needed:

$$H_2 + Cl_2 \rightarrow 2HCl$$

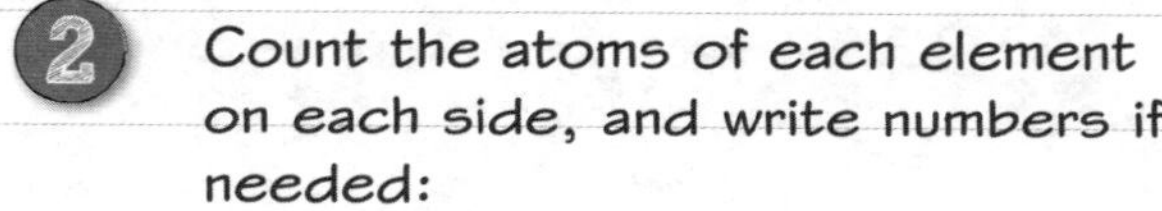

2HCl in the equation means two HCl molecules.

An example balanced equation

The balanced chemical equation for the reaction between iron(III) oxide and carbon is:

$$Fe_2O_3 + 3C \rightarrow 2Fe + 3CO$$

It shows that one unit of iron(III) oxide reacts with three carbon atoms. The products are two iron atoms and three carbon monoxide molecules:

Element	On left	On right
Fe	2	2
O	3	3
C	3	3

State symbols

State symbols show the physical state of each substance in a balanced equation:

- ✓ solid (s)
- ✓ gas (g)
- ✓ liquid (l)
- ✓ aqueous solution (aq). (dissolved in water)

Worked example

Aluminium reacts with hot dilute sulfuric acid, H_2SO_4. Aluminium sulfate, $Al_2(SO_4)_3$, and hydrogen, H_2, are formed.
Write a balanced equation for this reaction and include state symbols. **(2 marks)**

$2Al(s) + 3H_2SO_4(aq) \rightarrow Al_2(SO_4)_3(aq) + 3H_2(g)$

The unbalanced equation is:

$$Al + H_2SO_4 \rightarrow Al_2(SO_4)_3 + H_2$$

- There are two Al atoms on the right, so a **2** is placed in front of the Al on the left.
- There are three units of SO_4 on the right, so a **3** is placed in front of the H_2SO_4 on the left.
- There are now 3 × H_2 on the left, so a **3** is placed in front of the H_2 on the right.

Now try this

1 Balance these chemical equations:
(a) $Mg + O_2 \rightarrow MgO$ **(1 mark)**
(b) $N_2 + H_2 \rightarrow NH_3$ **(1 mark)**
(c) $CH_4 + O_2 \rightarrow CO_2 + H_2O$ **(1 mark)**

2 Balance these chemical equations and include state symbols.
(a) $CuO + HNO_3 \rightarrow Cu(NO_3)_2 + H_2O$ (the reaction between copper oxide powder and dilute nitric acid, forming copper nitrate solution and water). **(2 marks)**
(b) $Fe + Cl_2 \rightarrow FeCl_3$ (the reaction between iron filings and chlorine gas, forming iron(III) chloride powder). **(2 marks)**

Ionic equations

Ionic equations are used to model chemical changes involving ions.

Ions

An **ion** is an electrically charged particle. It is formed when an atom, or group of atoms, loses or gains electrons.

In general:

- hydrogen atoms and metal atoms lose electrons to form positively charged ions
- non-metal atoms gain electrons to form negatively charged ions.

You can revise how ions form on page 96.

Representing ions

You show the charge on an ion using a superscript + or –. For example:

Positive ion	Negative ion
H^+	Cl^-
Na^+	O^{2-}
Mg^{2+}	NO_3^-
Al^{3+}	CO_3^{2-}
NH_4^+	SO_4^{2-}

one S atom, four O atoms, two negative charges

Writing an ionic equation

Oppositely charged ions in solution may join to form an **insoluble** solid. This solid is called a **precipitate**. For example, silver ions and chloride ions in solution form solid silver chloride:

$$Ag^+(aq) + Cl^-(aq) \rightarrow AgCl(s)$$

In an ionic equation:

- all substances are shown by their formula
- the numbers of atoms of each element in the reactants and products are the same
- the total electrical charge is the same on both sides of the equation.

You may need to write a number in front of an ion or substance to balance an ionic equation.

Balancing an ionic equation

Aluminium ions react with hydroxide ions to form aluminium hydroxide.

Write all the symbols and formulae:

$$Al^{3+} + OH^- \rightarrow Al(OH)_3$$

Count the atoms of each element on each side, and write numbers if needed: $Al^{3+} + 3OH^- \rightarrow Al(OH)_3$

Using ratios

Simple **ratios** are used to balance equations. The ratio of the charges on the ions (Al^{3+} and OH^-) is 3 : 1. The ratio of the number of each ion (Al^{3+} and OH^-) in the product is 1 : 3.

Worked example

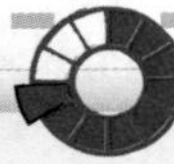

Magnesium reacts with copper sulfate solution to form copper and magnesium sulfate solution.

(a) Write a balanced equation for this reaction. **(1 mark)**

$Mg + CuSO_4 \rightarrow Cu + MgSO_4$

(b) Write a balanced ionic equation for this reaction. **(2 marks)**

$Mg + Cu^{2+} \rightarrow Cu + Mg^{2+}$

You might be asked to include state symbols:
$Mg(s) + CuSO_4(aq) \rightarrow Cu(s) + MgSO_4(aq)$

If you write all the ions involved, you get:
$Mg + Cu^{2+} + SO_4^{2-} \rightarrow Cu + Mg^{2+} + SO_4^{2-}$
The sulfate ions, SO_4^{2-}, are unchanged in the reaction. They are called **spectator ions**.
You can leave these out of an ionic equation.

Now try this

1 Lead(II) ions, Pb^{2+}, react with bromide ions, Br^-. Cream-coloured lead(II) bromide forms, $PbBr_2$. Write an ionic equation for the reaction. **(1 mark)**

2 Barium chloride solution, $BaCl_2(aq)$, reacts with sodium sulfate solution, $Na_2SO_4(aq)$. Solid barium sulfate, $BaSO_4(s)$, and sodium chloride solution, $NaCl(aq)$, form.

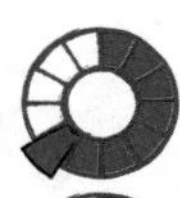

(a) Write the balanced equation. **(1 mark)**

(b) Write an ionic equation for this reaction. **(1 mark)**

Had a look ☐ Nearly there ☐ Nailed it! ☐

Hazards, risks and precautions

You should be able to evaluate the risks in a practical procedure. You should also be able to suggest suitable precautions.

Hazards

A **hazard** is something that could cause:

- damage or harm to someone or something
- adverse health effects, which may occur immediately or later on.

For example, ethanol is flammable. This is a hazard. If the ethanol ignited, it could cause burns or a fire.

Practical skills: Risks

A **risk** is the chance that someone or something will be harmed if exposed to a hazard. The amount of risk depends on factors such as:

- how much someone is exposed to a hazard
- the way in which exposure happens
- how serious the effects of exposure are.

The risk from heating ethanol using a hot water bath is less than when using a Bunsen burner.

Hazard symbols

The labels on containers of hazardous substances include **hazard symbols**. These are intended to:

- warn about the dangers associated with the substance in the container
- let people know about the precautions to take when they use the substance.

Some common hazard symbols

harmful or irritant

flammable

respiratory sensitiser

toxic

corrosive

oxidising

Precautions

A **precaution** is something that you can do to reduce the risk of harm from a hazard. Precautions include:

- using a less hazardous substance
- using protective clothing, such as gloves and eye protection
- using a different method or apparatus.

Worked example

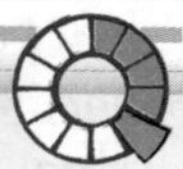

A student is preparing a dry sample of copper sulfate. She heats some copper sulfate solution in an evaporating basin. She then allows it to cool. Crystals of copper sulfate appear.

Describe and explain one safety precaution she should use. **(3 marks)**

She should heat the solution gently. This reduces the risk that it will spit out of the evaporating basin. The hot solution could cause skin burns or eye damage.

The answer is specific to this activity. It is not a general lab rule such as not running or not drinking the solution.

Other suitable precautions that could be mentioned, if linked to the activity, include:

- wearing gloves if toxic substances are used
- tying hair back or tucking in a tie if a Bunsen burner is used for heating.

Now try this

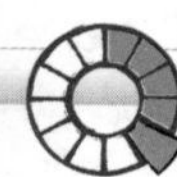

1. State **one** reason why hazard symbols are used. **(1 mark)**
2. A student carries out electrolysis on a concentrated sodium chloride solution. Toxic chlorine gas and flammable hydrogen gas are produced.
   Describe **two** precautions the student could take to reduce the risk of harm in this experiment. **(2 marks)**

Atomic structure

Dalton's simple model has changed over time because of the discovery of subatomic particles.

A brief atomic timeline

When	Who	What
1803	Dalton	Solid atom model: all atoms of an element are identical; different elements have different atoms
1897	Thomson	Discovers the electron
1904	Thomson	Plum pudding model: atoms are spheres of positive charge with negative electrons dotted around inside
1911	Rutherford	Solar system model: atoms have a positive nucleus surrounded by negative electrons in orbits
1913	Bohr	Electron shell model: electrons occupy shells or energy levels around the nucleus
1918	Rutherford	Discovers the proton
1932	Chadwick	Discovers the neutron

Atomic structure

An **atom** consists of a central **nucleus**, which:

- contains **protons** and **neutrons**
- is surrounded by electrons in shells.

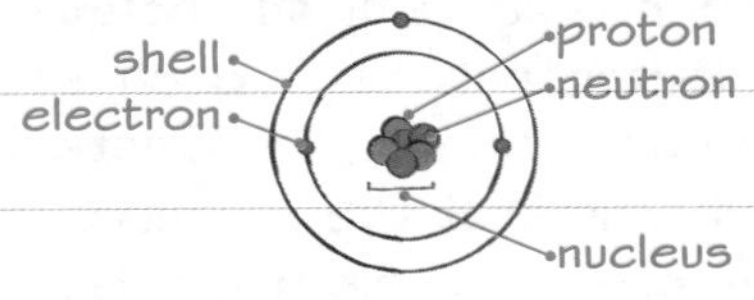

Maths skills: Relative values

Atoms and their subatomic particles are very small. Their diameters, masses and electrical charges are expressed as **relative values.** For example, the diameter of a hydrogen nucleus is 1.75×10^{-15} m. The diameter of a hydrogen atom is 1.06×10^{-10} m.

The diameter of the nucleus relative to the atom:

$$= \frac{1.75 \times 10^{-15}\,\text{m}}{1.06 \times 10^{-10}\,\text{m}} = \frac{1}{60\,600}$$

The nucleus is very small compared with the overall size of the atom.

Worked example

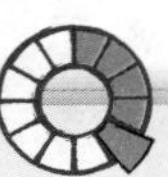

Complete the table with the names and properties of subatomic particles. **(3 marks)**

Particle	Relative charge	Relative mass
proton	+1	1
neutron	0	1
electron	−1	$\frac{1}{1836}$

- Atoms have equal numbers of protons and electrons. They have equal numbers of positive and negative charges, and so are neutral overall.
- Electrons have very little mass compared with protons and neutrons. Most of the mass of an atom is concentrated in its nucleus.

Now try this

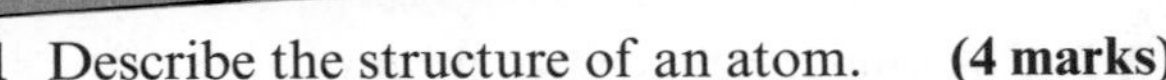

1 Describe the structure of an atom. **(4 marks)**

2 Explain why atoms are electrically neutral overall. **(2 marks)**

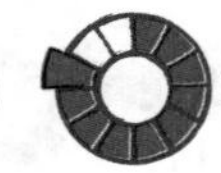

3 The mass of a proton is 1.6726×10^{-27} kg. The mass of an electron is 9.1094×10^{-31} kg. Calculate the mass of an electron relative to a proton. Give your answer to 4 significant figures. **(2 marks)**

Isotopes

The atoms of an element have identical chemical properties, but can exist as different isotopes.

Numbers of particles

Each atom can be described by its:

- **mass number**, the total number of protons and neutrons in the nucleus
- **atomic number**, the number of protons in the nucleus.

These are in full chemical symbols:

mass number → $^{23}_{11}Na$ ← atomic number

Atoms of a given element have the same number of protons in the nucleus:

- They have the same atomic number.
- This number is unique to that element.

Maths skills: Calculating particle numbers

Use the atomic number and mass number to calculate the number of subatomic particles in an atom.

For example, for sodium, Na:

1. atomic number = 11
   number of protons = 11
   number of electrons = 11
   (equal numbers of protons and electrons)

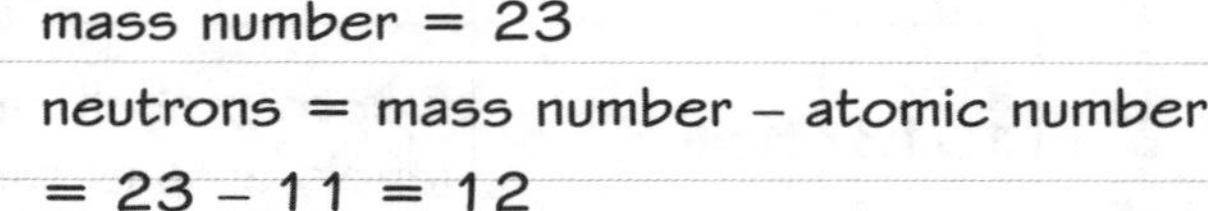

2. mass number = 23
   neutrons = mass number − atomic number
   = 23 − 11 = 12

Isotopes

Isotopes are atoms of an element with:

- the same number of protons
- different numbers of neutrons.

You can recognise isotopes of the same element because they will have the same atomic number, but different mass numbers.

Isotopes of an element have the same chemical properties because they have the same number of electrons.

You can revise electronic configurations on page 95.

Relative atomic mass

Take care not to confuse **relative atomic mass**, A_r, with mass number:

✓ A_r is the mean mass of the atoms of an element, relative to 1/12th the mass of a ^{12}C atom.

A_r values take into account the **relative abundance** or percentage of each isotope in a sample of an element. The existence of isotopes means that the A_r values of elements may not be whole numbers.

Worked example

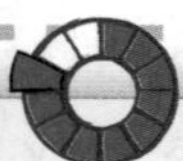

Gallium has two isotopes: $^{69}_{31}Ga$ and $^{71}_{31}Ga$ (sometimes written as gallium-69 and gallium-71). The relative abundance of $^{69}_{31}Ga$ is 60%. Calculate, to 1 decimal place, the relative atomic mass of gallium. **(3 marks)**

relative abundance of $^{71}_{31}Ga$ = 100% − 60%
= 40%

$$\text{relative atomic mass} = \frac{(69 \times 60) + (71 \times 40)}{100}$$

$$= \frac{4140 + 2840}{100}$$

$$= 69.8$$

Multiply the mass number of each isotope by its **relative abundance**, and then add them all together and divide by 100.

Now try this

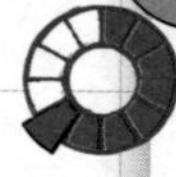

1 Bromine has two natural isotopes, $^{79}_{35}Br$ and $^{81}_{35}Br$. Explain, in terms of the numbers of subatomic particles, why these are isotopes of the same element. **(4 marks)**

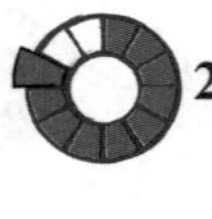

2 Chlorine has two natural isotopes, 75.8% $^{35}_{17}Cl$ and 24.2% $^{37}_{17}Cl$. Calculate the relative atomic mass, A_r, of chlorine. Give your answer to 1 decimal place. **(2 marks)**

Mendeleev's table

Dmitri Mendeleev's **periodic table** was successful and developed into the modern periodic table. In 1869, Mendeleev arranged all the elements known at the time into a table:

Mendeleev put the elements in order of the relative atomic masses.

He checked the properties of the elements and their compounds.

He swapped the places of some elements so that elements with similar properties lined up.

He left gaps where he thought there were other elements, and predicted their properties.

When these elements were discovered, Mendeleev's predictions fitted the properties very well.

Pair reversals

Mendeleev thought he had arranged elements in order of increasing relative atomic mass. This was not always true – the positions of elements in some pairs were reversed. For example, Mendeleev put tellurium in group 6 and iodine in group 7. This matched the properties of the elements and their compounds.

However, according to their relative atomic masses, iodine should be first:

A_r of Te = 128 A_r of I = 127

Iodine naturally exists only as ^{127}I. ^{126}Te exists, but ^{128}Te and ^{130}Te are its most abundant isotopes. Their high relative abundance gives tellurium a greater A_r than iodine.

Group 6 and 7 elements

Group 6 elements	Group 7 elements
do not react with water	all react with water
all react with oxygen (except oxygen itself)	do not react with oxygen
all form compounds with hydrogen with the general formula: H_2X	all form compounds with hydrogen with the general formula: HX
6 electrons in their atom's outer shell	7 electrons in their atom's outer shell

This information was not available in Mendeleev's time.

See page 95 to revise electronic configurations.

Worked example

Mendeleev published another table in 1871. Part of this is shown on the right. Mendeleev left gaps in his table, shown as * here.

Explain the importance of doing this. **(3 marks)**

The gaps were for elements not discovered then. Mendeleev used his table to predict the properties of these elements. When they were discovered later, their properties closely matched his predicted properties. This supported the ideas behind his table.

One of the gaps was for 'eka-silicon'. This was discovered in 1886 and named germanium.

Group

1	2	3	4	5	6	7
H						
Li	Be	B	C	N	O	F
Na	Mg	Al	Si	P	S	Cl
K Cu	Ca Zn	* *	Ti *	V As	Cr Se	Mn Br
Rb Ag	Sr Cd	Y In	Zr Sn	Nb Sb	Mo Te	* I

Now try this

1 Describe how Mendeleev arranged the elements known to him. **(2 marks)**

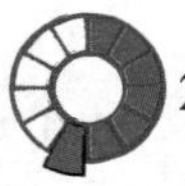

2 Suggest **one** reason why other scientists at the time thought that Mendeleev's table was not correct. **(1 mark)**

Had a look ☐ Nearly there ☐ Nailed it! ☐

The periodic table

The modern periodic table is useful for describing and predicting properties of elements.

Atomic number

In Mendeleev's periodic table, atomic number was just the position of an element in the table.

Later discoveries showed that:

- atomic number is actually the number of protons in the nucleus of an atom
- each element has a unique atomic number
- if the elements are arranged in order of increasing atomic number, Mendeleev's pair reversals are explained.

Explaining pair reversals

Iodine should be placed before tellurium according to its relative atomic mass, but after tellurium using its atomic number:

Relative atomic mass	128	127
Element symbol	Te	I
Atomic number	52	53

Mendeleev did not know about atomic structure. He could explain the pair reversal only in terms of the elements' properties.

Features of the modern periodic table

The elements in the modern periodic table are arranged in order of their atomic numbers.

Key: relative atomic mass / symbol / atomic number

Period	Group 1	2											3	4	5	6	7	0
1										1 H 1								4 He 2
2	7 Li 3	9 Be 4											11 B 5	12 C 6	14 N 7	16 O 8	19 F 9	20 Ne 10
3	23 Na 11	24 Mg 12											27 Al 13	28 Si 14	31 P 15	32 S 16	35.5 Cl 17	40 Ar 18
4	39 K 19	40 Ca 20	45 Sc 21	48 Ti 22	51 V 23	52 Cr 24	55 Mn 25	56 Fe 26	59 Co 27	59 Ni 28	63.5 Cu 29	65 Zn 30	70 Ga 31	73 Ge 32	75 As 33	79 Se 34	80 Br 35	84 Kr 36
5	85 Rb 37	88 Sr 38	89 Y 39	91 Zr 40	93 Nb 41	96 Mo 42	(98) Tc 43	101 Ru 44	103 Rh 45	106 Pd 46	108 Ag 47	112 Cd 48	115 In 49	119 Sn 50	122 Sb 51	128 Te 52	127 I 53	131 Xe 54
6	133 Cs 55	137 Ba 56	139 La 57	178 Hf 72	181 Ta 73	184 W 74	186 Re 75	190 Os 76	192 Ir 77	195 Pt 78	197 Au 79	201 Hg 80	204 Tl 81	207 Pb 82	209 Bi 83	(209) Po 84	(210) At 85	(222) Rn 86
7	(223) Fr 87	(226) Ra 88	(227) Ac 89	(261) Rf 104	(262) Db 105	(266) Sg 106	(264) Bh 107	(277) Hs 108	(268) Mt 109	(271) Ds 110	(272) Rg 111							

The horizontal rows are called **periods**.

Metals are on the left-hand side and in the centre.

Elements with similar properties are placed in the same vertical **groups**.

Non-metals are on the right-hand side.

Worked example

The table shows information about cobalt and nickel. Their positions in the periodic table were difficult to determine in Mendeleev's time.

Explain how knowledge of atomic structure helps to determine the positions of cobalt and nickel in the periodic table. **(2 marks)**

Relative atomic mass	58.9	58.7
Element symbol	Co	Ni
Atomic number	27	28

Elements are arranged in order of increasing atomic number (number of protons) in the modern table. The atomic number of cobalt is lower than that of nickel.

The answer uses the modern definition of the term **atomic number.** This is the number of protons in the nucleus of an atom. Mendeleev used a different definition.

Now try this

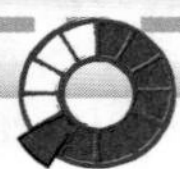

The relative atomic mass of an element is often given as a whole number.
Suggest why cobalt and nickel may **not** appear to be a pair reversal.
Use the information in the Worked example. **(1 mark)**

Electronic configurations

An **electronic configuration** describes the arrangement of electrons in shells in an atom or ion.

Modelling the arrangement of electrons

In an atom:

- electrons occupy electron **shells**
- shells are filled, starting with the innermost shell
- different shells hold different maximum numbers of electrons.

The diagram shows the electronic configuration of sodium, which has 11 electrons in its atoms. You can also show it in writing as 2.8.1 (each dot separates two occupied shells).

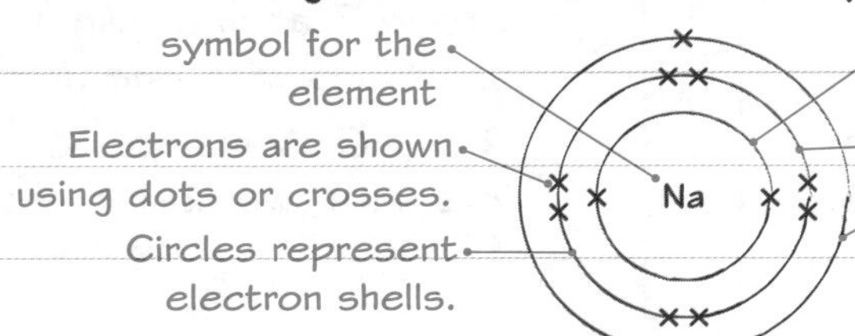
symbol for the element
Electrons are shown using dots or crosses.
Circles represent electron shells.
Na
1st (inner) shell holds 2 electrons.
2nd shell holds 8 electrons.
3rd shell holds 8 electrons (there is only 1 shown here, because sodium atoms only have 11 electrons).

Electronic configurations and the periodic table

The electronic configuration of hydrogen is 1. Below is a 'short form' of the periodic table for the remaining first 20 elements (He to Ca). In general, for these 20 elements:

- metal atoms have up to two electrons in their outer shell (the exception is Al which has three)
- non-metal atoms (except for group 0) have four or more electrons in their outer shell.

	1	2	3	4	5	6	7	0
Period 1								2 He 2
Period 2	2.1 Li 3	2.2 Be 4	2.3 B 5	2.4 C 6	2.5 N 7	2.6 O 8	2.7 F 9	2.8 Ne 10
Period 3	2.8.1 Na 11	2.8.2 Mg 12	2.8.3 Al 13	2.8.4 Si 14	2.8.5 P 15	2.8.6 S 16	2.8.7 Cl 17	2.8.8 Ar 18
Period 4	2.8.8.1 K 19	2.8.8.2 Ca 20						

The number of electrons in the outer shell is the same as the group number (except for elements in group 0, which have full outer shells).

The number of occupied shells is the same as the period number.

Elements in a group have the same number of electrons in their outer shell, except for helium in group 0.

Worked example

Flerovium is an element that was discovered at the end of the twentieth century.

Its electronic configuration is 2.8.18.32.32.18.4.

Explain where it should be placed in the periodic table. **(2 marks)**

Flerovium should be placed in group 4 because it has four electrons in its outer shell. It should be placed in period 7 because it has seven occupied shells.

You only need to be able to predict the electronic configurations of the first 20 elements. You should be able to do this in writing (as above) and as diagrams.

The total number of electrons in the atom is 114 (2 + 8 + 18 + 32 + 32 + 18 + 4).

This means that flerovium atoms also have 114 protons. Its atomic number is 114, so flerovium should go between elements 113 and 115.

Now try this

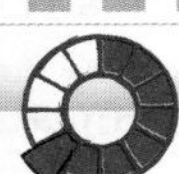

1 Draw the electronic configuration of chlorine, atomic number 17. **(2 marks)**

2 Describe the links between the electronic configuration of an element and its position in the periodic table. **(2 marks)**

Had a look ☐ Nearly there ☐ Nailed it! ☐

Ions

An **ion** is an atom or a group of atoms with a positive or negative charge.

Cations

A **cation** is:

- a positively charged ion
- formed when an atom or group of atoms loses one or more electrons.

Cations usually form from hydrogen or metals:

- Group 1 atoms lose 1 electron to form ions with one positive charge, +.
- Group 2 atoms lose 2 electrons to form ions with two positive charges, 2+.

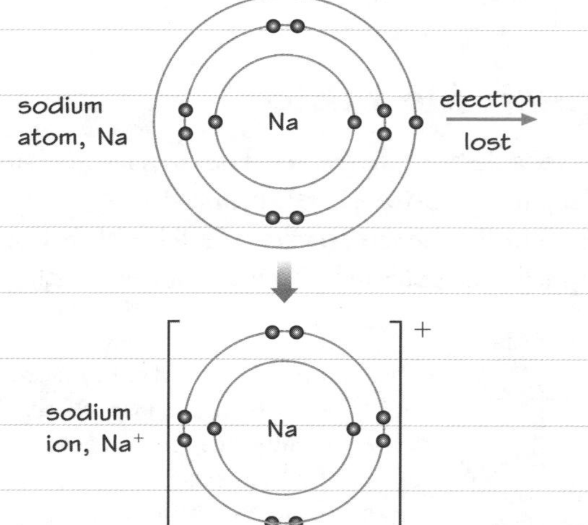

Anions

An **anion** is:

- a negatively charged ion
- formed when an atom or group of atoms gains one or more electrons.

Anions usually form from non-metals:

- Group 7 atoms gain 1 electron to form ions with one negative charge, –.
- Group 6 atoms gain 2 electrons to form ions with two negative charges, 2–.

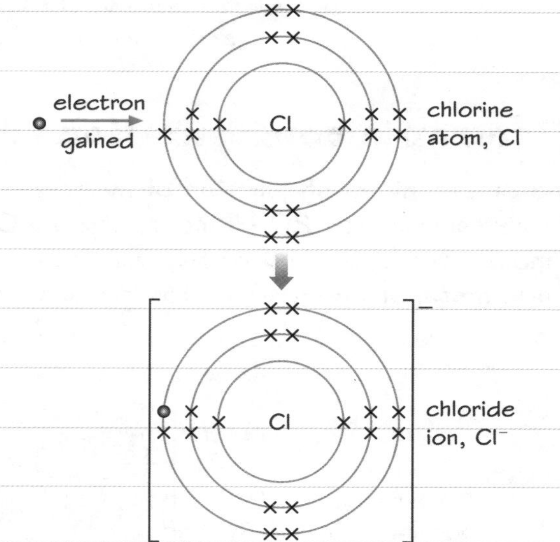

2D models of 3D objects

In these **dot-and-cross diagrams** each dot or cross represents an electron.

Worked example

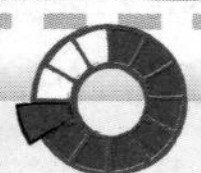

Calculate the number of protons, neutrons and electrons in a sodium ion, $^{23}_{11}Na^{+}$. **(3 marks)**

number of protons = 11

number of neutrons = (23 – 11) = 12

number of electrons = 11 – 1 = 10

Remember that neutral atoms have equal numbers of protons and electrons.

Reduce the number of electrons for positive ions and increase it for negative ions.

For example, for $^{16}_{8}O^{2-}$:

- number of protons = 8
- number of electrons = 8 + 2 = 10

Now try this

In which groups are these elements found?

Look at the periodic table on page 94.

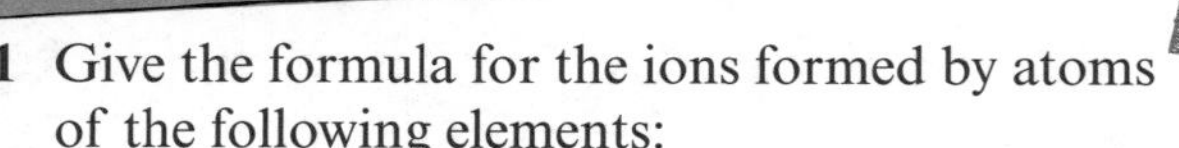

1 Give the formula for the ions formed by atoms of the following elements:

(a) lithium **(1 mark)**
(b) magnesium **(1 mark)**
(c) sulfur **(1 mark)**
(d) bromine. **(1 mark)**

2 Describe the meaning of the term **ion**. **(2 marks)**

3 Calculate the number of protons, neutrons and electrons in the following ions:

(a) $^{40}_{20}Ca^{2+}$ **(3 marks)**
(b) $^{19}_{9}F^{-}$ **(3 marks)**

Formulae of ionic compounds

You should be able to write the formulae of ionic compounds from the formulae of their ions.

Naming ions

An ion's name depends on the charge, and whether the ion also contains oxygen. **Compound ions** contain atoms of two different elements.

- Positively charged ions formed from hydrogen or metal atoms take the name of the element:

Formula	Name	
H^+	hydrogen	
Li^+	lithium	group 1
Na^+	sodium	
K^+	potassium	
Mg^{2+}	magnesium	group 2
Ca^{2+}	calcium	
Ba^{2+}	barium	
Al^{3+}	aluminium	group 3
Ag^+	silver	transition metals (the elements between groups 2 and 3)
Cu^{2+}	copper	
Zn^{2+}	zinc	
Fe^{2+}	iron(II)	
Fe^{3+}	iron(III)	
NH_4^+	ammonium	compound ion

- Negatively charged ions formed from single non-metal atoms take the name of the element, but end in **-ide**.
- Negatively charged ions in compounds containing three or more elements, one of which is oxygen, end in **-ate**:

Formula	Name	
F^-	fluoride	group 7
Cl^-	chloride	
Br^-	bromide	
I^-	iodide	
O^{2-}	oxide	group 6
S^{2-}	sulfide	
NO_3^-	nitrate	compound ions
CO_3^{2-}	carbonate	
SO_4^{2-}	sulfate	
OH^-	hydroxide	

OH^- does not follow the -ate rule.

Worked example

Write down the formulae of these ionic compounds:

(a) sodium fluoride **(1 mark)**

NaF

(b) magnesium oxide **(1 mark)**

MgO

(c) iron(III) oxide **(1 mark)**

Fe_2O_3

(d) sodium nitrate **(1 mark)**

$NaNO_3$

(e) barium nitrate. **(1 mark)**

$Ba(NO_3)_2$

If a compound ion is present **and** more than one of these ions is needed in the formula:

- put the compound ion in brackets, with the number after the closing bracket.

Maths skills: Balancing charges

This is simple if both ions have:

✓ 1 charge, as in NaF (Na^+ and F^-), or

✓ 2 charges, as in MgO (Mg^{2+} and O^{2-}).

If the ions have different numbers of charges, it may help to multiply them together. For example, for Fe^{3+} and O^{2-}, $(3 \times 2) = 6$. This means that you need:

✓ 2 Fe^{3+} ions to get 6 positive charges,

✓ 3 O^{2-} ions to get 6 negative charges.

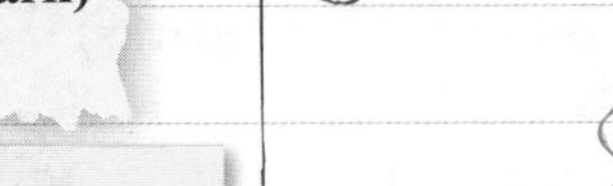

So the formula of iron(III) oxide is Fe_2O_3.

Now try this

You can use the tables on this page to help you.

Give the formulae of these ionic compounds:

(a) calcium sulfide **(1 mark)**

(b) iron(II) chloride **(1 mark)**

(c) ammonium hydroxide **(1 mark)**

(d) ammonium carbonate **(1 mark)**

(e) sodium sulfate. **(1 mark)**

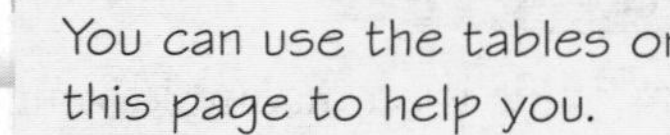

Had a look ☐ Nearly there ☐ Nailed it! ☐

Properties of ionic compounds

You can explain physical properties of ionic compounds in terms of bonding and structure.

Bonding

Ionic bonds are strong **electrostatic forces** of attraction between oppositely charged ions. For example, when sodium reacts with chlorine to form sodium chloride, NaCl:

- electrons transfer from sodium atoms to chlorine atoms
- Na^+ ions and Cl^- ions form
- Na^+ and Cl^- ions attract each other.

Structure

The ions in an ionic compound form a **lattice** structure which has:

- a regular arrangement of ions
- ionic bonds between oppositely charged ions.

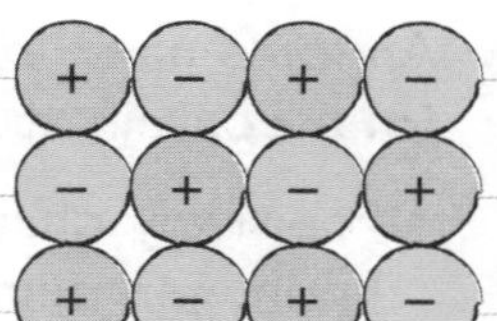

Melting and boiling points

Ionic compounds usually have:

- high melting points
- high boiling points.

As a result, they are in the **solid** state at room temperature. You can explain this in terms of bonding and structure:

- There are many strong ionic bonds.
- Large amounts of energy must be transferred to the lattice structure to break these bonds.

Solubility in water

Ionic compounds are often **soluble** in water. They **dissolve** to form **aqueous solutions**.

You can revise more about this on page 126.

Maths skills: 3D models

The diagram above models an ionic lattice in two dimensions. However, the lattice extends in three dimensions. It is called a **giant** lattice because it involves very many ions.

The structure of sodium chloride can be modelled in three dimensions as follows:

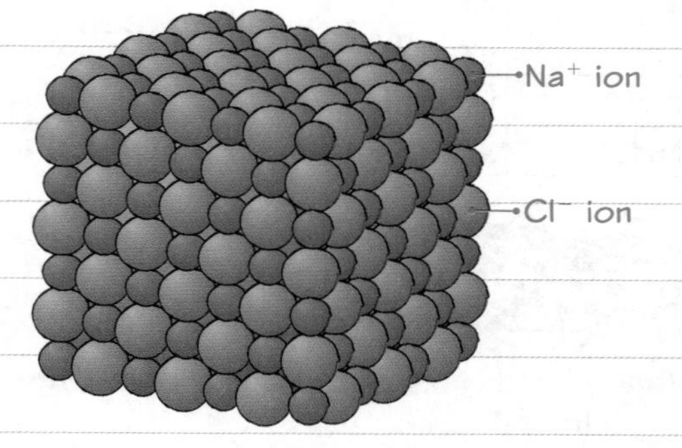

Worked example

(a) State why sodium chloride does **not** conduct electricity when it is in the solid state. **(1 mark)**

The ions are not free to move around in a solid.

> Although ions are electrically charged, they are held in fixed positions in the lattice structure.

(b) Explain why sodium chloride **does** conduct electricity when it is molten or in aqueous solution. **(2 marks)**

The ions are free to move around when sodium chloride is liquid or when it is dissolved in water. This means they can carry electric charge from place to place.

> An electric current is a flow of charge. A substance will conduct electricity if:
> - it contains charge carriers (such as ions).
>
> These charge carriers are free to move through the substance.

Now try this

Aluminium oxide is an insoluble ionic compound.

(a) State why solid aluminium oxide cannot conduct electricity. **(1 mark)**

(b) Describe how you could make aluminium oxide conduct electricity. **(2 marks)**

(c) Suggest why the method described in **(b)** may be expensive. **(2 marks)**

Covalent bonds

A **covalent bond** is formed when a pair of electrons is shared between two atoms.

A shared pair of electrons

Covalent bonds:

- are strong
- form between non-metal atoms
- often produce **molecules**, which can be elements or compounds.

A hydrogen atom can form one covalent bond. Usually, for atoms of other non-metals:

- number of bonds = (8 – group number).

Group	Example	Covalent bonds
4	carbon, C	4
5	nitrogen, N	3
6	oxygen, O	2
7	chlorine, Cl	1
0	helium, He	none

Helium and other elements in group 0 have full outer shells. They do not transfer or share electrons, so they are unreactive.

Modelling covalent bonds

There are three ways you can represent a covalent bond, e.g. hydrogen, H_2:

1. dot-and-cross (with shells)

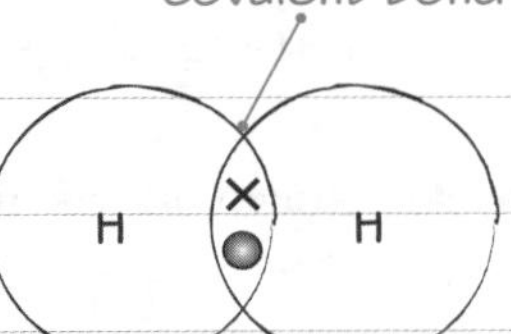

2. dot-and-cross (without shells)

H ● H
  ×

3. **structural formula**

H–H

The size of atoms is around 1×10^{-10} m and the size of simple molecules is around 1×10^{-9} m.

Some examples of dot-and-cross diagrams

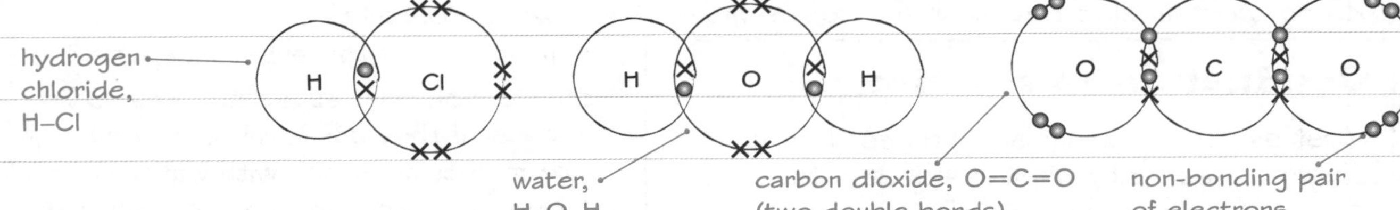

When you draw diagrams like these:

- draw overlapping circles to represent the outer shells
- draw a dot to represent an electron from one atom, and a cross to represent an electron from the other atom (make sure each bonding pair of electrons is between the two atoms)
- make sure that each atom (apart from hydrogen) has eight electrons in its outer shell.

Worked example

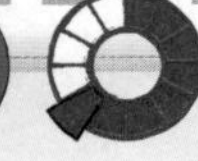

Methane, CH_4, can be represented by this structure:

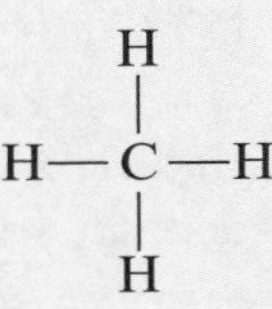

Draw a dot-and-cross diagram to represent methane. **(1 mark)**

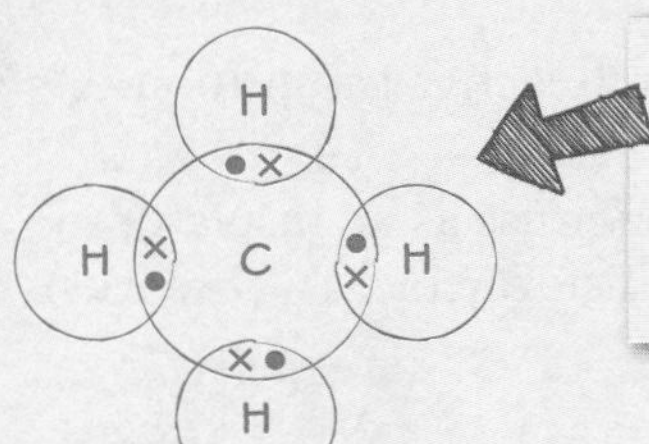

Hydrogen atoms do not have non-bonding pairs of electrons, but other atoms may do.

Now try this

1 Describe covalent bonding. **(2 marks)**

The molecule contains a double bond.

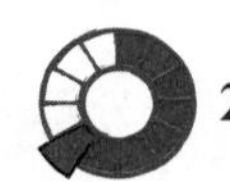

2 Draw a dot-and-cross diagram to represent a molecule of oxygen, O_2. **(2 marks)**

Had a look ☐ Nearly there ☐ Nailed it! ☐

Simple molecular substances

You can explain the physical properties of simple molecular substances in terms of their bonding and structure.

Bonding

A **simple molecule** consists of just a few atoms, joined to each other by strong covalent bonds. Simple molecular substances can be:

- non-metal elements, such as H_2, O_2, Cl_2
- compounds of non-metals, such as HCl, H_2O, CH_4.

Simple molecular substances usually have:

- low melting points
- low boiling points.

They are usually in the **gas** or **liquid** state at room temperature because of this.

Simple molecular substances with relatively large molecules, such as wax, are in the solid state at room temperature.

Melting and boiling points

There are weak attractive forces between molecules, called **intermolecular forces**.

When a simple molecular substance such as oxygen, O_2 or O=O, melts or boils:

- ✓ intermolecular forces are overcome
- ✓ covalent bonds do not break.

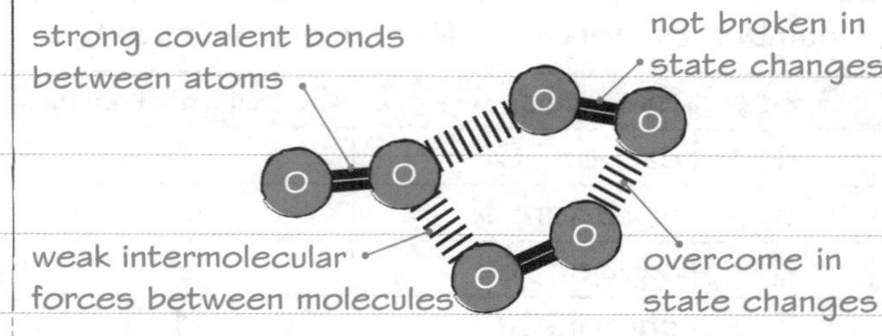

Non-conductors of electricity

Simple molecular substances do not conduct electricity when solid, liquid or gas. This is because their molecules:

- are not electrically charged, and
- do not contain electrons that are free to move.

Conducting in solution

Most simple molecular substances do not conduct electricity when in solution. However:

- ✓ some of them break down when they dissolve in water, forming ions
- ✓ these ions can move around, so the solution conducts electricity.

Solubility in water

Many simple molecular substances are **insoluble** in water. The intermolecular forces between water and these substances are weaker than those between:

- water molecules
- molecules of the substance itself.

Simple molecular substances dissolve in water if they can form strong enough intermolecular forces with water molecules:

- Hydrogen and oxygen are **sparingly soluble** (very little dissolves).
- Chlorine, carbon dioxide, sulfur dioxide and ammonia are soluble in water.
- Ethanol and ethanoic acid are soluble.
- Sugar is soluble in water.

Worked example

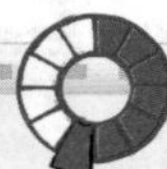

Nitrogen is a colourless, unreactive gas at room temperature.
Explain why it is suitable for use as an insulator in high-voltage electrical transformers. **(2 marks)**

Nitrogen is unreactive. So, it will not react with the materials used in the transformer. Nitrogen can insulate the parts in the transformer because it does not conduct electricity.

A nitrogen molecule consists of two nitrogen atoms joined together by a triple covalent bond: N≡N

A lot of energy must be transferred to break this very strong bond. This is why nitrogen is unreactive.

The answer includes only the properties of nitrogen that make it useful in transformers.

Now try this

1 Ammonia, NH_3, is a simple molecular substance. Explain, in terms of intermolecular forces, why it is in the gas state at room temperature. **(2 marks)**

2 Petrol and water are simple molecular substances. Explain why petrol does not dissolve in water. **(3 marks)**

Giant molecular substances

Giant molecules contain very many atoms, rather than just a few.

Bonding and structure

A **giant molecule** consists of many atoms. In giant molecules, the atoms are:

- joined by strong covalent bonds
- arranged in a regular lattice structure.

Giant molecular substances can be:

- non-metal elements, such as carbon
- compounds such as silica.

Giant molecular substances usually have:

- high melting points
- high boiling points.

They are solids at room temperature. A lot of energy must be transferred to break the many strong covalent bonds during melting and boiling. Giant molecular substances are insoluble in water.

Modelling giant molecules

There are many atoms in an entire giant molecule. You cannot represent an entire giant molecule using displayed formulae or dot-and-cross diagrams.

This is a **ball-and-stick model** of a small part of a silica molecule (found in sand):

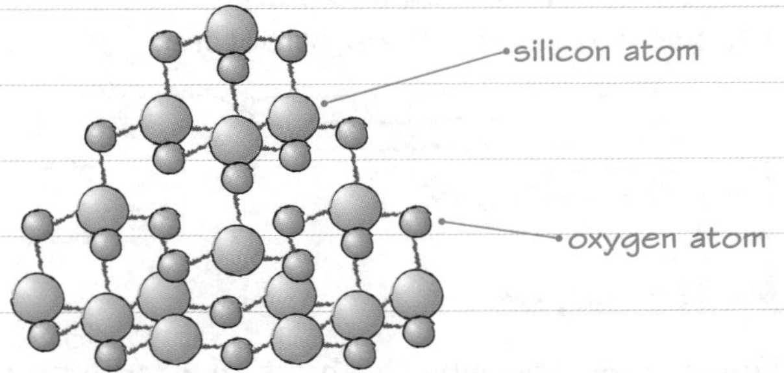

The formula of silica is SiO_2. As with other giant molecular substances, this is its **empirical formula** (see page 106).

Diamond

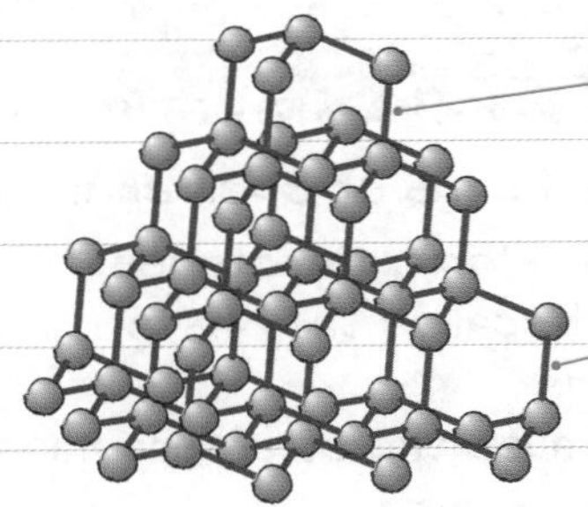

Graphite

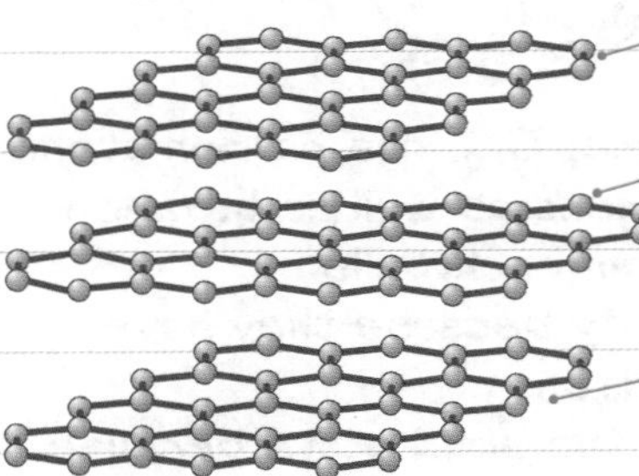

Diamond and graphite are both forms of carbon. They are giant molecular substances.

Worked example

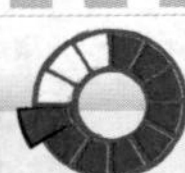

Graphite is used to make electrodes because it conducts electricity. Explain why it conducts electricity but diamond does not. **(4 marks)**

A carbon atom can form four covalent bonds. In graphite each carbon atom only forms three covalent bonds. The non-bonding outer electrons become delocalised. This means that they can move through the structure, so graphite conducts electricity. Diamond does not have delocalised electrons and does not conduct.

The rigid lattice structure and strong bonds of diamond make it very hard. This is why it is useful for cutting tools.

In graphite, the weak intermolecular forces let the layers slide over each other. This is why it is slippery and useful as a lubricant.

Metals also have delocalised electrons. This is why metals are good conductors of electricity. You can revise the structure and properties of metals on page 103.

Now try this

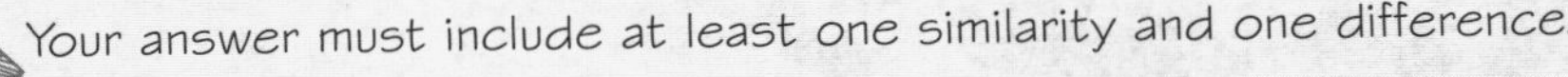

Your answer must include at least one similarity and one difference.

1 Compare and contrast the structure and bonding of diamond and graphite. **(4 marks)**

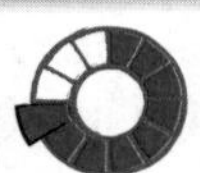

2 Explain, in terms of bonding and structure, why graphite is used as a lubricant. **(2 marks)**

Had a look ☐ Nearly there ☐ Nailed it! ☐

Other large molecules

Graphene and fullerenes are forms of carbon that exist as giant molecules.

Graphene

Graphene is a giant molecular substance. Its structure resembles a single layer of graphite:

- Each carbon atom is covalently bonded to three other carbon atoms.
- It has a regular lattice structure.

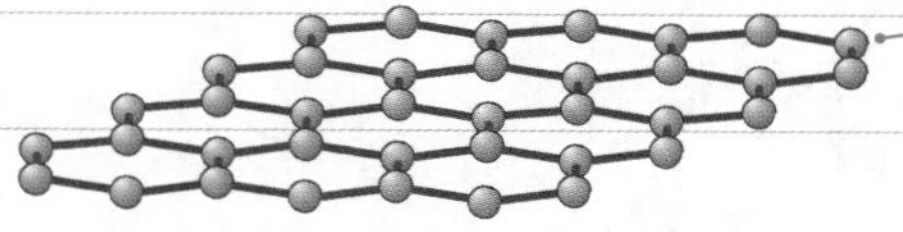

interlocking hexagonal rings of carbon atoms

Properties of graphene

Graphene conducts electricity:

- ✓ The non-bonding outer electrons become delocalised.
- ✓ They can move through the structure.

Graphene is very strong and flexible:

- ✓ It contains many strong covalent bonds.

Graphene is almost transparent:

- ✓ Its layers are just one atom thick.

Fullerenes

Fullerenes resemble a sheet of graphene rolled to form:

- hollow balls, often called **buckyballs**

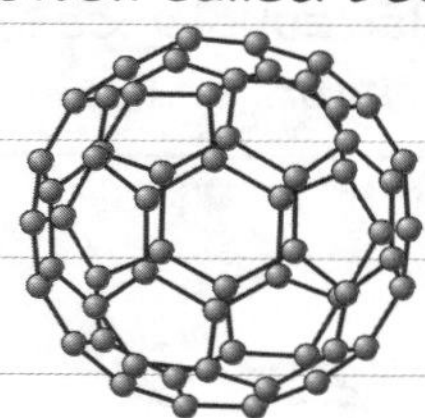

Buckminsterfullerene, C_{60}, has carbon atoms arranged in pentagons as well as hexagons. Materials made from buckyballs:

- conduct electricity because they have delocalised electrons
- are soft when in the solid state because they have weak intermolecular forces.

- hollow tubes, called carbon **nanotubes**.

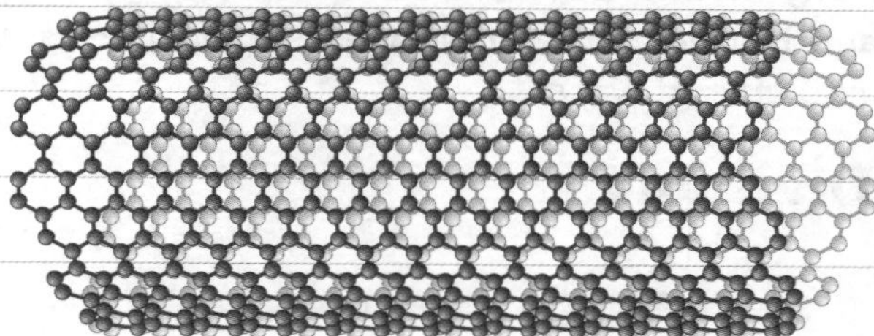

Nanotubes have closed ends or open ends. They can be several mm long. Nanotubes:

- conduct electricity because they have delocalised electrons
- are very strong because the structure has many strong covalent bonds.

Worked example

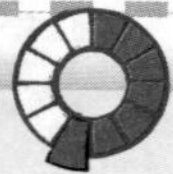

The diagram shows a section of a molecule of poly(ethene), a simple polymer.

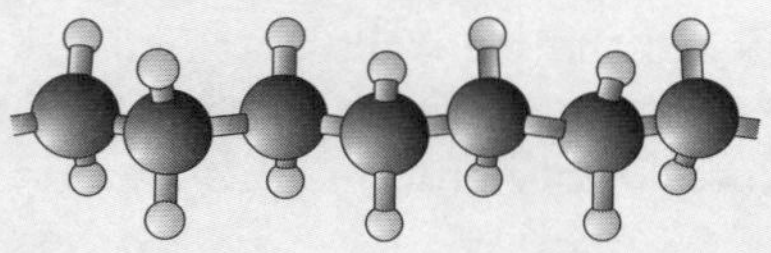

Describe the structure of poly(ethene). **(2 marks)**

Poly(ethene) consists of large molecules containing chains of carbon atoms. These atoms are joined to each other, and to hydrogen atoms, by covalent bonds.

Polymers are large molecules made from many smaller molecules, called **monomers**, joined together.

Poly(ethene) is not a fullerene:

- It is a **hydrocarbon** (a compound of carbon and hydrogen).

Polymer molecules are described as **macromolecules** rather than giant covalent molecules.

Now try this

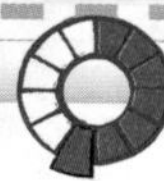

1 Describe the properties of graphene that make it suitable for use in flexible touch screens. **(2 marks)**

2 Explain why buckminsterfullerene has a much lower melting point than diamond. **(3 marks)**

Metals

Most elements are metals, and are placed on the left-hand side of the periodic table.

Metals versus non-metals

Some typical physical properties:

Property	Metals	Non-metals
appearance	shiny	dull
electrical conduction	good conductors	poor conductors
density	high	low
melting point	high	low

Mercury is liquid at room temperature.

Diamond and graphite have very high melting points. Graphite conducts electricity.

In addition:

- metals are **malleable** – they can be pressed into shape without shattering
- non-metals are **brittle** in the solid state – they shatter when bent or hit.

Metallic structure and bonding

A metal:

- consists of a giant lattice of positively charged metal ions
- has a 'sea' of delocalised electrons.

The delocalised electrons come from the outer shells of the atoms.

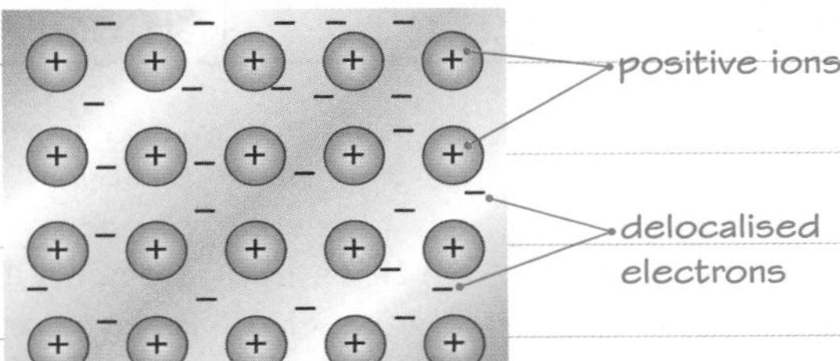

Metallic bonds are strong electrostatic forces of attraction between positive metal ions and delocalised electrons.

Malleable metals

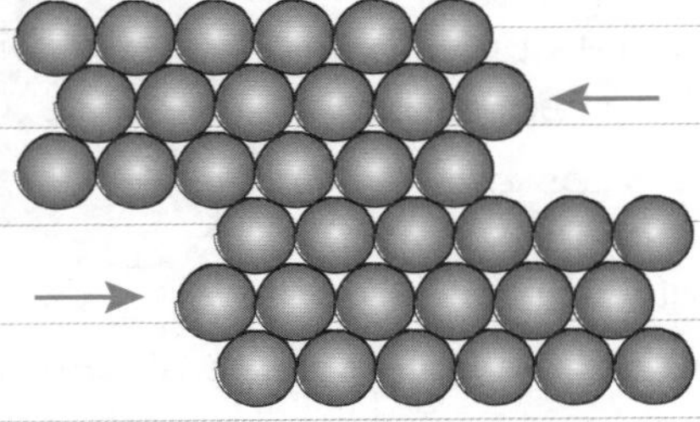

If a force is applied to a metal:

- layers of positive ions slide over each other
- the metal changes shape without shattering.

Insoluble metals

Metals are insoluble in water. However, some metals do seem to dissolve in water.

This is because they react with the water to produce soluble metal hydroxides. These dissolve, exposing more metal to the water.

For example, sodium reacts with water, forming sodium hydroxide solution and hydrogen:

$$2Na(s) + 2H_2O(l) \rightarrow 2NaOH(aq) + H_2(g)$$

You can revise the reactions of sodium and other group 1 metals on page 143.

Worked example

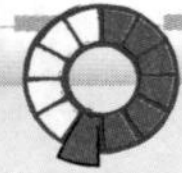

Copper is used in electricity cables. Explain, in terms of structure and bonding, why metals are good conductors of electricity. **(2 marks)**

Metals contain delocalised electrons which can move through the structure of the metal.

Copper is also **ductile**. It can be pulled to make wires without breaking. This is because its layers of positive ions can slide over each other.

The answer makes it clear that the delocalised electrons, not the metal ions, move through the structure.

Now try this

1 Aluminium is used to make overhead mains electricity cables. In terms of its structure and bonding, explain why aluminium is:
(a) ductile **(2 marks)**
(b) a good conductor of electricity. **(2 marks)**

2 Mercury is a liquid metal at room temperature. State whether it should conduct electricity, giving a reason for your answer. **(1 mark)**

Limitations of models

The structure and bonding of different substances are represented using **models**. Different models have different features and limitations. All the examples on this page refer to ethanoic acid.

Written formulae

The formula for a substance can be written as:

- an empirical formula
- a molecular formula
- a **structural** formula.

Empirical	Molecular	Structural
CH_2O	$C_2H_4O_2$	CH_3COOH

The simplest whole number ratio of atoms of each element. This does not show how the atoms are arranged, or (usually) the actual number of atoms.

The number of atoms of each element. This does not show how the atoms are arranged.

The number of atoms of each element. This does give an idea of how they are arranged.

Drawn structures

When you draw a structure, you should show all the covalent bonds in the molecule.

This carbon atom is covalently bonded to three hydrogen atoms. It is also covalently bonded to another carbon atom.

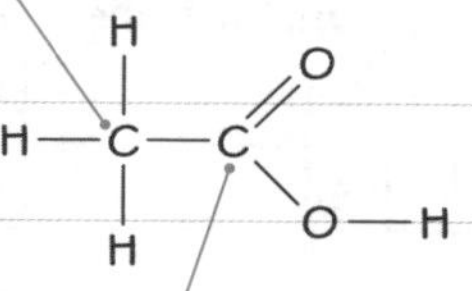

This carbon atom is covalently bonded to two oxygen atoms. It has a double bond to one of them.

This model does not show:

- the molecule's three-dimensional shape
- the bonding and non-bonding electrons.

Ball-and-stick models

You can draw **ball-and-stick models**. You can also make them using plastic modelling kits.

These models show:

- how each atom is bonded to other atoms
- the molecule's three-dimensional shape.

They do not show the bonding and non-bonding electrons, or each element's chemical symbol.

Space-filling models

Similar to ball-and-stick models but more accurately represent:

✓ the sizes of atoms relative to their bonds.

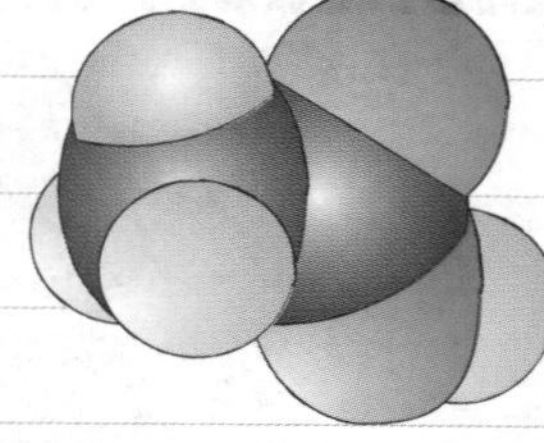

You may not be able to see all the atoms in a complex space-filling model.

Worked example

The diagram below is a dot-and-cross diagram.

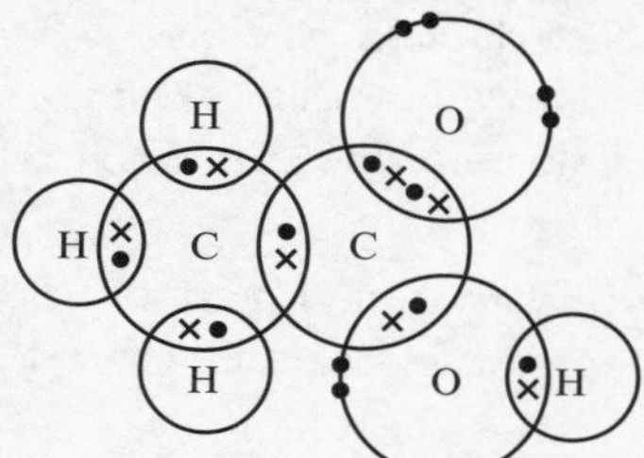

Describe the information it provides, and state a limitation of this model. **(4 marks)**

Dot-and-cross diagrams are two-dimensional models. This one may lead you to think incorrectly that ethanoic acid molecules are flat.

The diagram shows the symbol for each atom in the molecule. It also shows how each atom is bonded to other atoms. The pairs of electrons in each covalent bond are shown by dots and crosses. Non-bonding pairs of electrons in the outer shells are included.

It does not show the three-dimensional shape of the molecule.

Now try this

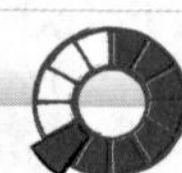

State **one** limitation of each model described on this page. **(7 marks)**

Relative formula mass

You should be able to calculate relative formula masses when given relative atomic masses.

Calculating relative formula mass

Relative formula mass has the symbol M_r.

To calculate the M_r of a substance, add together the relative atomic masses of all the atoms shown in its formula:

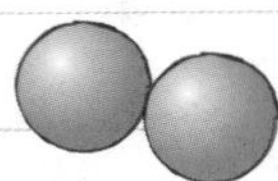

oxygen molecule – formula O_2
relative atomic mass of oxygen = 16
relative formula mass = 2 × 16
= 32

No units

M_r values are just numbers.

This is because an M_r value is the mass of a molecule or unit of a substance compared with 1/12th the mass of a ^{12}C atom. The M in M_r stands for 'molecular'.

You might see or hear the term 'relative molecular mass'. This really applies only to covalent substances.

Worked example

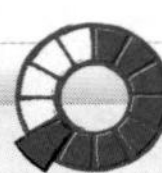

Calculate the relative formula mass of aluminium oxide, Al_2O_3. **(1 mark)**
(relative atomic masses: Al = 27, O = 16)

atoms in Al_2O_3:
(2 × Al) + (3 × O)
M_r = (2 × 27) + (3 × 16)
= 54 + 48
= 102

You do not need to learn any relative atomic masses. You will be given them in questions or you can find them on the periodic table.

This answer shows you the working out needed to obtain the answer.

If you show the working for steps in the calculation you may gain some marks even if your final answer is incorrect.

Worked example

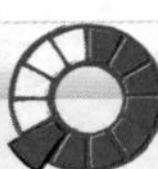

Calculate the relative formula mass of calcium nitrate, $Ca(NO_3)_2$. **(1 mark)**
(relative atomic masses: Ca = 40, N = 14, O = 16)

atoms in $Ca(NO_3)_2$:
(1 × Ca) + (2 × 1 × N) + (2 × 3 × O)
M_r = (1 × 40) + (2 × 14) + (6 × 16)
= 40 + 28 + 96
= 164

Maths skills You may find it easier if you first add up the A_r values for the atoms inside the brackets:

M_r of NO_3 = 14 + (3 × 16)
= 14 + 48
= 62

Then multiply your answer by the number outside, and add that to the remaining A_r values:

M_r of $Ca(NO_3)_2$ = (2 × 62) + 40
= 124 + 40
= 164

Now try this

relative atomic masses: H = 1, C = 12, O = 16, Na = 23, Al = 27, S = 32, Cl = 35.5, Cu = 63.5

Calculate the relative formula masses, M_r, of the following substances.

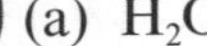

(a) H_2O **(1 mark)**
(b) CO_2 **(1 mark)**
(c) NaOH **(1 mark)**
(d) CCl_4 **(1 mark)**

(e) $CuCl_2$ **(1 mark)**

(f) Na_2SO_4 **(1 mark)**
(g) $Al(OH)_3$ **(1 mark)**
(h) $Al_2(CO_3)_3$ **(1 mark)**

Had a look ☐ Nearly there ☐ Nailed it! ☐

Empirical formulae

An **empirical formula** is the simplest whole number ratio of atoms of each element in a compound.

Calculating an empirical formula

A 10 g sample of a compound **X** contains 8 g of carbon and 2 g of hydrogen.

Step		C	H
1	Write the symbol of each element as a header.	C	H
2	Write down the mass of each element in g.	8	2
3	Write down the A_r of each element.	12	1
4	For each element, calculate: mass ÷ A_r	$\frac{8}{12} = 0.667$	$\frac{2}{1} = 2$
5	Divide each answer by the smallest answer (0.667 here).	$\frac{0.667}{0.667} = 1$	$\frac{2}{0.667} = 3$
6	You may then need to multiply all the numbers to remove fractions, then write out the empirical formula.	CH_3	

Finding a molecular formula

You can find the molecular formula of a compound from its empirical formula:

- if you know its relative formula mass, M_r.

The M_r of **X** in the example above is 30:

1. Calculate the M_r of the empirical formula:
   M_r of $CH_3 = 12 + (3 \times 1) = 15$
2. Divide the M_r of **X** by answer 1:
   $\frac{30}{15} = 2$
3. Multiply each number in the empirical formula by answer 2:

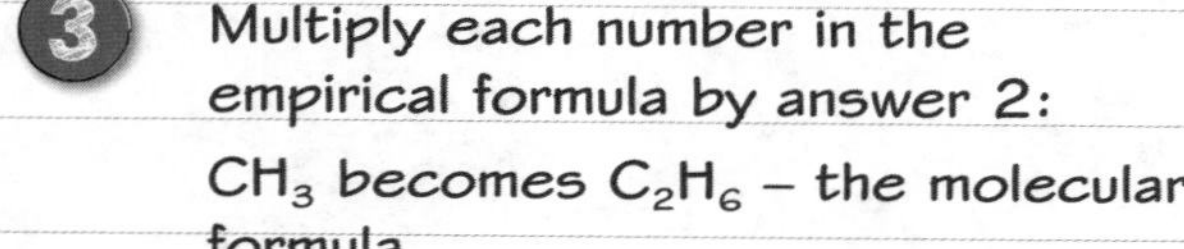

   CH_3 becomes C_2H_6 – the molecular formula

Practical skills: Determining empirical formula

You need to be able to describe an experiment to determine an empirical formula.

The apparatus below can be used to obtain results to do this for magnesium oxide.

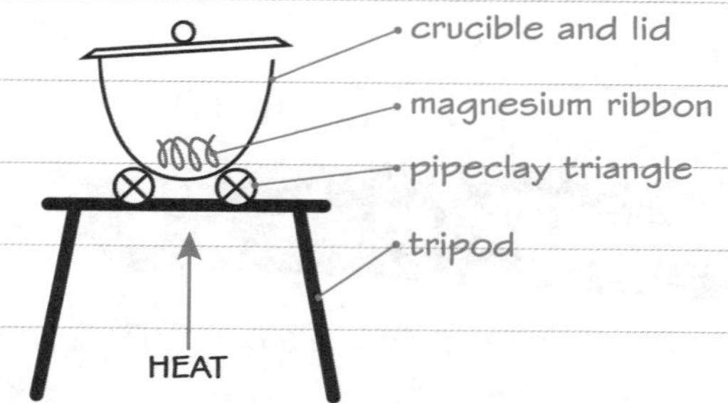

The crucible and its contents are weighed before and after heating the magnesium.

Worked example

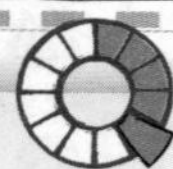

The table shows the results of an experiment to find the empirical formula of magnesium oxide.

Object	Mass (g)
empty crucible and lid	19.06
crucible, lid and Mg before heating	19.42
crucible, lid and Mg after heating	19.66

(a) Calculate the mass of magnesium used. **(1 mark)**

mass of magnesium = 19.42 – 19.06
= 0.36 g

(b) Calculate the mass of oxygen gained. **(1 mark)**

mass of oxygen = 19.66 – 19.42
= 0.24 g

Now try this

(a) Use the masses given in the Worked example to determine the empirical formula of magnesium oxide. (relative atomic masses: O = 16, Mg = 24) **(3 marks)**

(b) In the experiment described on this page, a lid is needed on the crucible. The lid must be kept slightly open during heating. Give **two** reasons why. **(2 marks)**

Conservation of mass

According to the **law of conservation of mass**, the total mass of reactants and products stays constant during a chemical reaction. The total mass before and after a reaction is the same.

Closed systems

A **closed system** is a situation in which no substances can enter or leave during a reaction.

Closed systems include:

- reactions in a sealed container, such as a flask fitted with a bung
- precipitation reactions in a beaker.

In a **precipitation reaction**, two **soluble** reactants form an **insoluble** product – the **precipitate**. For example:

$NaCl(aq) + AgNO_3(aq) \rightarrow NaNO_3(aq) + AgCl(s)$ ← forms a cloudy white precipitate

The total mass of the beaker and its contents stays the same during the formation of the silver chloride precipitate.

Non-enclosed systems

A **non-enclosed system** is a situation in which substances can enter or leave during a reaction.

Non-enclosed or 'open' systems include:

- reactions in an open flask, where a substance in the gas state may enter or leave.

Mass is conserved, but you will observe the following:

- The mass of a reactive metal **increases** if it is heated in air. This is because oxygen atoms combine with metal atoms to form a metal oxide.
- The mass of a reactive non-metal or a fuel **decreases** if it is heated in air. This is because products in the gas state escape from the container.
- The mass of a metal carbonate **decreases** if it is heated. This is because carbon dioxide gas is produced and this escapes from the container.

Worked example

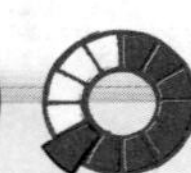

0.5 g of lithium metal is added to water. Calculate the mass of hydrogen gas produced. **(4 marks)**

(relative atomic masses: Li = 7, H = 1)

Always start with the balanced equation. You do not need state symbols.

$2Li + 2H_2O \rightarrow 2LiOH + H_2$

Work out the relative masses and multiply by the balancing numbers. You only need to do this for the substances asked for in the question.

$2 \times 7 \qquad 1 \times (1 + 1)$

$14 \qquad 2$

14 g of Li → 2 g of H_2

Divide by the number for lithium to find the mass of hydrogen produced for each gram of lithium.

1 g → $\frac{2}{14}$ g = 0.143 g

Multiply by the mass of lithum given in the question.

0.5 g → 0.143 g × 0.5 = 0.072

0.072 g of hydrogen is produced.

Now try this

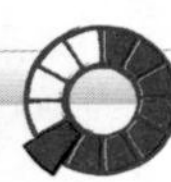

1 Methane, CH_4, burns in oxygen to form carbon dioxide and water.
$CH_4 + 2O_2 \rightarrow CO_2 + 2H_2O$
Calculate the mass of oxygen used when 10 g of methane burns.
(relative atomic masses: C = 12, H = 1, O = 16) **(4 marks)**

2 Potassium reacts with chlorine to form potassium chloride, KCl.
Calculate the mass of potassium needed to produce 20 g of potassium chloride.
(relative atomic masses: K = 39, Cl = 35.5) **(4 marks)**

Had a look ☐ Nearly there ☐ Nailed it! ☐

Reacting mass calculations

A balanced chemical equation is also called a **stoichiometric equation**.

Limiting or in excess?

A reactant is **in excess** if there is enough:

- to react with all the other reactant
- for some of it to be left over when the reaction stops.

The other reactant is the **limiting reactant**. For example, when magnesium reacts with hydrochloric acid the magnesium ribbon gradually becomes smaller:

Situation after reaction	Magnesium	Hydrochloric acid
some magnesium left	in excess	limiting
no magnesium left	limiting	in excess

The **stoichiometry** of a reaction is to do with the ratio of the amounts of reactants and products involved.

When you balance a chemical equation you are finding the 'stoichiometric equation'.

Limiting reactant and mass of product

The mass of product formed is controlled by the mass of the reactant that is not in excess:

✓ the reaction continues until all the particles of the limiting reactant have been used up.

Maths skills: Directly proportional

The mass of product is directly proportional to the mass of the limiting reactant, for example:

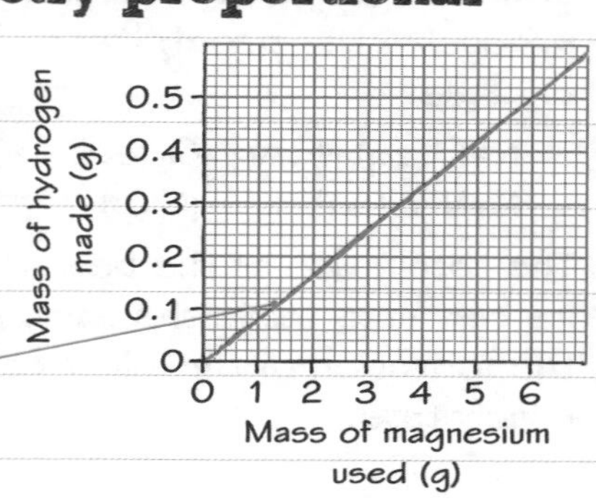

positive gradient and passes through the origin

Worked example

An oxide of copper is heated with excess hydrogen, H_2. In the reaction, 1.27 g of copper and 0.18 g of water, H_2O, form.
Use this information to determine the stoichiometry of the reaction. (relative atomic masses: H = 1, O = 16, Cu = 63.5) **(4 marks)**

amount of Cu $= \frac{1.27}{63.5} = 0.02$

M_r of $H_2O = 1 + 1 + 16 = 18$

amount of $H_2O = \frac{0.18}{18} = 0.01$

ratio of Cu:H_2O = 0.02:0.01
= 2:1

Right-hand side must be $\rightarrow 2Cu + H_2O$

So the left-hand side must be $Cu_2O + H_2 \rightarrow$

Equation is: $Cu_2O + H_2 \rightarrow 2Cu + H_2O$

Hydrogen is in excess, so the oxide of copper must be the limiting reactant. This means that the amount of the oxide of copper controls the amounts of copper and water produced.

The calculated amounts of each product are in moles (mol). You can revise mole calculations on page 110.

The oxide of copper must contain two copper atoms for every oxygen atom. This is why its formula is Cu_2O (and not CuO, the black copper(II) oxide you normally see).

Now try this

1 Explain why, in a chemical reaction, the mass of the reactant not in excess determines the mass of product formed. **(2 marks)**

2 Powdered carbon is heated with excess steam, producing 3.3 g of carbon dioxide and 0.30 g of hydrogen. Determine the stoichiometry of the reaction. (relative formula masses: $CO_2 = 44$, $H_2 = 2$) **(4 marks)**

Concentration of solution

You need to be able to calculate the concentration of solutions in g dm^{-3}.

Solute, solvent and solution

A **solution** is a mixture of a solute in a solvent:

- The **solute** is the substance that dissolves.
- The **solvent** is the substance that the solute dissolves in.

Water is the solvent in an **aqueous solution**. The state symbol for an aqueous solution in balanced equations is (aq). The symbol (l) is for substances in the liquid state.

dm^3 and cm^3

Measuring cylinders and other lab apparatus show volumes in cubic centimetres, cm^3. You need to convert these measurements into cubic decimetres, dm^3, when you calculate concentrations. It helps to know that:

- ✓ 1 dm^3 = 10 × 10 × 10 = 1000 cm^3
- ✓ To convert cm^3 to dm^3, divide by 1000.

Mass and volume

To calculate the **concentration** of a solution, you need to know:

- the mass of solute in **grams**, g, and
- the volume of solution in **cubic decimetres**, dm^3.

If you are making a solution, you can use the volume of the solvent instead.

Mass, volume and concentration

You use this equation to calculate the concentration of a solution in g dm^{-3}:

$$\text{concentration (g dm}^{-3}\text{)} = \frac{\text{mass of solute (g)}}{\text{volume of solution (dm}^3\text{)}}$$

Units

The unit g dm^{-3} means 'grams per cubic decimetre'. You may also see it written as g/dm^3.

Rearranging equations

You need to be able to change the subject of an equation. For example:

- ✓ mass of solute = concentration × volume
- ✓ $\text{volume} = \frac{\text{mass of solute}}{\text{concentration}}$

Worked example

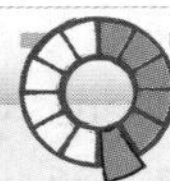

2.50 g of sodium hydroxide is dissolved in 250 cm^3 of water. Calculate the concentration of the solution formed in g dm^{-3}. **(2 marks)**

$250\ cm^3 = \frac{250}{1000} = 0.250\ dm^3$

$\text{concentration} = \frac{2.50\ g}{0.250\ dm^3} = 10\ g\ dm^{-3}$

Remember to convert the volume to dm^3 if it is given to you in cm^3.

The units are shown in the concentration calculation here. This makes it easier for you to see how it is done. You do not need to show units in your working out, but you must show the units in your final answer.

Now try this

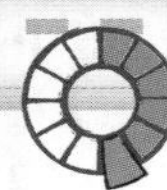

1 Calculate the concentrations of the following solutions formed:

(a) 0.40 g of glucose dissolved in 0.50 dm^3 of water. **(1 mark)**

(b) 1.25 g of copper chloride dissolved in 100 cm^3 of water. **(2 marks)**

2 Calculate the mass of sodium hydroxide needed to make 150 cm^3 of a 40 g dm^{-3} solution. **(2 marks)**

Avogadro's constant and moles

You need to be able to carry out calculations involving Avogadro's constant and the mole.

The mole

In chemistry, the 'amount' of a substance does not refer to its volume or mass.

The **mole** is the unit for amount of substance. It's shown as **mol** in calculations and values. One mole (1 mol) of particles of a substance is:

- Avogadro's constant number of particles (atoms, ions or molecules) of that substance.

The mass of 1 mol of particles is the 'relative particle mass' in grams.

Maths skills: Avogadro's constant

Avogadro's constant is 6.02×10^{23} mol^{-1} (to three significant figures).

The number 6.02×10^{23} is a number in its **standard form**. In general, you write such numbers as:

$a \times 10^n$

- $1 \leq a < 10$ (a is between 1 and 10)
- n: an integer (whole number)

6×10^{23} means 6 followed by 23 zeros:

600 000 000 000 000 000 000 000

Masses from moles

The mass of 1 mol of a substance is its A_r or M_r in grams. The table shows you three examples.

Substance	Mass of 1 mol
carbon, C, $A_r = 12$	12 g
oxygen, O_2, $M_r = 32$	32 g
carbon dioxide, CO_2, $M_r = 44$	44 g

To calculate the mass of an amount of substance:

$$\text{mass (g)} = M_r \times \text{amount (mol)}$$

(A_r for atoms)

Moles from masses

To calculate the amount of substance from a given mass:

$$\text{amount (mol)} = \frac{\text{mass (g)}}{A_r \text{ or } M_r}$$

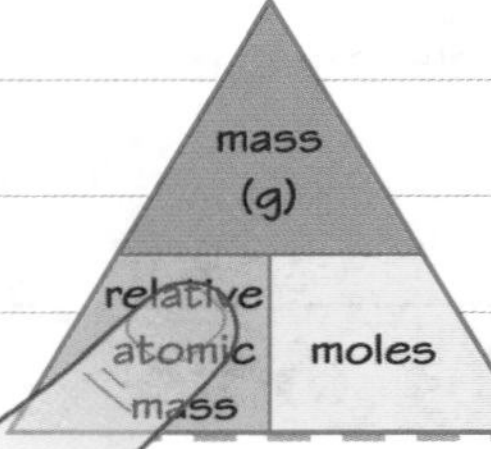

Worked example

(a) Calculate the amount, in mol, of ethane molecules in 45 g of ethane, C_2H_6. (relative atomic masses: C = 12, H = 1) **(2 marks)**

M_r of $C_2H_6 = (2 \times 12) + (6 \times 1) = 30$

amount $= \frac{45}{30} = 1.5$ mol

(b) Calculate the number of ethane molecules in 45 g of ethane. Give your answer to two significant figures. **(1 mark)**

number $= 1.5 \times 6.02 \times 10^{23}$
$= 9.0 \times 10^{23}$ ethane molecules

You also need to be able to calculate the mass of a substance if you are given its amount. For example, the mass of 2.0 mol of ethane $= 30 \times 2.0 = 60$ g

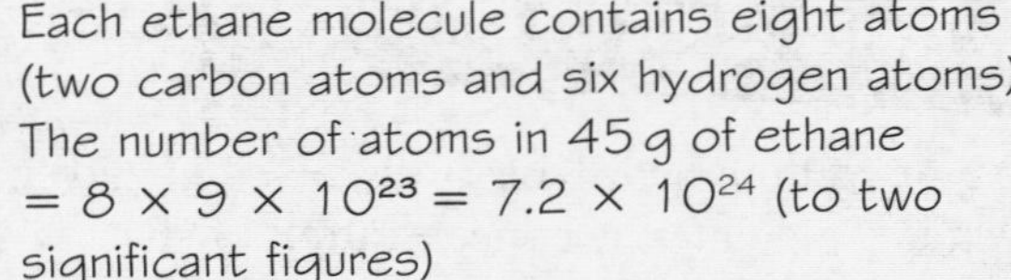

Each ethane molecule contains eight atoms (two carbon atoms and six hydrogen atoms). The number of atoms in 45 g of ethane $= 8 \times 9 \times 10^{23} = 7.2 \times 10^{24}$ (to two significant figures)

Now try this

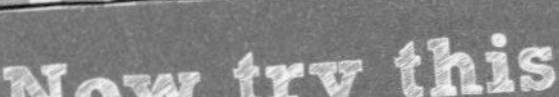

You will need to use Avogadro's constant and the A_r of carbon.

1 (a) Calculate the amount, in mol, of molecules in 22.5 g of water. (M_r of $H_2O = 18$) **(1 mark)**

(b) Calculate the amount, in mol, of atoms in 22.5 g of water. **(1 mark)**

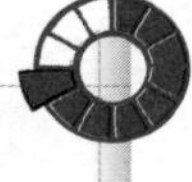

(c) Use your answer to (b) to calculate the number of atoms in 22.5 g of water. **(1 mark)**

2 (a) Calculate the number of atoms in 6.0 g of diamond. Give your answer to two significant figures. (A_r of C = 12) **(2 marks)**

(b) Calculate the mass, in grams, of 1.00×10^{12} carbon atoms. Give your answer to two significant figures. **(2 marks)**

Extended response – Types of substance

There will be at least one 6-mark question on your exam paper. For these questions, you will need to think scientifically and structure your answer logically, showing how the points you make are related to each other.

You can revise the topics for this question, which is about the **structure and bonding** of different types of substance, on pages 98–103.

Worked example

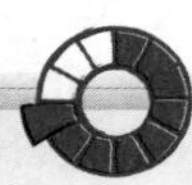

Sodium chloride is an ionic compound produced in the reaction between sodium and chlorine. The table shows information about how well these three substances conduct electricity at room temperature.

Substance	Ability to conduct
sodium	conducts
chlorine	does not conduct
sodium chloride	does not conduct

Explain, in terms of structure and bonding, the differences in the ability to conduct electricity.

(6 marks)

Sodium is a metal. It consists of a regular lattice of positively charged sodium ions attracted to a sea of delocalised electrons. These electrons are free to move, so solid sodium conducts electricity.

On the other hand, chlorine is a simple molecular substance. Its molecules are attracted to each other by weak intermolecular forces. Even though its molecules are free to move they are uncharged, so chlorine does not conduct.

Sodium chloride is an ionic compound. In the solid state, it contains oppositely charged ions held together strongly in a regular lattice. Therefore, even though it contains charged particles, these are not free to move. This is why solid sodium chloride does not conduct electricity.

Command word: Explain

When you are asked to **explain** something, it is not enough just to state or describe it. Your answer **must** contain some reasoning or justification of the points you make. Your explanation **can** include mathematical explanations, if calculations are needed.

In this example Extended response question, no calculations are needed.

You should give a description of the structure of a solid metal, making it clear that solid sodium contains sodium ions. You should also give the reason why sodium conducts electricity.

You should use **connectives** as you move from one idea to the next. Here the use of 'on the other hand' shows that this part of the answer is in contrast to the first part of the answer.

This part of the answer includes another connective: 'therefore' indicates that the sentence before has a consequence. In this case the lattice structure prevents the conduction of electricity.

You need to show comprehensive knowledge and understanding using relevant scientific ideas to support your explanations. You should consider each substance in turn, giving clear lines of reasoning. Leave out any information or ideas that are unnecessary to answer the question.

Now try this

Mercury is a metal and paraffin oil is a simple molecular compound. Both exist as liquids at room temperature. Zinc chloride exists as a liquid above 290 °C. Paraffin oil does not conduct electricity but the other two liquids do.

Explain, in terms of structure and bonding, the differences in the ability to conduct electricity.

(6 marks)

States of matter

The **particle theory** models the states of matter, with particles described as hard spheres.

	Solid	Liquid	Gas
Particle diagram			
Arrangement of particles	close together regular pattern	close together random	far apart random
Movement of particles	vibrate about fixed positions	move around each other	fast in all directions
Relative energy of particles	least stored energy	→	most stored energy

State changes

The interconversions between states of matter are called **state changes**.

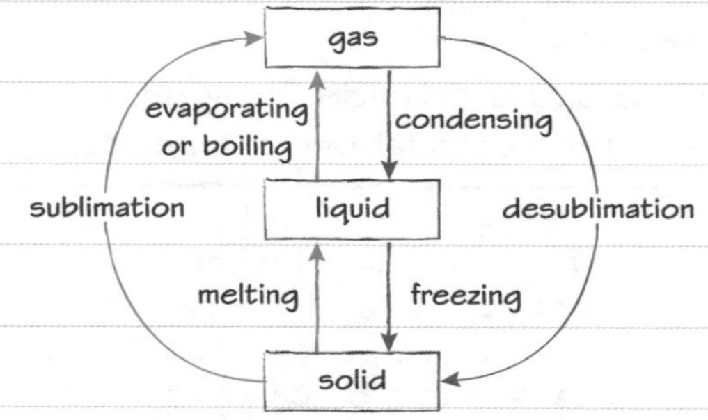

During a state change:

- energy is transferred to or from the particles
- the arrangement of particles changes
- the movement of particles changes.

Melting and boiling points

At its **melting point**, a substance begins to:

- melt if energy is transferred to the particles
- freeze if energy is transferred to the surroundings.

At its **boiling point**, a substance begins to:

- boil if energy is transferred to the particles
- condense if energy is transferred to the surroundings.

A substance evaporates if it changes from a liquid to a gas below its boiling point:

- Particles with high enough energy leave the surface of the liquid.
- The remaining particles have less energy.
- The liquid cools down unless it is heated.

Worked example

The table shows some data about bromine.

Melting point (°C)	Boiling point (°C)
−7	59

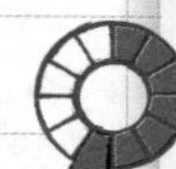

(a) Predict the state of bromine at 25 °C. **(1 mark)**

liquid

(b) Describe, in terms of the particle theory, what happens when bromine boils. **(3 marks)**

Forces of attraction between bromine particles are overcome and the particles move far apart. The particles no longer just move around each other, but move quickly in all directions.

A substance is a:

- solid below its melting point
- gas above its boiling point
- liquid between its melting and boiling points.

See pages 96–103 to revise the different types of bonds and forces. Some forces of attraction are overcome during melting, and the remainder during boiling.

State changes are **physical changes** because the particles themselves are unchanged.

Chemical reactions cause **chemical changes** – the particles are changed by rearranging atoms.

Now try this

Naphthalene is a solid that can change directly from a solid to a gas at room temperature. Name this change of state, and describe what happens to the particles. **(5 marks)**

Pure substances and mixtures

Elements and compounds

An **element** is a substance that consists only of atoms with the same atomic number (the same number of protons in their nucleus), e.g.

- Hydrogen is an element because its atoms all have one proton in their nucleus.
- Oxygen is a different element because its atoms all have eight protons.

A **compound** is a substance that consists of atoms of two or more different elements, chemically joined together.

For example, water is a compound. This is because it consists of hydrogen atoms and oxygen atoms chemically joined together.

You can revise **ionic bonds** on page 96 and **covalent bonds** on page 99.

Atoms and molecules

Elements exist as atoms or molecules.

Argon is in group 0. It has no tendency to gain, lose or share electrons. It exists as single atoms (that is, it is **monatomic**).

Hydrogen and oxygen exist as simple molecules.

Carbon exists as giant molecules (diamond, graphite and graphene).

Compounds exist as:

- ✓ molecules, such as water, H_2O
- ✓ ionic structures.

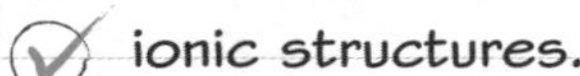

Pure substances and mixtures

In everyday use, the word 'pure' usually means that nothing has been added to a substance.

In chemistry, a **pure** substance contains only one element or compound, e.g.

- Pure hydrogen contains only hydrogen molecules.
- Pure water contains only water molecules.

Most substances are **mixtures** because they contain different elements and/or compounds. Mixtures are impure substances.

The components of a mixture are not chemically joined together.

Air

Air is a mixture of:

- ✓ elements, such as nitrogen, N_2, oxygen, O_2, and argon, Ar
- ✓ compounds, such as water, H_2O, and carbon dioxide, CO_2.

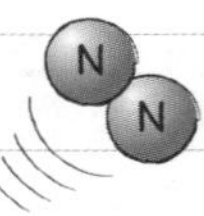

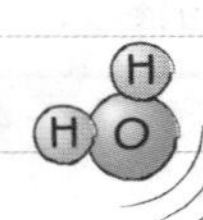

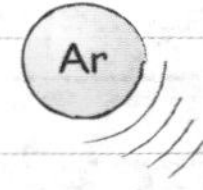

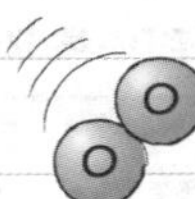

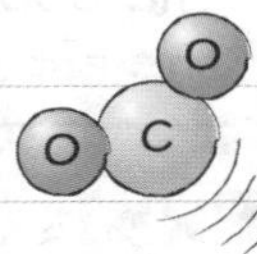

Worked example

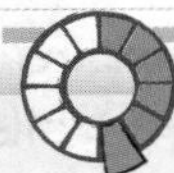

A student heats a sample of solid hexadecanol in a hot water bath. She measures its temperature at regular intervals until after it melts. The graph shows her results.

Explain how the results show whether the hexadecanol is pure or impure. **(2 marks)**

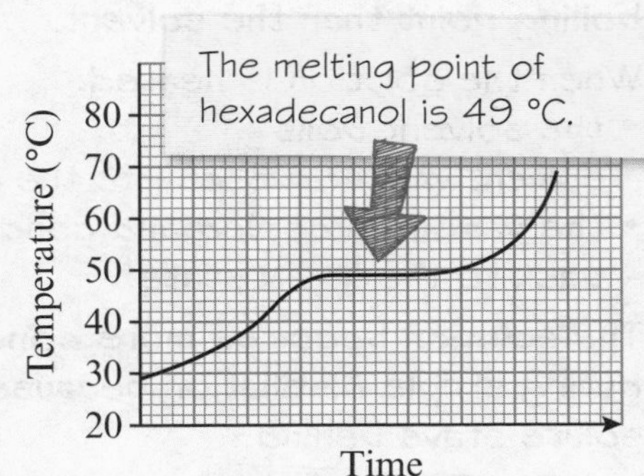

The hexadecanol is pure because it has a sharp melting point, shown by the horizontal part of the heating curve. A mixture would melt over a range of temperatures instead.

Now try this

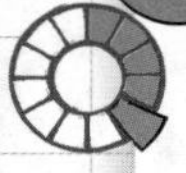

1 The label on a bottle describes the contents as 'Pure mineral water'.
Explain why the water is not pure in the scientific sense. **(3 marks)**

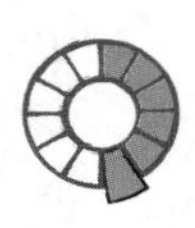

2 When it is heated, 18 carat gold melts between 915 °C and 963 °C.
Explain why its melting point is not a single temperature. **(2 marks)**

Had a look ☐ Nearly there ☐ Nailed it! ☐

Distillation

You can separate liquids from mixtures using **distillation**.

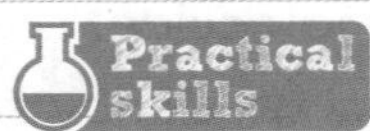

Simple distillation

You use **simple distillation** to separate:

- a solvent from a solution.

For example, you can separate water from sea water (a mixture of water and dissolved compounds) using this method.

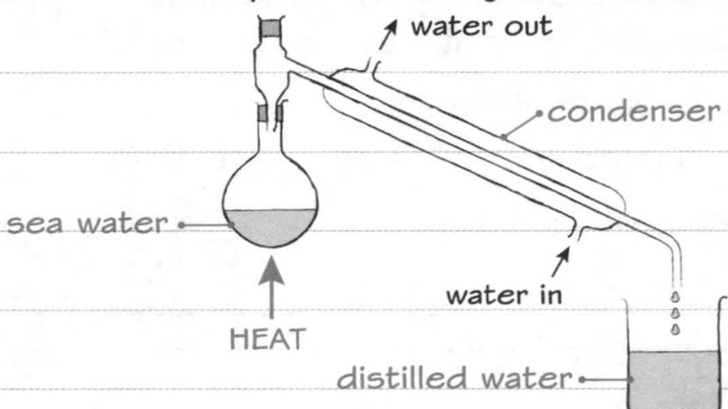

The **condenser** has two tubes, one inside the other:

- Cold water runs through the space between the two tubes, keeping the condenser cold.

The cooling water does not mix with the substance being separated.

Fractional distillation

You use **fractional distillation** to separate:

- a liquid from a mixture of **miscible** liquids (liquids that mix completely with each other).

For example, you can separate ethanol from a mixture of ethanol and water using this method.

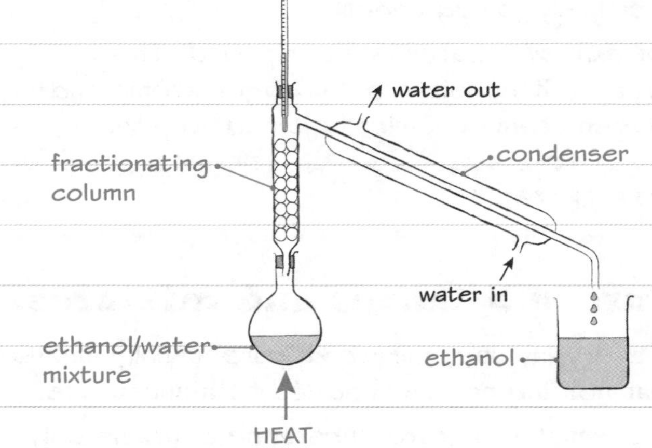

The **fractionating column** used in fractional distillation has a **temperature gradient**:

- hottest at the bottom, coldest at the top.

Explaining simple distillation

Simple distillation works because the solute in the solution has a much higher boiling point than the solvent.

When the solution is heated:

- the solvent boils
- solvent vapour passes into the condenser
- the vapour is cooled and condensed back to the liquid state.

The solution becomes more **concentrated** during simple distillation because the solute stays behind.

Explaining fractional distillation

Fractional distillation works because the liquids in the mixture have different boiling points.

When the mixture is heated:

- the mixture boils
- hot vapour rises up the fractionating column
- vapour condenses when it hits the cool surface of the column and drips back
- the fraction with the lowest boiling point reaches the top of the column first
- its vapour passes into the condenser.

If you carry on heating, vapours from fractions with higher boiling points pass to the condenser.

Now try this

(a) Explain the function of a condenser in distillation. **(2 marks)**

(b) State the physical property that allows substances to be separated by distillation. **(1 mark)**

(c) Describe the temperature gradient in a fractionating column. **(1 mark)**

Filtration and crystallisation

Filtration and **crystallisation** are two important techniques for separating mixtures.

Practical skills Filtration

You use **filtration** to separate an insoluble substance from a liquid or a solution. There are two reasons for doing this:

1. to purify a liquid or a solution by removing solid impurities from it, e.g. sand from sea water
2. to separate the solid you want from the liquid it is mixed with, e.g. to separate crystals from a solution after crystallisation.

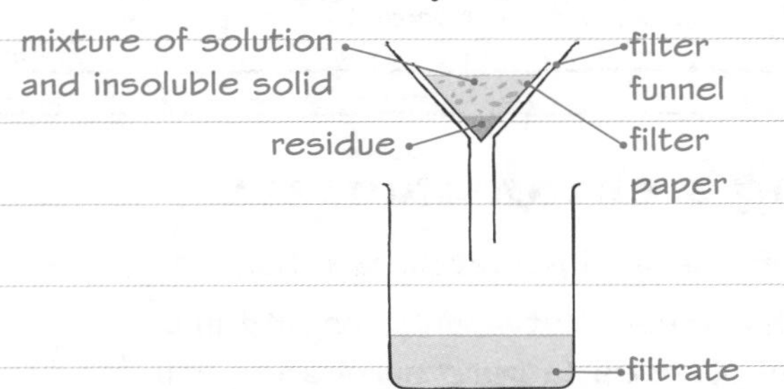

Practical skills Crystallisation

You use **crystallisation** to produce:

- ✓ solid crystals from a solution.

In crystallisation:

- ✓ The solution is heated to remove enough solvent to produce a **saturated** solution (one that cannot hold any more solute).
- ✓ The saturated solution is allowed to cool.
- ✓ **Crystals** form in the solution.
- ✓ The crystals are separated from the liquid and dried.

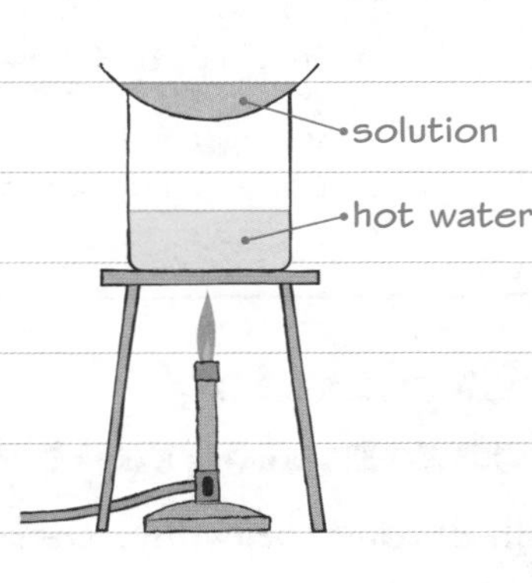

A hot water bath gives you more control over heating than using a Bunsen burner flame directly on the evaporating basin.

Explaining filtration

Filtration works because the filter paper has tiny pores. These are:

- large enough to let water molecules and dissolved substances through
- small enough to stop insoluble solid particles going through.

Explaining crystallisation

Solubility is the mass of solute that dissolves in a given volume of solvent at a given temperature.

Crystallisation works because:

- the solubility of the solute decreases as the saturated solution cools
- crystals form from the excess solute.

Worked example

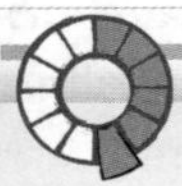

A student wants to produce some copper sulfate crystals from copper sulfate solution. She heats the solution in an evaporating basin to remove some of the water.

Describe the steps that she should then take to obtain dry crystals of copper sulfate. **(3 marks)**

She should let the solution cool down so that crystals form. She should then decant the remaining liquid and dry the crystals in an oven.

It is safer to make a concentrated or saturated solution than to remove all the water. The crystals will be a better size and shape, and will not spit out of the evaporating basin.

Decanting means pouring the liquid away carefully so that the solid stays behind. The student could use filtration instead to separate the crystals from the liquid. She could pat them dry using filter paper.

Now try this

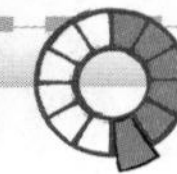

Glass and dust are insoluble in water but copper chloride is soluble.

Outline a method to produce dry copper chloride from a mixture of broken glass, dust and copper chloride powder. **(4 marks)**

Paper chromatography

Paper **chromatography** is used to separate mixtures of soluble substances.

Practical skills: Paper chromatography

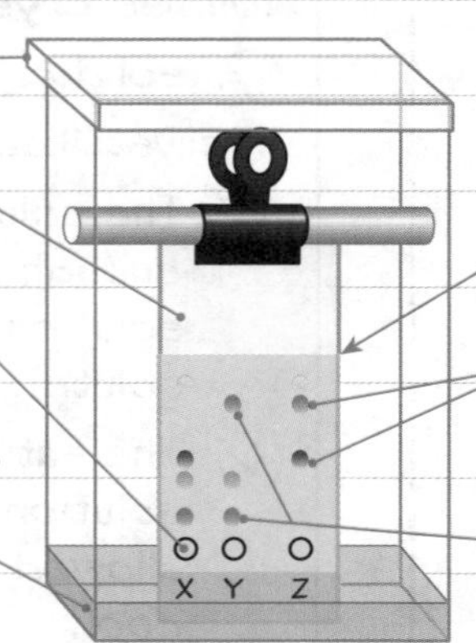
lid (to stop evaporation of solvent)
paper
Drops of the different samples are put onto the paper and allowed to dry. The bottom of the paper is then dipped into a solvent.
solvent (this can be water or some other liquid that the samples will dissolve in)
solvent front (the solvent has reached this level)
The different compounds in a sample dissolve to different extents in the solvent.
More soluble compounds are carried up the paper faster than less soluble ones, so the compounds separate out.
X Y Z

Explaining chromatography

In chromatography, there are two phases:

 a **stationary phase** (a substance that does not move)

 a **mobile phase** (a substance that moves through the stationary phase – the solvent).

During chromatography:

- each soluble substance in the mixture forms bonds with the two phases
- substances that form stronger attractive forces with the stationary phase stay near the bottom
- substances that form stronger attractive forces with the mobile phase move towards the top.

Using a chromatogram

You can use a **chromatogram** to:

- ✓ distinguish between pure and impure substances (a pure substance will produce only one spot)
- ✓ identify a substance by comparing its pattern of spots with those of a known substance
- ✓ identify substances using R_f values.

Maths skills: R_f values

$$R_f = \frac{\text{distance travelled by spot}}{\text{distance travelled by solvent}}$$

R_f values have no units. They vary from 0 (spot stays on baseline) to 1 (spot travels with solvent front).

Worked example

What is the R_f value of the lowest spot for sample C in this chromatogram? **(1 mark)**

☐ **A** 0.20
☒ **B** 0.40
☐ **C** 0.44
☐ **D** 0.90

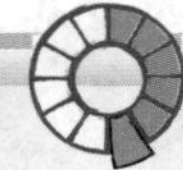

$$R_f = \frac{4\text{ cm}}{10\text{ cm}} = 0.4$$

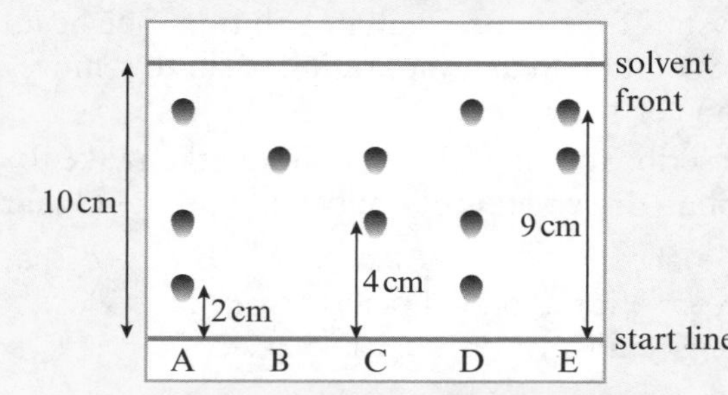
solvent front
10 cm
9 cm
4 cm
2 cm
start line
A B C D E

Now try this

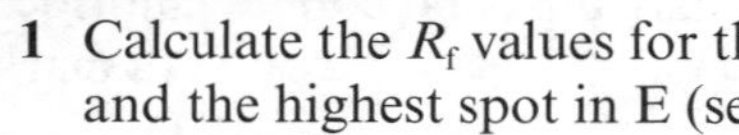

1 Calculate the R_f values for the lowest spot in A and the highest spot in E (see above). **(2 marks)**

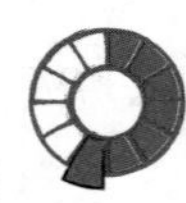

2 Explain, in terms of attractive forces, why a spot travels far up the paper. **(2 marks)**

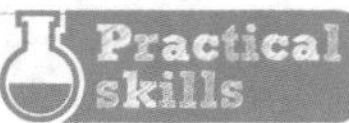

Investigating inks

You can use **simple distillation** and **paper chromatography** to investigate the composition of inks.

Core practical

Paper chromatography

Aim

To investigate the composition of inks using paper chromatography.

Apparatus

- eye protection
- chromatography paper
- boiling tube with bung
- dropping pipette
- pencil and ruler
- solvent for mobile phase

Method

Draw a pencil line near the bottom of the chromatography paper. Apply a small spot of ink, and then place the paper into a boiling tube containing a little solvent. Replace the bung and allow the solvent to travel through the paper.

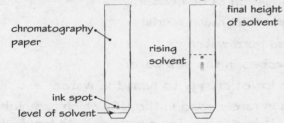

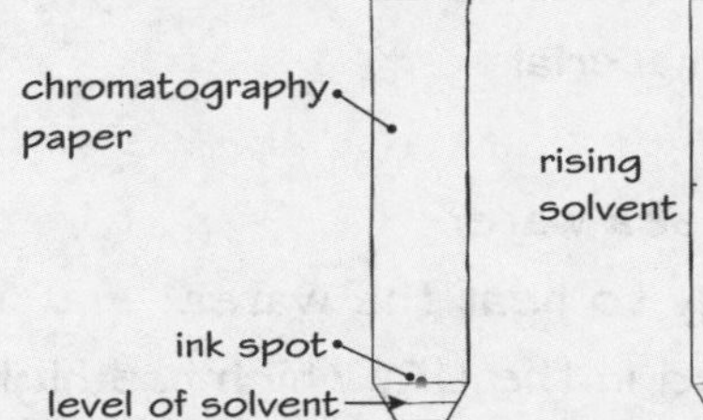

Results

Measure:

- the distance from the pencil line to the solvent front
- the distance from the pencil line to the centre of each spot.

Record the colour of each spot and its R_f value.

You can revise paper chromatography on page 116.

Ink is a mixture of coloured substances dissolved in a solvent.

Eye protection is important because you should treat all substances in the lab (apart from water straight from the tap) as potentially harmful. You could use a narrow glass **capillary tube** instead of a pipette to obtain tiny drops of ink.

The solvent used depends on the type of ink. For ballpoint pen ink, you may need to use propanone rather than water.

If you write a **risk assessment**, remember to include hazardous substances and procedures. For example, propanone is flammable and it has a harmful vapour, so you would need to keep it away from Bunsen burner flames and make sure that the lab is well ventilated.

Simple distillation

You can use simple distillation to separate the solvent in the ink from the coloured substances. The method you use will depend on the apparatus available to you in school.

Revise simple distillation on page 114.

Maths skills You could use a suitable table to record the results. Remember:

$$R_f = \frac{\text{distance travelled by spot}}{\text{distance travelled by solvent}}$$

Measure distances in millimetres rather than centimetres.

Now try this

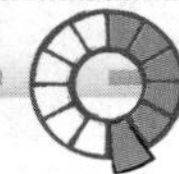

A student uses simple distillation to purify the solvent in a sample of ink. The method that he uses is as follows:

- Add some ink to a boiling tube, then fit a delivery tube and bung.
- Hold the boiling tube with a test-tube holder.
- Heat the ink using a roaring Bunsen burner flame until it is dry.
- Collect the vapour in a test tube.

State **two** ways of improving his method using the same essential equipment. **(2 marks)**

You could suggest different ways of working. You could also suggest additional apparatus to improve the safety or efficiency of the practical activity.

Had a look ☐ Nearly there ☐ Nailed it! ☐

Drinking water

Waste water and ground water must be treated to make the water **potable** or safe to drink.

Drinking water

Potable drinking water must have:
- low levels of contaminating substances
- low levels of microbes.

Fresh water can come from reservoirs, lakes and rivers. It is likely to contain:
- objects such as leaves and twigs
- insoluble solids such as particles of soil
- microbes, which may cause disease.

The water must be treated to remove these.

Safe but not pure

Tap water is potable but it is not a pure substance. It contains:

 dissolved salts

 dissolved chlorine.

Stages in water treatment

The main stages are as follows:

 **sedimentation** – large insoluble particles sink to the bottom of a tank

 **filtration** – small insoluble particles are removed by filtering through beds of sand

 **chlorination** – chlorine gas is bubbled from the water to kill microbes.

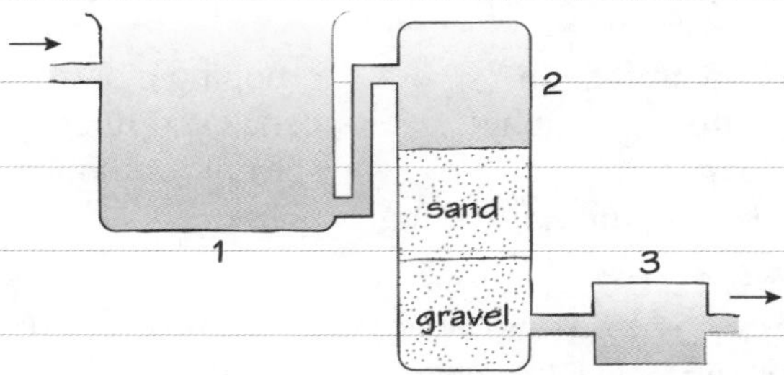

Sea water and distillation

Sea water contains dissolved salts. Their concentration is too high to drink it safely. Sea water can be made potable by simple distillation:
- Filtered sea water is boiled.
- The water vapour is cooled and condensed to form **distilled water**.

Distilled water does not contain dissolved salts, but it contains dissolved gases from the air.

Evaluating distillation

Simple distillation of sea water:
- 👍 uses a plentiful raw material
- 👍 produces pure water
- 👍 kills microbes in the sea water
- 👎 needs a lot of energy to heat the water.

Distillation is rarely used in the UK, which has high fuel costs. It is suitable for countries with low fuel costs, little fresh water, or plentiful sunshine to evaporate the water without using a fuel.

Worked example

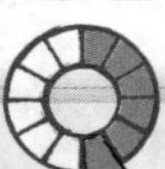

Explain why water used for a chemical analysis must not contain any dissolved salts. **(3 marks)**

Dissolved salts could react with the substances used in the analysis. A product formed in the reaction could interfere with the analysis, giving a false result. If the water used does not contain any dissolved salts, this will not happen.

Many substances used for chemical analysis are used in aqueous solution (dissolved in water).

Dissolved salts may react with a sample, forming insoluble precipitates. Analytical instruments may detect the dissolved salts, giving a false result.

Now try this

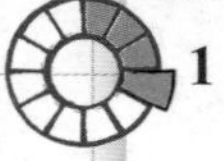

1 Describe what is meant by 'potable'. **(1 mark)**

2 In the UK, fresh water is piped from reservoirs so that it can be treated to make it safe to drink. Name and describe the three main stages in water treatment. **(3 marks)**

Extended response – Separating mixtures

There will be at least one 6-mark question on your exam paper. For these questions, you will need to think scientifically and structure your answer logically, showing how the points you make relate to each other.

You can revise the topics for this question, which is about **states of matter** and **separation techniques**, on pages 112–117.

Worked example

Nitrogen and oxygen are useful substances in air. They can be separated from air by:

- cooling air until water vapour and carbon dioxide solidify
- filtering the chilled air
- further cooling the filtered air to –200 °C
- carrying out fractional distillation.

Substance	Melting point (°C)	Boiling point (°C)
nitrogen	–210	–196
oxygen	–219	–183
water vapour	0	100
carbon dioxide	–78.5	–78.5

The table gives information about four substances found in air. Use the information above, and your knowledge and understanding of separation methods, to explain how nitrogen and oxygen may be separated in this way. **(6 marks)**

If the air is cooled to below –78.5 °C but above –183 °C, the water vapour and carbon dioxide will become solid. The nitrogen and oxygen will stay as gases, so the solid water and carbon dioxide could then be filtered to remove them. This will stop them contaminating the oxygen and nitrogen later in the process.

Use data from the table to suggest the range of temperatures needed to solidify water and carbon dioxide. Use your knowledge and understanding of separation techniques to suggest how they could be separated from oxygen and nitrogen.

When the filtered air is cooled to –200 °C, the oxygen and nitrogen will condense to form a mixture of liquids. This mixture can then be piped into a fractionating column. The column should be warmer at the bottom and colder towards the top.

Use data from the table to describe what happens to the oxygen and nitrogen at –200 °C. Use your knowledge and understanding of fractional distillation to describe the temperature gradient needed in the column.

In the fractionating column, the nitrogen will boil more easily than the oxygen, because it has the lower boiling point. Nitrogen gas can leave from the top of the column and oxygen from the bottom of the column. Depending on the temperature at the bottom, the oxygen could leave as a liquid or a gas.

Use data from the table to explain why oxygen and nitrogen can be separated by fractional distillation. Use your knowledge and understanding to describe where each substance will leave the fractionating column.

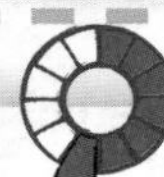

You need to show comprehensive knowledge and understanding. Be prepared to apply what you know to a new context. Use the data given to you in a structured way with clear lines of reasoning.

Now try this

Substance	Solubility at 25 °C ($g\,dm^{-3}$ water)
sodium carbonate	307
calcium carbonate	0.013

Sodium carbonate and calcium carbonate are both white powders at room temperature. The table shows some information about them. A student accidentally makes a mixture of the two powders. Use the information in the table, and your knowledge and understanding of separation methods, to explain how the student could produce separate, dry samples of sodium carbonate and calcium carbonate from this mixture. **(6 marks)**

Had a look ☐ Nearly there ☐ Nailed it! ☐

Acids and alkalis

You can recognise acids and alkalis by their effects on indicator solutions.

Acids

Acids have these properties:

- The pH of their aqueous solutions is less than 7.
- They are a source of **hydrogen ions**, $H^+(aq)$ in solution.

Hydrochloric acid releases H^+ ions:

$$HCl(aq) \rightarrow H^+(aq) + Cl^-(aq)$$

Ethanoic acid releases H^+ ions:

$$CH_3COOH(aq) \rightarrow CH_3COO^-(aq) + H^+(aq)$$

- The higher the concentration of $H^+(aq)$ ions, the lower the pH of the **acidic** solution.

Alkalis

Alkalis have these properties:

- The pH of their aqueous solutions is more than 7.
- They are a source of **hydroxide ions**, $OH^-(aq)$ in solution.

Sodium hydroxide releases OH^- ions:

$$NaOH(aq) \rightarrow Na^+(aq) + OH^-(aq)$$

Ammonia produces OH^- ions in solution:

$$NH_3(g) + H_2O(l) \rightarrow NH_4^+(aq) + OH^-(aq)$$

- The higher the concentration of OH–(aq) ions, the higher the pH of the **alkaline** solution.

The pH scale

The **pH scale** is a measure of the **acidity** or **alkalinity** of a solution (how acidic or alkaline it is). It goes from 0 to 14. **Neutral** solutions are pH 7.

Indicators are substances that have different colours, depending on their pH. Universal indicator solution or paper is often used to estimate the pH of a solution.

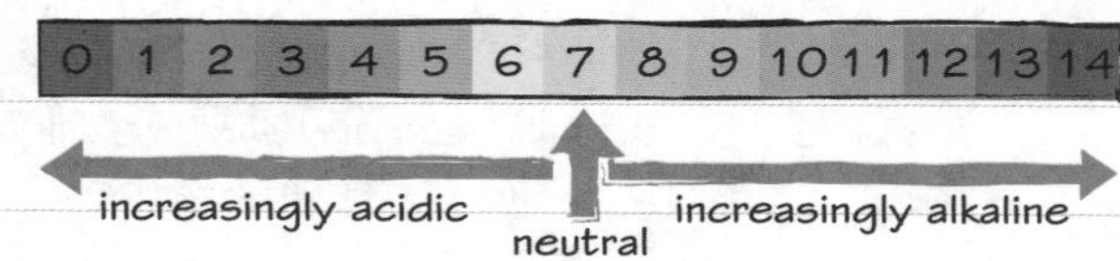

The indicator colour is matched to a colour chart, which shows the pH of each colour.

Worked example

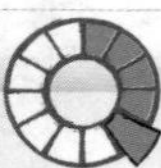

Litmus and phenolphthalein are single indicators. Complete the table to show their colours in acidic and alkaline solutions. **(4 marks)**

Indicator	In acidic solution	In alkaline solution
litmus	red	blue
phenolphthalein	colourless	pink

Methyl orange is another single indicator (unlike universal indicator, which is a mixed indicator). Single indicators are useful in **titrations**.
You can revise titrations on page 125.

Methyl orange is:

- red in acidic solutions
- yellow in alkaline solutions.

You will see an orange colour at the end-point of a titration.

Now try this

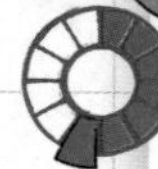

1 Explain why sodium chloride, NaCl, dissolves in water to produce a solution with a pH of 7, but hydrogen chloride, HCl, dissolves in water to produce a solution with a pH below 7. **(3 marks)**

2 Potassium reacts with water:

$$2K(s) + 2H_2O(l) \rightarrow 2KOH(aq) + H_2(g)$$

State and explain the effect that the solution formed has on phenolphthalein. **(2 marks)**

The paper can change colour (as litmus solution does) or stay the same colour.

	Acidic solution	Alkaline solution
Red litmus		
Blue litmus		

3 Copy and complete this table to show the colours of litmus paper in different solutions. **(2 marks)**

Strong and weak acids

The strength and concentration of an acid determines its pH.

Concentrated versus dilute

For a given volume:

- a **concentrated** solution has a greater amount of dissolved solute particles than a **dilute** solution.

You can change a concentrated solution into a dilute solution by adding more water to it. You can change a dilute solution to a more concentrated solution by:

- dissolving more solute in it
- evaporating some of the water.

Maths skills: Ratio and proportion

Diluting a solution and working out its new concentration involves understanding **ratios**.

For example, 1 cm^3 of 2 $g\,dm^{-3}$ acid is added to 4 cm^3 of water. This 'bar model' shows the mixture, where A is acid and W is water:

A	W	W	W	W

The ratio of acid to water is 1:4.

The proportion of acid is $\frac{1}{5}$ of the mixture.

The new concentration is:

$$\frac{1}{5} \times 2 = 0.4\,g\,dm^{-3}$$

Strong acids

Hydrochloric acid and sulfuric acid are **strong acids** because they **fully dissociate** into ions in solution. All their molecules release $H^+(aq)$ ions:

$$HCl(aq) \rightarrow H^+(aq) + Cl^-(aq)$$

$$H_2SO_4(aq) \rightarrow 2H^+(aq) + SO_4^{2-}(aq)$$

The concentration of hydrogen ions determines the pH of the solution. The table shows examples:

Concentration of $H^+(aq)$ (g/dm³)	pH of solution
1	0
0.1	1
0.01	2

- The pH increases by 1 when the H^+ ion concentration decreases by a factor of 10.

You can revise concentration on page 109.

Weak acids

Ethanoic acid is a **weak acid** because it **partially dissociates** into ions in solution. Only a few molecules release $H^+(aq)$ ions:

$$CH_3COOH(aq) \rightleftharpoons CH_3COO^-(aq) + H^+(aq)$$

Reaction is **reversible**.

At a given concentration of acid:

- a strong acid has a higher concentration of H^+ ions than a weak acid
- a strong acid has a lower pH than a weak acid at the same concentration (see table below).

Acidic solution	pH
hydrochloric acid	0.3
ethanoic acid	2.5

You can revise reversible reactions and use of the ⇌ symbol on page 141.

Worked example

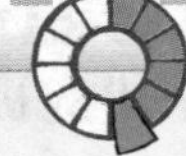

50 $g\,dm^{-3}$ nitric acid has a pH of 0.125. State the pH when 2 cm^3 of this acid is mixed with 198 cm^3 of water. **(1 mark)**

pH 2.125

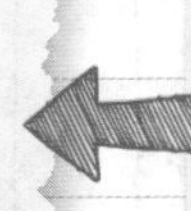

The new total volume of acid = (2 + 198) = 200 cm^3

So the concentration of the acid decreases by (200/2) = 100 times.

If the concentration decreases by two factors of 10 (10^2 = 100), the pH increases by 2.

Now try this

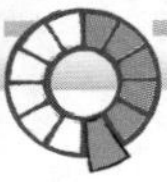

1 Explain why a sample of concentrated ethanoic acid may have the same pH as a sample of dilute nitric acid. **(2 marks)**

2 The pH of 7.3 $g\,dm^{-3}$ hydrochloric acid is 0.70. Deduce the concentration of hydrochloric acid that has a pH of 2.70. **(1 mark)**

Had a look ☐ Nearly there ☐ Nailed it! ☐

Bases and alkalis

An **alkali** is a soluble **base** – one that will dissolve in water.

Reactions of acids with bases

A **base** is:

- any substance that reacts with an acid to form a salt and water **only**.

Bases are metal oxides and metal hydroxides. In general:

base + acid → salt + water

For example:

$NaOH(aq) + HCl(aq) \rightarrow NaCl(aq) + H_2O(l)$

$CuO(s) + 2HNO_3(aq) \rightarrow Cu(NO_3)_2(aq) + H_2O(l)$

Naming salts

A **salt** forms when hydrogen ions in an acid are replaced by metal ions or ammonium ions. The name of a salt consists of two parts:

- first part – the metal in the base
- second part – from the acid used.

Acid used	Second part
hydrochloric acid, $HCl(aq)$	chloride
nitric acid, $HNO_3(aq)$	nitrate
sulfuric acid, $H_2SO_4(aq)$	sulfate

Reactions of acids with metals

Reactive metals react with acids to produce a salt and hydrogen **only**. In general:

metal + acid → salt + hydrogen

For example:

$Mg(s) + 2HCl(aq) \rightarrow MgCl_2(aq) + H_2(g)$

$2Al(s) + 3H_2SO_4(aq) \rightarrow Al_2(SO_4)_3(aq) + 3H_2(g)$

Practical skills: Hydrogen

A **lighted splint** ignites hydrogen with a 'pop'.

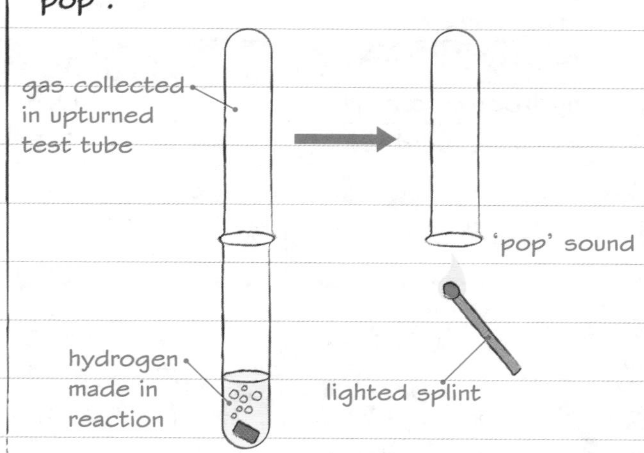

Reactions of acids with metal carbonates

Metal carbonates react with acids to produce a salt, water **and** carbon dioxide. In general:

metal carbonate + acid → salt + water + carbon dioxide

For example:

$CaCO_3(s) + 2HCl(aq) \rightarrow CaCl_2(aq) + H_2O(l) + CO_2(g)$

Practical skills: Carbon dioxide

Carbon dioxide turns **limewater** milky or cloudy white.

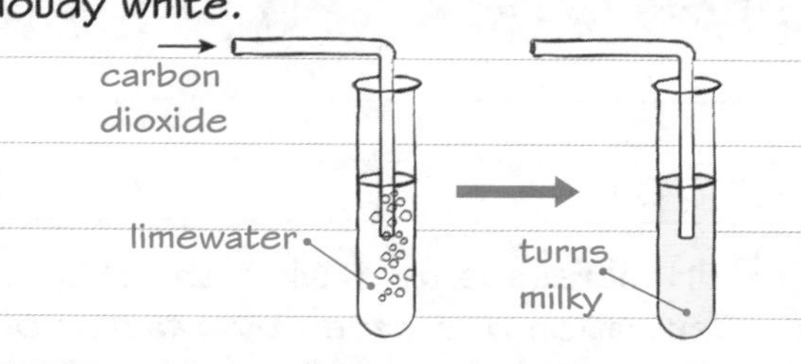

Neutralisation

Neutralisation is the reaction between an acid and a base. In an acid–alkali neutralisation, hydrogen ions from the acid react with hydroxide ions from the alkali to form water:

$H^+(aq) + OH^-(aq) \rightleftharpoons H_2O(l)$

Now try this

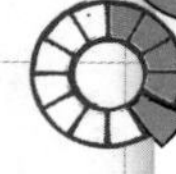

1 Explain why all alkalis are bases but not all bases are alkalis. **(2 marks)**

2 Give the formulae, and names, for the five salts formed in the reactions shown on this page. **(5 marks)**

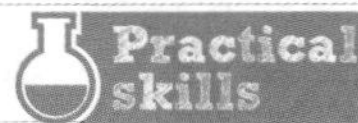

Neutralisation

Core practical

Investigating neutralisation

Aim

To investigate the change in pH when powdered calcium hydroxide, a base, is added to a fixed volume of dilute hydrochloric acid.

> Calcium hydroxide is an irritant but calcium oxide is corrosive, so calcium hydroxide is safer to use.

Apparatus

- eye protection
- beaker
- measuring cylinder
- balance
- spatula
- stirring rod
- white tile
- universal indicator paper and pH colour chart
- dilute hydrochloric acid
- calcium hydroxide powder.

> You should choose appropriate apparatus for your practical work. If you are using a measuring cylinder, choose one that is closest to the volume of liquid you need. For example, to measure 50 cm^3, choose a 50 cm^3 measuring cylinder. Do not choose a 25 cm^3 or 100 cm^3 one.

Method

Add some dilute hydrochloric acid to the beaker. Measure and record the pH of the contents of the beaker. Add a small mass of calcium hydroxide powder, stir, and then measure and record the pH again. Repeat until the pH no longer changes.

> You could use a pH meter instead of universal indicator paper. This produces a more precise reading. It will be accurate only if you calibrate it first using standard pH solution.

Results

Use a table to record your results. For example:

Mass of $Ca(OH)_2$ added (g)	pH of mixture
0	
0.3	

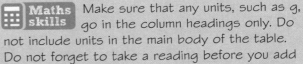

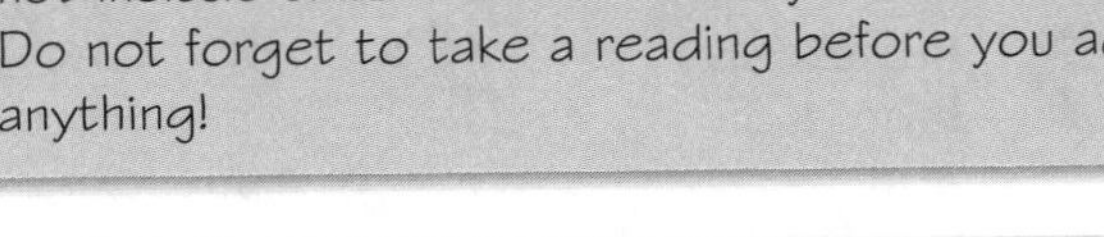

> Make sure that any units, such as g, go in the column headings only. Do not include units in the main body of the table. Do not forget to take a reading before you add anything!

Analysis

Plot a titration curve – a line graph with:

- mass (or volume) added on the horizontal axis
- pH on the vertical axis.

Draw a smooth curve best fit, ignoring anomalous results (results that are too high or low).

> ✓ Label each axis with the quantity and unit (if there is one), e.g. 'Mass of $Ca(OH)_2$ (g)'.
>
> ✓ Choose scales so that the plotted points occupy at least 50% of the graph area.
>
> ✓ Use a sharp pencil for your graph.
>
> ✓ Plot each point with × (but you might find it more accurate to use + instead).

Now try this

> What pH values do you expect at the start, middle and end? Why?

A student adds portions of a powdered soluble base to dilute hydrochloric acid. She estimates the pH each time using universal indicator paper.

(a) Describe **two** ways in which she can ensure that her measurements are precise. **(2 marks)**

(b) Describe how she should use the universal indicator paper to estimate the pH. **(3 marks)**

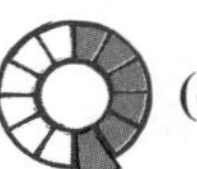

(c) The student added the base until it was in excess.
Explain the changes in pH during her experiment. **(6 marks)**

Salts from insoluble bases

Practical skills

You can make pure, dry, hydrated copper sulfate crystals, starting from copper oxide.

Core practical

Preparing copper sulfate

Aim

To make a sample of pure, dry, hydrated copper sulfate crystals using copper oxide and sulfuric acid:

$$CuO(s) + H_2SO_4(aq) \rightarrow CuSO_4(aq) + H_2O(l)$$

'Hydrated' copper sulfate crystals are the familiar blue ones. Their structure includes water molecules as **water of crystallisation**.

Apparatus

- eye protection
- beakers
- spatula
- stirring rod
- Bunsen burner
- heat-resistant mat
- tripod and gauze mat
- filter funnel and filter paper
- evaporating basin
- dilute sulfuric acid
- copper oxide powder.

The apparatus is used in the three main stages of this practical:

stage 1: reacting warm sulfuric acid with excess copper oxide powder

stage 2: filtering the reaction mixture to remove excess (unreacted) copper oxide

stage 3: producing hydrated copper sulfate crystals by crystallisation with the help of a hot water bath.

You can revise filtration and crystallisation on page 115.

Method

The diagram describes the main steps needed.

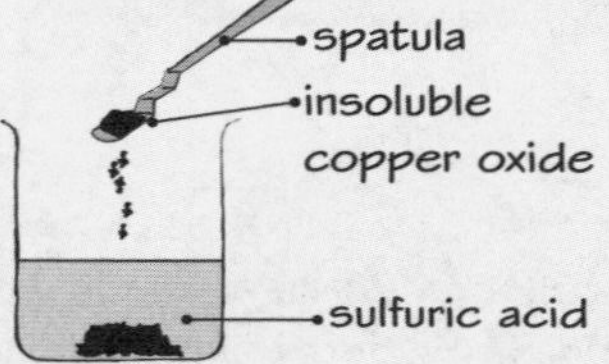

1 Add excess base to the acid.

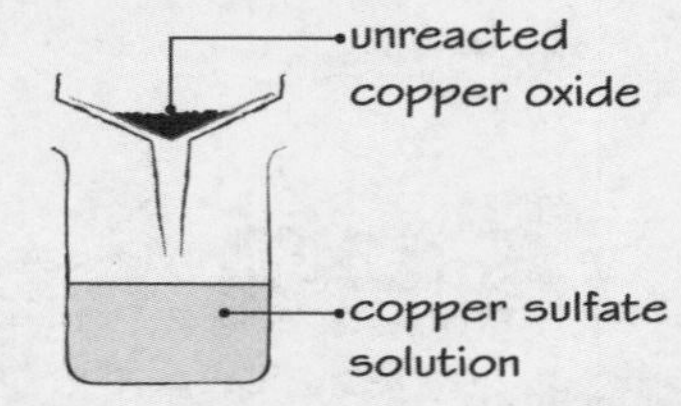

2 Filter to remove unreacted copper oxide.

copper sulfate crystals formed by evaporating the water

3 Crystallise the copper sulfate solution by heating it or leaving it to stand in a warm place.

The same method works using a metal but hydrogen, $H_2(g)$, is made rather than water. However, copper does not react with sulfuric acid.

Results

Record your observations at each stage and the appearance of the crystals at the end. This could include the dry mass of the crystals, and their colour, sizes and shapes.

Record your observations as you go, rather than trying to remember them afterwards.

Now try this

Calcium carbonate is an insoluble white solid. It reacts with dilute nitric acid to form calcium nitrate solution:

$$CaCO_3(s) + 2HNO_3(aq) \rightarrow Ca(NO_3)_2(aq) + H_2O(l) + CO_2(g)$$

(a) Describe what you will see when excess calcium carbonate is used in the reaction. **(3 marks)**

(b) Devise a method to prepare a sample of pure, dry, hydrated calcium nitrate crystals. **(6 marks)**

Devise means that you need to plan or invent a procedure from existing ideas.

Salts from soluble bases

You should be able to describe how to carry out a **titration** to prepare a pure, dry salt.

Using a soluble base

The flow chart shows the three main steps needed to make a soluble salt from a soluble base.

Use an acid–base titration to find the exact volume of the soluble base that reacts with the acid. → Mix the acid and soluble base in the correct proportions, producing a solution of the salt and water. → Warm the salt solution to evaporate the water – this will leave crystals of the salt behind.

Practical skills: Doing a titration

To carry out a typical titration:

- Put acid into a **burette**.
- Use a **pipette** to put a known volume of alkali into a conical flask.
- Put a few drops of a suitable **indicator** solution, such as phenolphthalein or methyl orange, into the alkali.
- Record the burette start reading.
- Add acid to the alkali until the colour changes – the **end-point**.
- Record the burette end reading.

Maths skills: Mean titres

The **titre** is the volume of acid added to exactly neutralise the alkali:

titre = (end reading) – (start reading)

Concordant titres are identical to each other, or very close together (usually within $0.10\,cm^3$).

You normally calculate the **mean** titre using your concordant results only:

$$\text{mean titre} = \frac{\text{sum of concordant titres}}{\text{number of concordant titres}}$$

Worked example

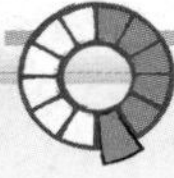

(a) Explain why titration must be used to produce a salt from dilute hydrochloric acid and potassium hydroxide solution. **(2 marks)**

Titration lets you find the correct proportions of acid and alkali to mix together to produce a solution that contains only a salt and water.

(b) Describe **three** steps needed to obtain an accurate titre during a titration. **(3 marks)**

Swirl the flask continuously during the titration to mix the acid and alkali thoroughly. Near the end-point, add the acid drop by drop, pausing between each addition. Rinse the inside of the flask to make sure all the acid mixes with the alkali.

Potassium hydroxide is a soluble base (an alkali) and the salt (potassium chloride) is also soluble. You can revise solubility rules on page 126.

When you are making a salt, you find the mean titre, then use the burette to add this volume of acid to the alkali without the indicator.

Obtain accurate readings by clamping the burette vertically and reading it at eye level.

Apparatus

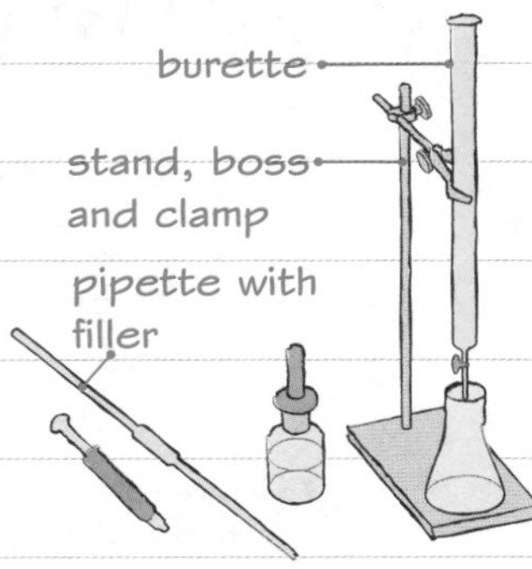

Now try this

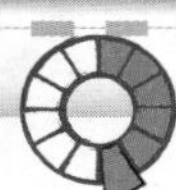

The charcoal is still a powder.

Instead of carrying out a final titration without the indicator, powdered charcoal can be added at the end-point. This forms strong chemical bonds with the indicator.
Explain how the indicator is then removed from the salt solution. **(2 marks)**

Had a look ☐ Nearly there ☐ Nailed it! ☐

Making insoluble salts

You need to be able to predict, using solubility rules, whether a precipitate will be formed when named solutions are mixed together. If a precipitate is formed, you should be able to name it.

General rules

This is why solutions in practicals are often sodium compounds.

This is very useful to remember.

This follows the first rule at the top.

Soluble	Insoluble
all common sodium, potassium and ammonium salts	
all nitrates	
common chlorides	silver chloride lead chloride
common sulfates	lead sulfate barium sulfate calcium sulfate
sodium hydroxide potassium hydroxide ammonium hydroxide	common hydroxides
sodium carbonate potassium carbonate ammonium carbonate	common carbonates

Worked example

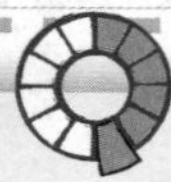

(a) Predict whether a precipitate forms when silver nitrate and sodium chloride solutions are mixed. Name any precipitate formed. **(2 marks)**

A precipitate will form – silver chloride.

Two products form, sodium nitrate and silver chloride, but only silver chloride is insoluble. It forms a cloudy **precipitate** in the mixture:

$AgNO_3(aq) + NaCl(aq) \rightarrow NaNO_3(aq) + AgCl(s)$

(b) Name **two** substances that could be used to produce insoluble barium sulfate. **(2 marks)**

barium nitrate and sodium sulfate

All nitrates, and all sodium salts, are soluble. So, to make an insoluble salt **XY**, choose:

- **X** nitrate, and
- sodium **Y**

as your two soluble reactants.

Making insoluble salts

1 Mix solutions of two substances that will form the insoluble salt.

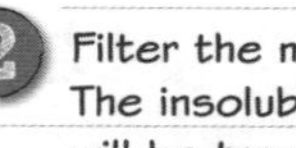

2 Filter the mixture. The insoluble salt will be trapped in the filter paper.

3 Wash the salt with distilled water.

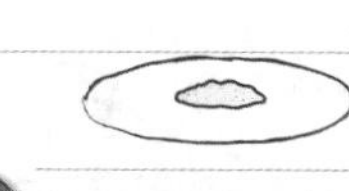

4 Leave the salt to dry on the filter paper. It could be dried in an oven.

Now try this

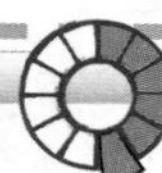

Predict whether a precipitate forms when these solutions are mixed, naming any precipitate formed.

(a) sodium carbonate and calcium chloride **(1 mark)**

(b) ammonium nitrate and potassium hydroxide **(1 mark)**

(c) lead nitrate and sodium sulfate. **(1 mark)**

Extended response – Making salts

There will be at least one 6-mark question on your exam paper. For these questions, you will need to think scientifically and structure your answer logically, showing how the points you make are related to each other.

You can revise the topics for this question, which is about **acids, alkalis and salts**, on pages 120–126.

Worked example

Soluble salts can be made by the reactions of acids with insoluble metal compounds. The salt produced depends on the reactants chosen.

Devise a method to prepare pure, dry crystals of zinc sulfate, $ZnSO_4$, from a zinc compound and a suitable acid. Begin your answer by choosing suitable reactants.

You should also write a balanced equation as part of your plan. **(6 marks)**

I would use zinc carbonate and sulfuric acid. This is the equation for the reaction:

$ZnCO_3 + H_2SO_4 \rightarrow ZnSO_4 + H_2O + CO_2$

I would put some of the acid into a beaker and warm it with a Bunsen burner, tripod and gauze.

I would then use a spatula to add a little zinc carbonate to the warm acid and stir it with a stirring rod. I would repeat this until all the bubbling stopped and some solid was left in the bottom of the beaker.

Next, I would filter the mixture to remove the excess zinc carbonate, using a filter funnel and filter paper. I would collect the zinc sulfate solution in a conical flask.

I would pour the solution into an evaporating basin and leave it on a windowsill for water to evaporate and crystals to form. I would carefully pour away the excess liquid and dry the crystals with filter paper.

Command word: Devise

When you are asked to **devise** something, you need to plan or invent a procedure from existing principles or ideas. These may include familiar chemistry in an unfamiliar context, so make sure that you study carefully any information given to you in the question.

Zinc oxide would also work:

$ZnO + H_2SO_4 \rightarrow ZnSO_4 + H_2O$

Zinc hydroxide would work too:

$Zn(OH)_2 + H_2SO_4 \rightarrow ZnSO_4 + 2H_2O$

You should describe how to make sure that all the acid has reacted. You would not see bubbling if you chose zinc oxide or zinc hydroxide, but you would see unreacted solid left in the beaker when all the acid had reacted.

If you were asked to prepare an insoluble salt by precipitation, you could wash the salt in the filter paper with distilled water at this stage.

In this part of the answer, you could mention heating the solution to evaporate most of the water instead. You would then leave the crystals in a warm place to dry, such as an oven.

You should make sure that your plan is thought out well, with a clear and logical structure. Aim to support it with scientific information and ideas. You could write a method in continuous prose, as here, or you could write it in numbered steps instead. You should apply your knowledge and understanding of the necessary techniques, and use this to answer the question even if you have not made the required salt in school. Include appropriate apparatus without going into irrelevant detail.

Now try this

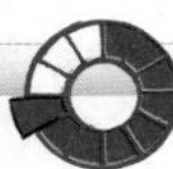

An insoluble salt can be made by mixing two suitable solutions together to form a precipitate.

Devise a method to prepare a pure, dry sample of insoluble lead(II) chloride, $PbCl_2$.

Begin your answer by choosing suitable solutions and precautions needed for safe working.

You may also write a balanced equation, including state symbols, as part of your plan. **(6 marks)**

Had a look ☐ Nearly there ☐ Nailed it! ☐

Electrolysis

Electrolysis is used to decompose ionic compounds in the molten state or dissolved in water.

Some key words

An **electrolyte** is:

- an ionic compound in the **molten** state (liquid) or **dissolved** in water.

Electrolysis is:

- a process in which **electrical energy**, from a **direct current** (d.c.) supply, decomposes an electrolyte.

Anions are:

- negatively charged ions that move to the positive **electrode** (anode).

Cations are:

- positively charged ions that move to the negative electrode (cathode).

Moving charges

The ions in an electrolyte are charged particles. An electric current will pass through the electrolyte only if the ions are free to move from place to place.

Remember that ions:

- can move about in liquids
- can move about in solutions
- cannot move about in solids.

You can revise the arrangement and movement of particles in solids and liquids on page 112.

Electrolysis of molten lead bromide

The electrolysis of hot, molten lead bromide produces lead and bromine: $PbBr_2(l) \rightarrow Pb(l) + Br_2(g)$

During electrolysis positive charged ions are attracted to the negative electrode and move to it.

Negatively charged ions are attracted to the positive electrode and move to it during electrolysis.

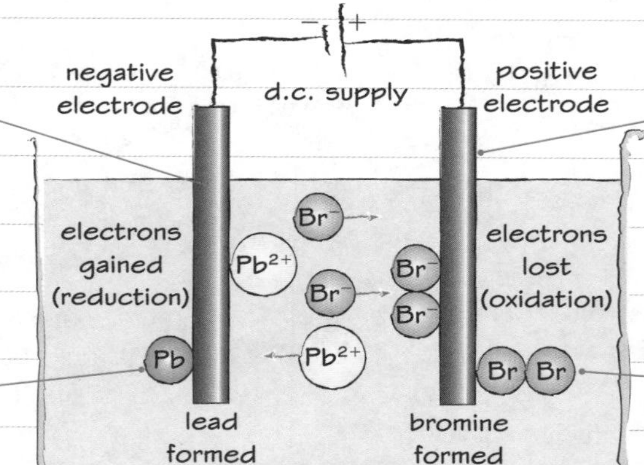

Remember 'oil rig': **O**xidation **I**s **L**oss of electrons, and **R**eduction **I**s **G**ain of electrons.

Positively charged ions gain electrons and are **reduced**.

Negatively charged ions lose electrons and are **oxidised**.

Worked example

Molten zinc chloride is electrolysed.

(a) Predict the products formed at each electrode. **(2 marks)**

Zinc forms at the negative electrode and chlorine at the positive electrode.

(b) Write half equations for the reactions occurring at each electrode. **(2 marks)**

negative electrode: $Zn^{2+} + 2e^- \rightarrow Zn$

positive electrode: $2Cl^- \rightarrow Cl_2 + 2e^-$

Zinc chloride (and lead bromide) is a **binary** ionic compound. This means that it consists of two elements only. You must be able to predict the products of electrolysis of such compounds in the molten state.

Electrons are shown as e^- in these equations:

- **Reduction** (gain of electrons) happens at the negative electrode.
- **Oxidation** (loss of electrons) happens at the positive electrode.

Now try this

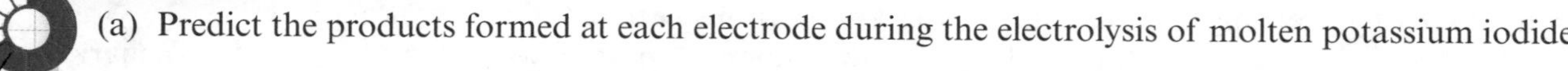

(a) Predict the products formed at each electrode during the electrolysis of molten potassium iodide. **(2 marks)**

(b) Write half-equations for the reactions occurring at each electrode. **(2 marks)**

(c) Explain at which electrode oxidation occurs. **(2 marks)**

Electrolysing solutions

You need to be able to explain products formed by the electrolysis of compounds in solution.

Ions in a solution

Water is a covalent compound. Some of its molecules naturally form ions:

$$H_2O(l) \rightleftharpoons H^+(aq) + OH^-(aq)$$

You can revise reversible reactions and the use of the $\rightleftharpoons$ symbol on page 141.

The presence of these ions means that a solution of an ionic compound contains:

- cations and anions from the dissolved ionic compound, and
- H^+ and OH^- ions from the water.

Competing ions

During electrolysis of an aqueous solution, all the ions in the electrolyte compete to be discharged and form products.

1. Hydrogen gas is produced if H^+ ions are discharged:
   $2H^+(aq) + 2e^- \rightarrow H_2(g)$
2. Oxygen gas is produced if OH^- ions are discharged:
   $4OH^-(aq) \rightarrow 2H_2O(l) + O_2(g) + 4e^-$

At the cathode

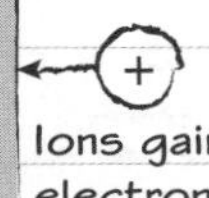

At the negative electrode:

- ✓ hydrogen is produced, **unless**
- ✓ the compound contains ions from a metal less reactive than hydrogen.

In that case:

- ✓ the metal is produced instead.

Copper and silver are below hydrogen in the reactivity series (revise this series on page 132).

At the anode

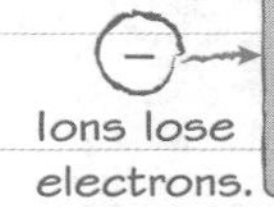

At the positive electrode:

- ✓ oxygen is produced (from OH^- ions) **unless**
- ✓ the compound contains halide ions (Cl^-, Br^- or I^-).

In that case:

- ✓ chlorides produce chlorine, Cl_2
- ✓ bromides produce bromine, Br_2
- ✓ iodides produce iodine, I_2.

Worked example

Predict the products formed at each electrode during the electrolysis of the following concentrated aqueous solutions.

(a) copper chloride solution **(1 mark)**

copper at the cathode, chlorine at the anode

(b) sodium chloride solution **(1 mark)**

hydrogen at the cathode, chlorine at the anode

(c) sodium sulfate solution **(1 mark)**

hydrogen at the cathode, oxygen at the anode

Copper is less reactive than hydrogen, so copper is produced. Chloride ions are present, so chlorine is produced.

Sodium is more reactive than hydrogen, so hydrogen is produced instead of sodium.

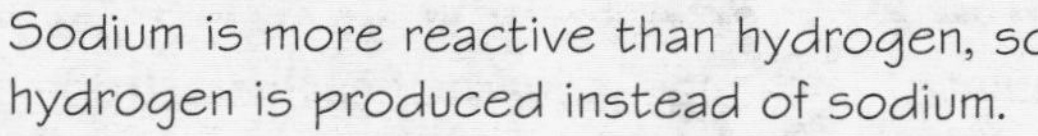

Halide ions are not present, so oxygen is produced when OH^- ions are discharged.

Now try this

Which four ions are present, which are discharged and which stay in solution?

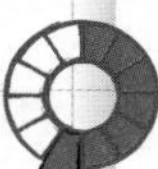

Dilute sulfuric acid, $H_2SO_4(aq)$, is an electrolyte.

(a) Give the formulae of all the ions in the electrolyte, and where each comes from. **(3 marks)**

(b) Predict the product formed at each electrode, and write a balanced half-equation for its formation. **(4 marks)**

Investigating electrolysis

Copper can be purified by the electrolysis of copper sulfate solution using copper electrodes.

Core practical

Electrolysis of copper sulfate

Aim

To investigate the change in mass of the anode and the cathode when copper sulfate solution is electrolysed using copper electrodes.

The cathode gains mass as copper ions gain electrons:

$Cu^{2+}(aq) + 2e^- \rightarrow Cu(s)$ (reduction)

The anode loses mass as the copper atoms lose electrons:

$Cu(s) \rightarrow Cu^{2+}(aq) + 2e^-$ (oxidation)

Apparatus

- eye protection
- beaker
- two strips of copper
- crocodile clips and leads
- d.c. power supply
- copper sulfate solution
- filter paper
- ±0.01 g balance.

You could bend the strips of copper over the edge of the beaker. Hold them in place using crocodile clips. You can measure the current if you connect an ammeter in series with one of the electrodes.

Method

Measure and record the mass of each electrode. Connect the apparatus as shown below.

Allow the experiment to run as directed by your teacher, or by your own preliminary experiments. Disconnect the two electrodes, carefully dry them with filter paper, and then weigh them again.

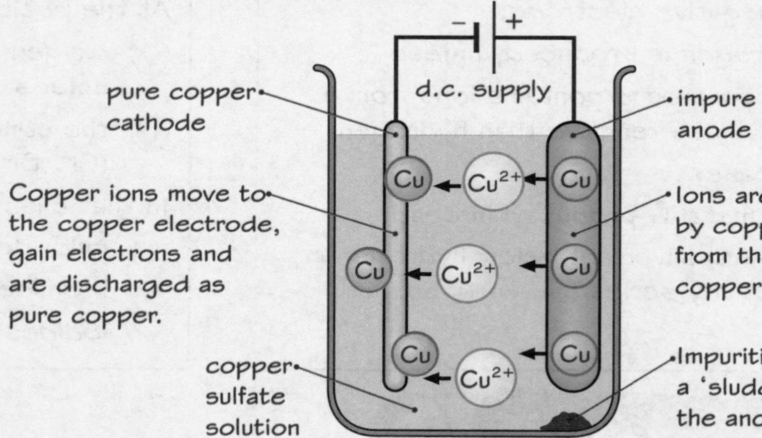

−
+
d.c. supply
pure copper cathode
impure copper anode
Copper ions move to the copper electrode, gain electrons and are discharged as pure copper.
Ions are replaced by copper ions from the impure copper anode.
copper sulfate solution
Impurities form a 'sludge' below the anode.
Cu
Cu2+

Results

Record the masses of each electrode at the start and end. A table is one way to do this neatly.

Make sure your table has columns to record which electrode is being weighed (at the start and the end) and the mass recorded.

Analysis

Calculate the change in mass of each electrode. Compare these two values and discuss reasons for any difference observed.

The gain in mass of the cathode should equal the loss in mass of the anode.

Now try this

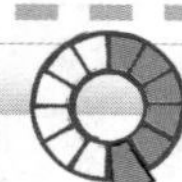

During the electrolysis of copper sulfate solution with copper electrodes, the cathode gains 0.19 g and the anode loses 0.43 g.

(a) The electrodes are rinsed with ethanol, which boils at 78 °C, before being dried and weighed. Suggest why they dry more quickly than if water is used. **(1 mark)**

(b) Give **two** reasons why the change in mass of the cathode is less than expected. **(2 marks)**

Extended response – Electrolysis

There will be at least one 6-mark question on your exam paper. For these questions, you will need to think scientifically and structure your answer logically, showing how the points you make are related to each other.

You can revise the topics for this question, which is about **electrolysis**, on pages 128–130.

Worked example

Potassium iodide is an ionic compound. It contains potassium ions, K^+, and iodide ions, I^-. When molten potassium iodide is electrolysed, potassium metal and iodine vapour are formed. Explain how the potassium ions and iodide ions in solid potassium iodide are converted into potassium and iodine by electrolysis. You may include suitable half-equations in your answer. **(6 marks)**

The potassium iodide must be heated. This is so that its ionic bonds break and it becomes molten. Electrolysis cannot happen in the solid state because the ions are not free to move. It happens in the liquid state because the ions are free to move.

Potassium ions are attracted to the oppositely charged electrode, the negatively charged cathode. At the cathode, the potassium ions gain electrons and become potassium atoms:

$K^+ + e^- \rightarrow K$

This is reduction because the ions gain electrons.

Iodide ions are attracted to the positively charged electrode, the anode. Here they lose electrons to form iodine atoms. This is oxidation because the ions lose electrons. Pairs of iodine atoms join together to form an iodine molecule. Overall:

$2I^- \rightarrow I_2 + 2e^-$

Command word: Explain

When you are asked to **explain** something, it is not enough just to state or describe it. Your answer **must** contain some reasoning or justification of the points you make. Your explanation **can** include mathematical explanations, if calculations are needed.

In this example Extended response question, no calculations are needed.

You need to remember what the starting point in the question is. In this case it is solid potassium iodide, not molten potassium iodide. Remember to explain how to melt the compound, and why this is needed.

You should include knowledge and understanding about what happens at the cathode. Include a half-equation and reduction as the gain of electrons.

You should include knowledge and understanding about what happens at the anode. Include a half-equation and oxidation as the loss of electrons. Also mention that a covalent bond forms between two iodine atoms when they share a pair of electrons.

Your answer should show comprehensive knowledge and understanding. Use relevant scientific ideas to support your explanations, including appropriate balanced equations. You should include why potassium iodide must be melted first, then what happens at each electrode. You should leave out information or ideas that are unnecessary to answer the question.

Now try this

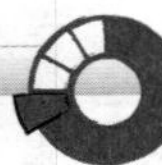

Remember to explain whether the lithium and chloride ions will be free to move in the molten potassium chloride. Also explain why this matters.

Lithium-ion batteries are found in many consumer products, including smartphones and electric cars. Lithium is manufactured by the electrolysis of lithium chloride dissolved in molten potassium chloride. During this process, molten lithium rises to the surface and chlorine gas bubbles off.
Explain how the lithium ions and chloride ions in solid lithium chloride are converted into lithium and chlorine by electrolysis. You may include suitable half-equations in your answer. **(6 marks)**

Had a look ☐ Nearly there ☐ Nailed it! ☐

The reactivity series

You can find the relative **reactivity** of a metal by comparing its reactions with other metals.

Reactions with water

Hydrogen is produced if a metal reacts with water:

metal + water → metal hydroxide + hydrogen

For example:

sodium + water → sodium hydroxide + hydrogen

$2Na(s) + 2H_2O(l) \rightarrow 2NaOH(aq) + H_2(g)$

In general, the more reactive the metal, the greater the rate of bubbling.

Reactions with acids

Hydrogen is produced if a metal reacts with a dilute acid:

metal + acid → salt + hydrogen

These reactions, and the lab test for hydrogen, are explained in more detail on page 122.

- The rate of reaction is greater in warm acid than in cold acid.
- In general, the more reactive the metal, the greater the rate of bubbling.

Some exceptions

Some reactive metals react unexpectedly slowly with water or acids.

1. Aluminium has a layer of aluminium oxide that stops water reaching the metal below.
2. In the reaction of calcium with dilute sulfuric acid, a layer of insoluble calcium sulfate forms, slowing the reaction.
3. In the reaction of magnesium with water, a layer of sparingly soluble magnesium hydroxide forms, slowing the reaction.

Magnesium reacts vigorously with steam:

$Mg(s) + H_2O(g) \rightarrow MgO(s) + H_2(g)$

Metals less reactive than hydrogen do not react with water or dilute acids.

A reactivity series

A **reactivity series** shows elements arranged in order of their reactivity.

	Reaction with	
	Water	Dilute acid
potassium	reacts quickly with cold water	violent reaction
sodium		
calcium		
magnesium	very slow	reaction becoming less vigorous ↓
aluminium	none	
zinc	reacts with steam	
iron		
hydrogen		
copper	no reaction with water or steam	no reaction with dilute acids
silver		
gold		

Remember that hydrogen is not a metal.

Worked example

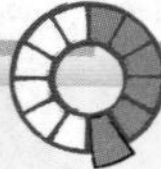

The diagram shows how five metals react with cold dilute hydrochloric acid and with cold water. Deduce the order of reactivity, starting with the least reactive. **(2 marks)**

copper, iron, zinc, magnesium, calcium

The rate of bubbling from both diagrams is used to list the metals in the order asked for.

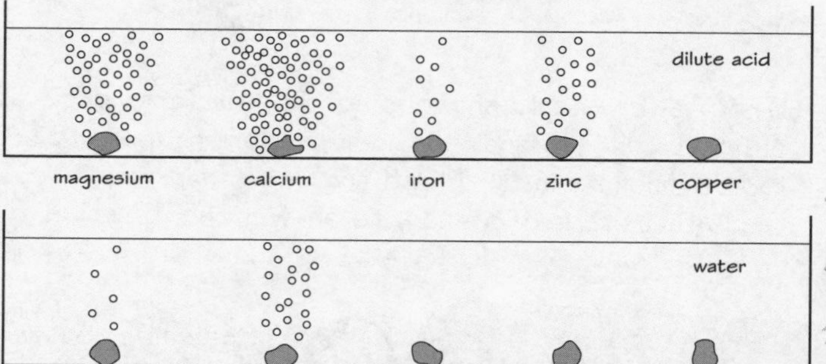

Now try this

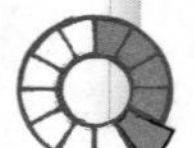

Using the information in the Worked example:

(a) Name **one** metal that reacts with cold water. **(1 mark)**

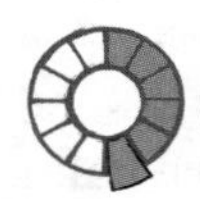

(b) Name **two** metals that react with dilute hydrochloric acid but do not appear to react with cold water. **(2 marks)**

Metal displacement reactions

You can work out the relative reactivity of metals using displacement reactions.

Displacement

A more reactive metal will **displace** a less reactive metal from its salts in solution. Magnesium is more reactive than copper. It can displace copper from copper sulfate solution:

$Mg(s) + CuSO_4(aq) \rightarrow MgSO_4(aq) + Cu(s)$

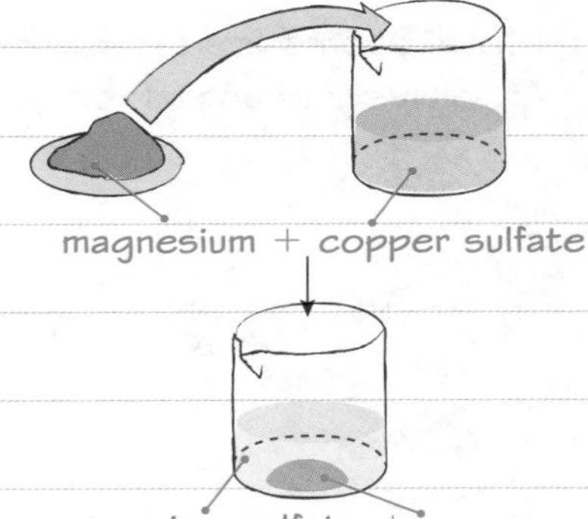

In this reaction, you observe:

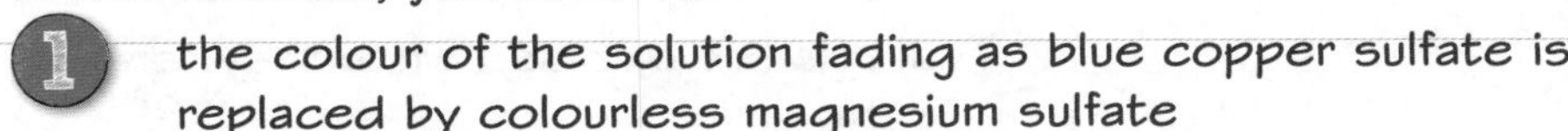

1. the colour of the solution fading as blue copper sulfate is replaced by colourless magnesium sulfate

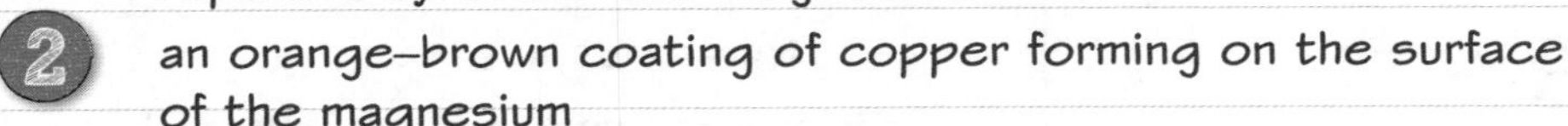

2. an orange–brown coating of copper forming on the surface of the magnesium
3. an increase in temperature because the reaction is **exothermic** (see page 152).

Practical skills Method for deducing a reactivity series

You can work out a reactivity series for metals by observing what happens when you mix different combinations of metals and their salt solutions. To do this:

- ✓ add a powdered metal to a test tube or beaker of a metal salt solution (as above right), or
- ✓ dip a small piece of metal into a metal salt solution on a spotting tile.

Then look for evidence of a reaction, such as a change in colour or temperature.

Some example results

Here are the results of an experiment with three metals and their salt solutions. A ✓ shows where a coating forms on the metal.

	$MgSO_4(aq)$	$ZnSO_4(aq)$	$CuSO_4(aq)$
Mg(s)	✗	✓	✓
Zn(s)	✗	✗	✓
Cu(s)	✗	✗	✗

From the results:

- ✓ magnesium is the most reactive
- ✓ copper is the least reactive
- ✓ a metal does not displace itself.

Worked example

A student adds a spatula of powdered metal to 10 cm³ of copper sulfate solution in boiling tubes. He stirs the mixtures and records the maximum increases in temperature. The table shows his results.

Metal	Temperature increase (°C)	Metal	Temperature increase (°C)
copper	0.0	magnesium	12.5
iron	5.0	zinc	9.5

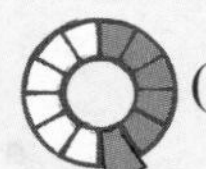

(a) Explain, using the results, which metal was the most reactive of the four metals tested. **(2 marks)**

Magnesium was the most reactive because its reaction gave the greatest temperature rise.

(b) Explain the results for copper. **(2 marks)**

There was no temperature change because there was no reaction. A metal does not displace itself from its compounds.

Metals less reactive than copper would also have no reaction with copper sulfate solution.

Now try this

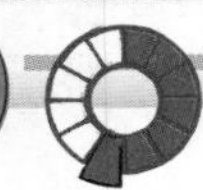

Using the table in the Worked example:

(a) Deduce a reactivity series for the four metals and explain your answer. **(2 marks)**

(b) Silver is less reactive than copper. Explain the expected temperature change if silver is used in the experiment. **(2 marks)**

Had a look ☐ Nearly there ☐ Nailed it! ☐

Explaining metal reactivity

You can explain the reactivity of metals in terms of how easily they form cations.

Forming cations

Metal atoms lose electrons to form **cations** (positively charged ions). For example:

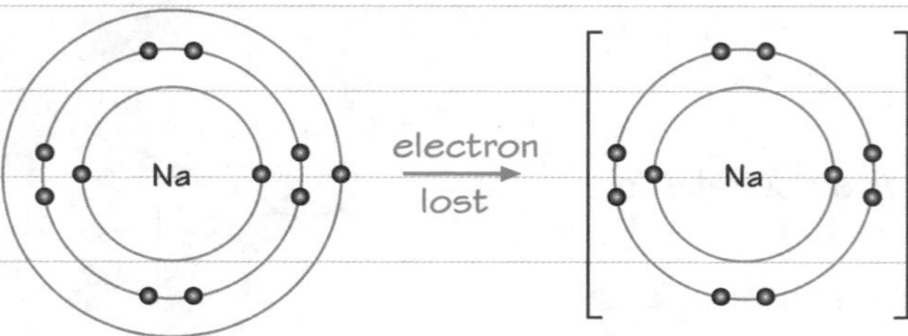

sodium atom, Na sodium ion, Na^+

A full outer shell is obtained when a metal atom loses electrons. The electronic configuration of Na is 2.8.1 but 2.8 for Na^+.

OIL RIG

Remember that:

- **O**xidation **i**s **l**oss of electrons.
- **R**eduction **i**s **g**ain of electrons.

This means that:

- atoms are oxidised when they form cations:

$$Na \rightarrow Na^+ + e^-$$

- cations are reduced when they form atoms:

$$Na^+ + e^- \rightarrow Na$$

You can revise forming ions on page 96.

Metals in reactions

Metal atoms form cations when metals react with water or dilute acids.

1. Metals react with water to form metal hydroxides, e.g.

$$2Na + 2H_2O \rightarrow 2NaOH + H_2$$

NaOH contains Na^+ and OH^- ions.

2. Metals react with dilute acids to form salts, e.g.

$$2Na + 2HCl \rightarrow 2NaCl + H_2$$

NaCl contains Na^+ and Cl^- ions.

Metal displacement reactions

Displacement reactions are **redox reactions**:

- The atoms of the more reactive metal are **oxidised** – they lose electrons.
- The metal cations of the less reactive metal are **reduced** – they gain electrons.

For example, magnesium displaces copper:

$$Mg + CuSO_4 \rightarrow MgSO_4 + Cu$$

You can show this as two **half-equations**:

$Mg \rightarrow Mg^{2+} + 2e^-$ oxidation

$Cu^{2+} + 2e^- \rightarrow Cu$ reduction

The sulfate ions, SO_4^{2-}, are unchanged.

Worked example

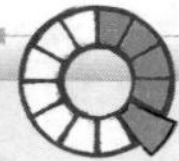

The list shows four metals arranged in order of decreasing reactivity from left to right.

sodium, aluminium, iron, copper

Which atoms form cations most easily? **(1 mark)**

☒ **A** sodium atoms
☐ **B** aluminium atoms
☐ **C** iron atoms
☐ **D** copper atoms

The reactions of metals (with water, dilute acids and salt solutions) show the **relative tendency** of metal atoms to form cations. The more easily a metal atom loses electrons, the more reactive the metal is.

Remember that sodium and iron react with water but aluminium, placed between them in the reactivity series, does not. It is protected by a layer of aluminium oxide.

Now try this

One way of joining two pieces of railway line together involves the 'thermite reaction'. A mixture of powdered aluminium and iron(III) oxide, Fe_2O_3, is heated. Aluminium oxide, Al_2O_3, and molten iron are produced. Use the description of the thermite reaction to help you to explain the difference in reactivity of aluminium and iron. **(3 marks)**

Metal ores

Most metals are extracted by the reduction of ores found in the Earth's crust.

Ores

Rocks contain metals or their compounds. An **ore** is a rock that contains enough of a metal to make its extraction economical.

Rocks may contain too little metal to make extraction worthwhile (if the cost of extracting the metal is greater than the value of the metal itself). Over time, metal prices may rise and these **low-grade ores** may become useful.

Unreactive metals

Unreactive metals such as gold are placed at the bottom of the reactivity series. They are found in the Earth's crust. They are in their 'native state' uncombined with other elements:

✓ They are not found as **compounds**

but

✓ they may occur naturally as **alloys** (mixtures of metals).

Oxidation

Oxidation is:

the gain of oxygen by a substance, e.g. magnesium is **oxidised** to magnesium oxide in air:

$$2Mg + O_2 \rightarrow 2MgO$$

the loss of electrons by a substance, e.g.

$$Mg \rightarrow Mg^{2+} + 2e^-$$

Reduction

Reduction is:

the loss of oxygen by a substance, e.g. zinc oxide is **reduced** to zinc when it is heated with carbon:

$$ZnO + C \rightarrow Zn + CO$$

the gain of electrons by a substance, e.g.

$$Zn^{2+} + 2e^- \rightarrow Zn$$

Worked example

Copper can be produced from copper oxide, CuO, by heating it with hydrogen gas.

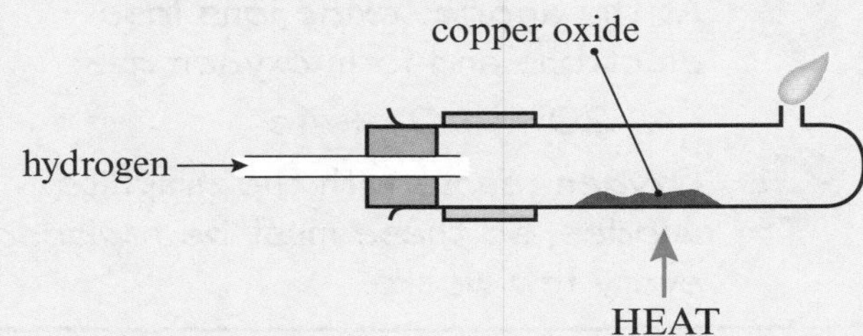

Explain, with the help of a balanced equation, why this is a redox reaction. **(3 marks)**

Copper oxide loses oxygen so it is reduced. At the same time, hydrogen gains oxygen so it is oxidised: $CuO + H_2 \rightarrow Cu + H_2O$

Resistance to corrosion

Metals **corrode** when they react with substances around them, such as air and water. How easily a metal corrodes depends upon how reactive it is:

easily corrodes	potassium
	sodium
	calcium
	magnesium
	(aluminium)
	zinc
	iron
	copper
corrodes slowly	silver
does not corrode	gold

Aluminium is protected by an oxide layer.

Now try this

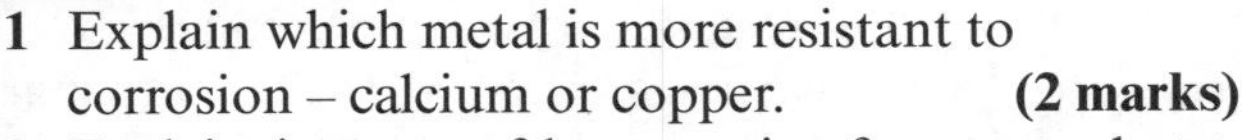

1 Explain which metal is more resistant to corrosion – calcium or copper. **(2 marks)**

2 Explain, in terms of loss or gain of oxygen, what happens during oxidation and reduction. **(2 marks)**

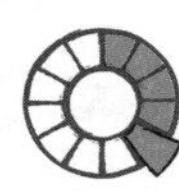

3 Explain why platinum is found in its native state in the Earth's crust. **(2 marks)**

Iron and aluminium

Most metals are extracted by the reduction of ores found in the Earth's crust.

Metal extraction

The method used to extract a metal from its ore is related to:

- the **cost** of the extraction process
- the metal's position in the **reactivity series** (shown on the right).

In principle, all metals can be extracted from their compounds using electrolysis. But:

- electricity is needed, which is expensive
- reduction by heating with carbon can be used if a metal is less reactive than carbon
- chemical reactions may be needed to separate silver and gold from other metals.

	Metal	Extraction method
most reactive	potassium	electrolysis of a molten compound
	sodium	
	calcium	
	magnesium	
	aluminium	
carbon →	zinc	reduction of its oxide using carbon
	iron	
hydrogen →	copper	
	silver	found as the element
least reactive	gold	

Hydrogen can reduce copper oxide to copper, but this is hazardous.

Iron extraction

Iron is less reactive than carbon, so it is produced by reducing iron oxide using carbon. This happens in industrial equipment called a **blast furnace**. You do not need to know any details about the furnace itself, but at the high temperatures inside it:

Iron oxide is reduced by carbon:

iron oxide + carbon → iron + carbon monoxide

$$Fe_2O_3(s) + 3C(s) \rightarrow 2Fe(l) + 3CO(g)$$

It is also reduced by carbon monoxide:

iron oxide + carbon monoxide → iron + carbon dioxide

$$Fe_2O_3(s) + 3CO(g) \rightarrow 2Fe(l) + 3CO_2(g)$$

3. Molten iron (iron in the liquid state) is produced.

Aluminium extraction

Aluminium is more reactive than carbon. It is produced by reducing aluminium oxide in an **electrolytic cell**.

Aluminium oxide is dissolved in molten **cryolite**. This reduces the temperature needed for electrolysis to happen, when an electric current passes through the mixture.

At the **cathode**, aluminium ions gain electrons and are reduced to aluminium atoms:

$$Al^{3+} + 3e^- \rightarrow Al$$

At the **anode**, oxide ions lose electrons and form oxygen gas:

$$2O^{2-} \rightarrow O_2 + 4e^-$$

4. Oxygen reacts with the graphite anodes, so these must be replaced every few weeks.

Worked example

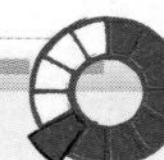

A student tries to extract the metals from magnesium oxide and lead oxide. He heats each oxide with powdered carbon. Explain why he can extract lead but not magnesium. **(3 marks)**

Carbon is above lead in the reactivity series but below magnesium. This means that carbon can reduce lead oxide but not magnesium oxide.

Lead and its compounds are toxic. Using a fume cupboard reduces the chance of breathing these in during the experiment.

It is not enough to write that magnesium is more reactive than lead. You need to include the role of carbon as a **reducing agent**.

Now try this

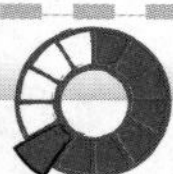

1. Suggest why tin oxide, but not calcium oxide, can be reduced by carbon. **(3 marks)**
2. Explain why electrolysis is not used to extract iron from iron oxide. **(2 marks)**

Biological metal extraction

There are biological methods for extracting metals, including use of plants or bacteria.

Using scrap iron

Copper is a valuable metal but we are running out of **high-grade** copper ores. It is expensive to extract copper from **low-grade** ores using traditional methods. Therefore, other methods are being researched.

Scrap iron can be used to produce copper from solutions of copper salts:

iron + copper sulfate → iron(II) sulfate + copper

$$Fe(s) + CuSO_4(aq) \rightarrow FeSO_4(aq) + Cu(s)$$

Iron is more reactive than copper. It can displace copper from copper compounds.

You can revise metal displacement on page 133.

Grades of ore

Grade of ore	Proportion of metal or metal compound
high	high
↓	↓
low	low

Compared with high-grade ores, low-grade ores:

- 👍 are more common because most high-grade ores have already been used
- 👎 are less profitable
- 👎 use more energy
- 👎 produce more waste when used.

Phytoextraction

Phytoextraction is a biological method of metal extraction that uses plants.

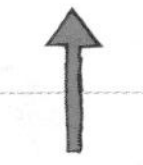

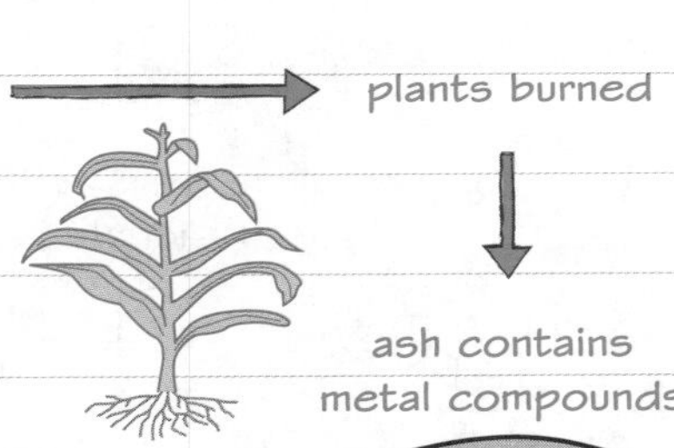

Bioleaching

Bioleaching is a biological method of metal extraction that uses bacteria. Copper can be extracted from copper sulfide, CuS, in the following way:

- ✓ The bacteria oxidise sulfide ions, S^{2-}.
- ✓ Copper sulfide ores break down.
- ✓ Cu^{2+} ions are released.

The solution that the bacteria produce is called a **leachate**. This leachate contains:

- ✓ a high concentration of metal ions.

Scrap iron can be used to obtain copper from the leachate.

Worked example

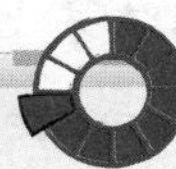

Copper can be produced from ores by two methods:
Method A involves mining the ore, purifying it and heating the purified ore with carbon.
Method B is phytoextraction using plants.
Evaluate these two extraction methods. **(6 marks)**

When you are asked to evaluate something, make sure that you give its advantages *and* disadvantages. Include a conclusion.

Method A is quicker but supplies of high-grade ores are limited. It uses more energy and produces large amounts of waste rock.

Method B produces ash with a high concentration of copper from low-grade ore. It does not cause pollution as a result of mining. On the other hand, it takes a lot longer because the plants have to grow and it uses more land.

Overall I think that method B is better because it conserves supplies of high-grade ores.

Now try this

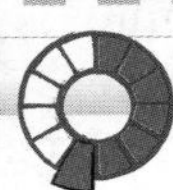

1 Describe how phytoextraction works. **(4 marks)**
2 Describe **one** advantage and **one** disadvantage of extracting metals using bacteria. **(2 marks)**

Recycling metals

Recycling metals rather than extracting them from ores has economic implications. It can also preserve the environment and the supply of limited raw materials.

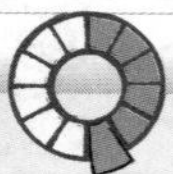

Extracting metals

Extracting metals from their ores:

- uses up limited resources
- uses a lot of energy
- damages the environment.

Recycling reduces these disadvantages. Used metal items are collected. Rather than throwing them away, these are taken apart. The metal is melted down to make new items.

Recycling

Recycling metals means:

- metal ores will last longer
- less energy is needed
- fewer quarries and mines are needed
- less noise and dust are produced
- less land is needed.

Worked example

The flow chart shows the main stages in extracting aluminium from its ore.
Use it to suggest the benefits of recycling aluminium. **(3 marks)**

Less waste rock will be produced from mining.

Aluminium oxide will not need separating and purifying from aluminium ore, which will save energy. Less carbon dioxide will be emitted because less fuel will be needed for heat and electricity and because carbon dioxide is produced from the electrolysis.

You need to be specific in your answer. To write that recycling is 'better for the environment' does not give enough detail.

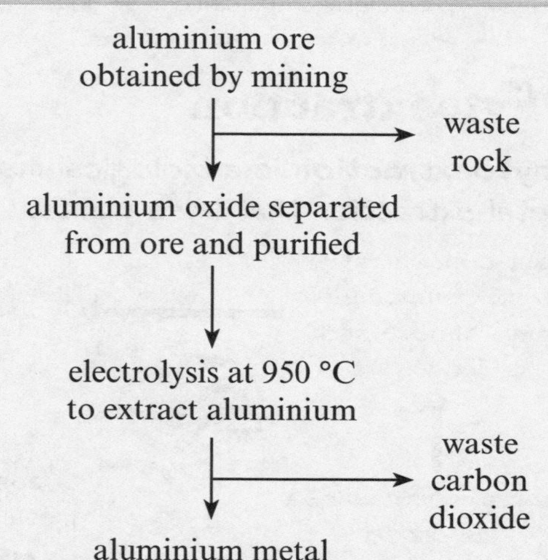

Drawbacks of recycling

- Used metal items must be collected and transported to the recycling centre.
- Different metals must be removed from used items and sorted.
- Recycling saves different amounts of energy, depending on the metal involved.

Recycling metals saves energy

Different amounts of energy are saved by recycling metals compared with extracting them from ores:

Metal	Percentage energy saved
aluminium	94
copper	86
iron and steel	70

Now try this

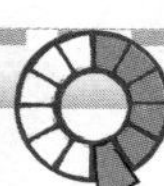

1 Describe **two** ways in which recycling copper, rather than extracting it from copper ores, can reduce pollution. **(2 marks)**

2 Suggest **two** reasons why more energy is saved when aluminium is recycled than when steel is recycled. **(2 marks)**

3 Explain why recycling metals conserves the supplies of metal ores. **(2 marks)**

Life-cycle assessments

You need to be able to describe the basic principles involved in carrying out a life-cycle assessment.

Cradle to grave

A **life-cycle assessment** (LCA) of a product is a 'cradle-to-grave' analysis of its impact on the environment. It includes these stages:

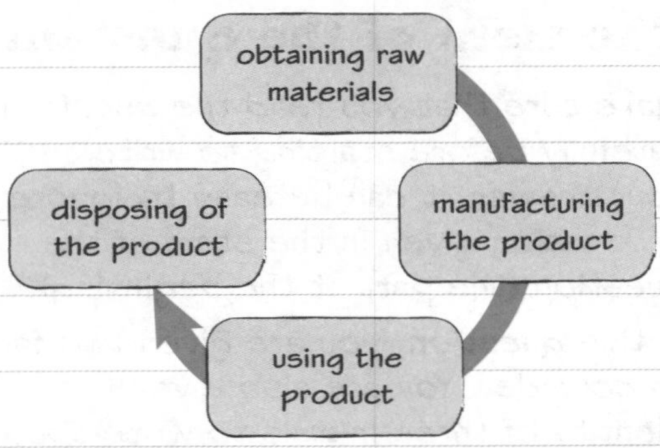

Data for an LCA

An LCA is likely to need data on these factors at most or all stages:

- the use of energy
- the release of waste materials
- transport and storage.

An LCA is also likely to need data on:

- whether the raw materials needed are renewable or non-renewable
- whether any of the product can be recycled or re-used
- how the product is disposed of.

An example of a life cycle

There are many detailed stages in the life cycle of a product, e.g. for a car.

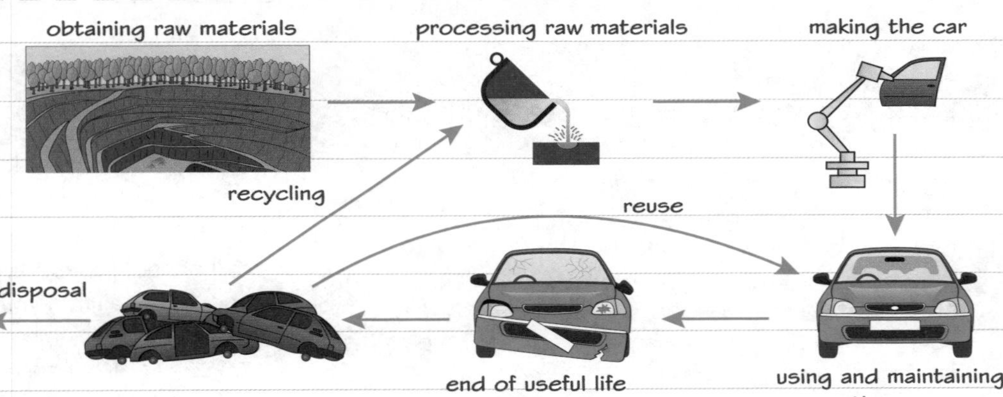

Worked example

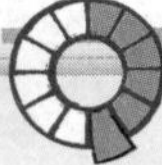

The table shows information about the energy needed in the life cycle of a polyester shirt.

Stage	Percentage of lifetime energy used
obtaining raw materials	5
manufacturing the shirt	14
using the shirt	80
disposing of the shirt	1

Discuss the use of energy in the life cycle of the shirt, and suggest how it may be improved. **(4 marks)**

Almost one-fifth of the energy used is to do with making the shirt. Very little is to do with disposing of it at the end of its life. Most of the energy used is to do with using the shirt, probably due to washing and ironing it. Energy use could be reduced by washing at low temperatures and drying it outside.

The question shows only part of the data that should be available to make a life-cycle assessment, e.g. no information is given about the environmental impact of each stage, or the lifespan of the shirt.

Now try this

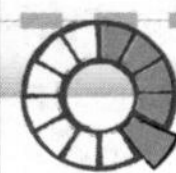

1 State the four main stages in a life-cycle assessment. **(1 mark)**

2 Life-cycle assessments may identify alternative materials that can be used with less impact. Give **one** other reason for carrying out a life-cycle assessment. **(1 mark)**

Had a look ☐ Nearly there ☐ Nailed it! ☐

Extended response – Reactivity of metals

There will be at least one 6-mark question on your exam paper. For these questions, you will need to think scientifically and structure your answer logically, showing how the points you make are related to each other.

You can revise the topics for this question, which is about **metal reactivity** and **metal extraction**, on pages 132–136.

Worked example

Some metals are found as uncombined elements, but most metals are extracted from ores found in the Earth's crust. Different metals are extracted using different methods.

Explain how the method used to obtain a metal is related to its position in the reactivity series and to the cost of the extraction process. In your answer, refer to aluminium, iron and gold as examples. **(6 marks)**

Gold is placed at the bottom of the reactivity series because it is an unreactive metal. It is found in the Earth's crust uncombined with other elements so the cost of extracting gold is low.

Aluminium is placed near the top of the reactivity series, because it is a very reactive metal. It is found combined with other elements, such as oxygen. Aluminium is more reactive than carbon, so it must be extracted using electrolysis. A lot of electricity is used in electrolysis, which makes it expensive to extract aluminium from its ore.

Iron is placed in the middle of the reactivity series, because it is less reactive than aluminium but more reactive than gold. Iron is also less reactive than carbon. Iron can be extracted by heating its ore with carbon. Although electrolysis could be used instead, carbon is cheaper than electricity.

The stem of the question

Make sure that you read the question carefully before starting to write your answer. It can be easy to ignore information given in the stem of the question (the part at the beginning).

In this question, you are given two factors to consider. You are also given the identity of three metals to write about.

You could begin your answer with any of the metals. The extraction of gold is simple to explain so it is covered first here. In your answer, refer clearly to the position of gold in the reactivity series, and also to the cost of extracting it.

You should include knowledge and understanding about why aluminium must be extracted using electrolysis. You should also include why this is expensive. You need to explain why heating with carbon is unsuitable. You can do this by referring to the position of carbon in the reactivity series.

You might point out that electrolysis could still be used to extract iron. You can then explain that heating with carbon is cheaper.

You should think clearly about how to present your ideas before you start. Aim for a clear and logical structure. You can support this with scientific information and ideas. You should leave out information or ideas that are unnecessary to answer the question.

Now try this

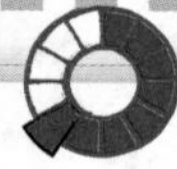

You are provided with these metal powders and solutions:

Metal powder	Solution
copper	copper sulfate
magnesium	magnesium sulfate
zinc	zinc sulfate

Explain how you could use these substances in displacement reactions to find the order of reactivity of the three metals. In your answer, include the expected results and how you would use them to show the correct order of reactivity. You should use equations as part of your answer. **(6 marks)**

The Haber process

Reversible reactions

Chemical reactions are **reversible**. The direction of some reversible reactions can be altered by changing the reaction conditions.

Ammonium chloride decomposes when heated: $NH_4Cl(s) \rightarrow NH_3(g) + HCl(g)$

Ammonia and hydrogen chloride combine when cool:
$NH_3(g) + HCl(g) \rightarrow NH_4Cl(s)$

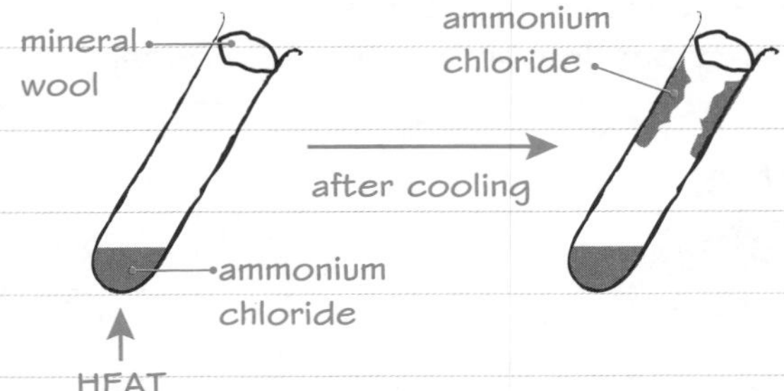

These changes can be modelled using ⇌, the **reversible symbol**:

$NH_4Cl(s) \rightleftharpoons NH_3(g) + HCl(g)$

Dynamic equilibrium

In a **closed system**, a container where no reacting substances can enter or leave, a reversible reaction can reach **equilibrium**. At equilibrium:

- rate of forward reaction = rate of backward reaction
- the concentrations of the reacting substances stay constant (do not change).

A chemical equilibrium is a **dynamic** equilibrium:

- the forward and backward reactions keep going – they do not stop at equilibrium.

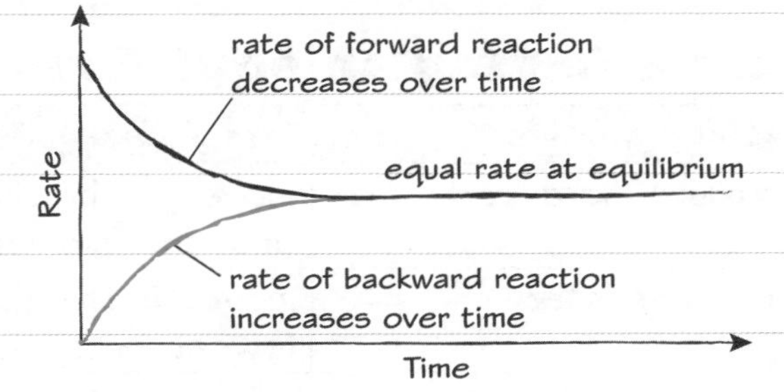

The Haber process

The **Haber process** is a reversible reaction between nitrogen and hydrogen to form ammonia.

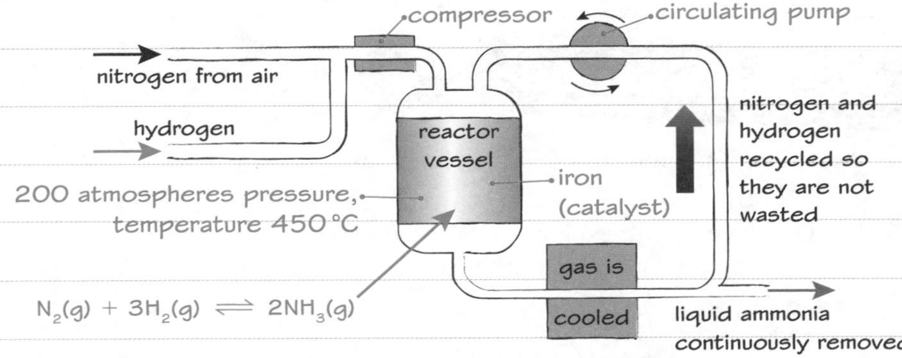

Nitrogen can be obtained by fractional distillation of liquefied air.

Methane in natural gas reacts with steam to produce carbon dioxide and hydrogen:

$CH_4(g) + 2H_2O(g) \rightarrow CO_2(g) + 4H_2(g)$

Worked example

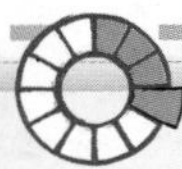

The Haber process requires nitrogen and hydrogen. State the raw materials needed to obtain these two reactants. **(2 marks)**

Nitrogen is extracted from the air and hydrogen is obtained from natural gas.

Now try this

Hydrated copper sulfate crystals turn from blue to white when they are heated:

hydrated copper sulfate (blue) ⇌ anhydrous copper sulfate (white) + water

(a) State why the word equation contains the symbol ⇌. **(1 mark)**

(b) Describe how anhydrous copper sulfate could be used to test for water. **(2 marks)**

More about equilibria

Position of a dynamic equilibrium

Equilibrium position gives you an idea of the relative concentrations of the substances in a reaction mixture at equilibrium. For example, bismuth chloride reacts reversibly with water:

	$BiCl_3$ + H_2O	$\rightleftharpoons$	$BiOCl$ + $2HCl$
concentrations	low		high
equilibrium position			to the right
concentrations	high		low
equilibrium position	to the left		

Going to completion

Some reactions do not reach equilibrium. Instead, they go to **completion**:

- all the limiting reactants are used up
- you can calculate the mass of product formed using the mass of the limiting reactant (see pages 107 and 108 for details).

Reactions that go to completion include:

- precipitation reactions
- reactions in which gases can escape.

Factors affecting an equilibrium position

Change in conditions	Effect on equilibrium position
temperature increased	moves in the direction of the endothermic reaction
pressure increased	moves in the direction of the fewest molecules of gas
concentration of a reacting substance increased	moves in the direction away from the reacting substance
catalyst added	no change

as seen in the balanced equation for the reaction

equilibrium reached sooner

Worked example

Ammonia is manufactured in the Haber process:

$N_2(g) + 3H_2(g) \rightleftharpoons 2NH_3(g)$

(a) Use the graph to explain why the temperature chosen, 450 °C, is a compromise. **(3 marks)**

As the temperature increases, the equilibrium yield of ammonia decreases, but the rate of reaction increases. 450 °C is low enough to obtain an acceptable yield of ammonia, but high enough to obtain it in an acceptable time.

(b) Use the graph to explain why the pressure used, 200 atmospheres, is a compromise. **(2 marks)**

As the pressure increases, the equilibrium yield of ammonia and the rate of reaction increase. However, very high pressures need stronger and more expensive equipment which uses more energy.

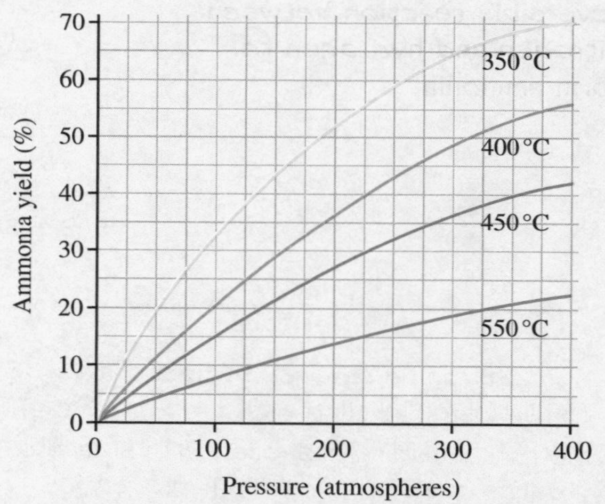

The equilibrium yield of ammonia at different temperatures and pressures.

Now try this

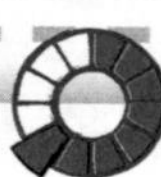

The forward reaction is exothermic.

Methanol is manufactured using this reaction:

$CO(g) + 2H_2(g) \rightleftharpoons CH_3OH(g)$

$\Delta H = -91\ kJ\ mol^{-1}$

Predict and explain the effects on the equilibrium concentration of methanol vapour when:

(a) The pressure is decreased. **(2 marks)**

(b) The temperature is decreased. **(2 marks)**

(c) The concentration of hydrogen is increased. **(2 marks)**

The alkali metals

The elements in group 1 of the periodic table are known as the **alkali metals**.

Physical properties

The alkali metals have some properties typical of metals. For example, they are:

- good conductors of heat and electricity
- shiny when freshly cut.

However, compared with most other metals, they:

- are soft (you can cut them with a knife)
- have relatively low melting points (but all are solid).

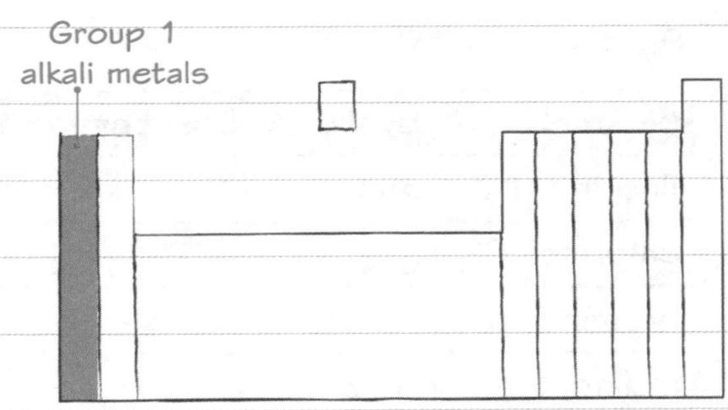

Reactions with water

The alkali metals react with water, producing an alkaline metal hydroxide and hydrogen, e.g.

sodium + water → sodium hydroxide + hydrogen

$2Na(s) + 2H_2O(l) \rightarrow 2NaOH(aq) + H_2(g)$

Reactivity

The **reactivity** of the alkali metals increases down the group:

- Lithium fizzes steadily.
- Sodium melts into a ball from the heat released in the reaction and fizzes rapidly.
- Potassium gives off sparks and the hydrogen produced burns with a lilac-coloured flame.

Density

Lithium, sodium and potassium are less dense than water, so they float.

Storage

Lithium, sodium and potassium are stored in oil. This is to keep air and water away.

Worked example

(a) Rubidium is placed below potassium in the periodic table.
Predict what is seen in the reaction of rubidium with water. **(2 marks)**

Rubidium will react very vigorously with water, producing sparks and bursting into flames explosively.

You need to be able to:

- describe the pattern of reactivity of lithium, sodium and potassium, and
- use it to predict the reactivity of other alkali metals (rubidium, caesium and francium).

(b) Explain why sodium is more reactive than lithium. **(2 marks)**

Sodium atoms are larger than lithium atoms. So the outer electron in a sodium atom is further from the nucleus than the outer electron in a lithium atom. This means that the force of attraction is weaker. So the outer electron is lost more easily from sodium than from lithium.

The alkali metals all have one electron in the outer shell of their atoms. They lose this electron in reactions to form ions with a 1+ charge, e.g. Li^+ and Na^+.The more easily the outer electron is lost, the more reactive the metal.

Going down the group, the number of occupied shells in an atom increases. The electrons in the inner shells **shield** the outer electron, reducing its attraction for the nucleus.

Now try this

Caesium, Cs, is placed below rubidium in group 1 of the periodic table.

(a) Write a balanced equation for the reaction of caesium with water. **(2 marks)**

(b) Predict what is seen in the reaction of caesium with water. **(2 marks)**

(c) State **two** physical properties that rubidium and caesium have in common. **(2 marks)**

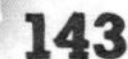

Had a look ☐ Nearly there ☐ Nailed it! ☐

The halogens

The elements in group 7 of the periodic table are non-metals known as the **halogens**.

Appearance

Element	State at room temperature	Colour
fluorine, F_2	gas	pale yellow
chlorine, Cl_2	gas	yellow–green
bromine, Br_2	liquid	red–brown
iodine, I_2	solid	dark grey

covalent molecules, each with two atoms

forms a purple vapour when warmed

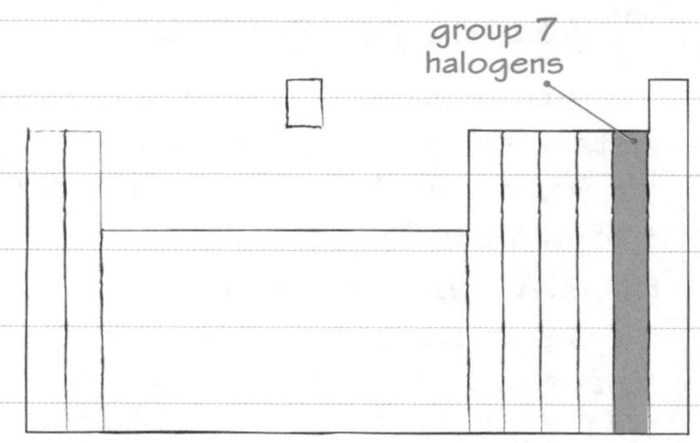

Melting and boiling points

Going down group 7:
- melting points increase
- boiling points increase.

When simple molecular substances melt or boil:
- weak intermolecular forces are overcome
- the strong covalent bonds joining atoms together in each molecule do not break.

Going down group 7:
- the intermolecular forces between molecules become stronger
- more heat energy is needed to overcome these forces.

Temperature (°C)

350, 250, 150, 50, −50, −150, −250

fluorine, chlorine, bromine, iodine

melting point, boiling point

You can revise simple molecular substances on page 100 and physical states on page 112.

Worked example

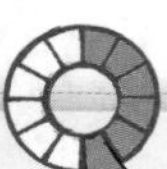

Chlorine is manufactured on an industrial scale by the electrolysis of concentrated sodium chloride solution.
Describe a simple laboratory test for the presence of chlorine. **(2 marks)**

Put a piece of damp blue litmus paper into the container. If chlorine is present, the litmus paper turns red, then is bleached white.

Practical skills When you describe a laboratory test, say what you would do and what you would observe.

In a different test, damp **starch iodide paper** turns blue–black in the presence of chlorine. The colour change happens because chlorine displaces iodine, which then reacts with the starch. You can revise halogen displacement reactions on page 146.

Now try this

Use evidence from the bar chart to explain your answer.

Astatine is a rare, radioactive element placed immediately below iodine in group 7.

(a) Suggest why astatine is thought to be dark in colour, even though no one has seen it. **(1 mark)**

(b) Predict the melting point of astatine, and justify your answer. **(2 marks)**

Reactions of halogens

The reactivity of the elements decreases down group 7 (the opposite trend to group 1).

Reactions with metals

The halogens react with metals to produce compounds called metal halides, e.g.

sodium + chlorine → sodium chloride

$2Na(s) + Cl_2(g) \rightarrow 2NaCl(s)$

The sodium burns with an orange flame, forming white sodium chloride.

Iron wool reacts with the halogens when it is heated, e.g.

iron + chlorine → iron(III) chloride

$2Fe(s) + 3Cl_2(g) \rightarrow 2FeCl_3(s)$

The iron burns to form a dark-purple solid.

Halide ions, halides

The elements in group 7 are called **halogens**. In reactions with metals, halogen atoms gain electrons and are reduced, e.g.

$Cl_2 + 2e^- \rightarrow 2Cl^-$

The ions formed:

- ✓ have a 1– charge, e.g. Cl^-, Br^- and I^-
- ✓ are called **halide ions**.

Halides are compounds of metals or hydrogen with halogens, such as sodium chloride.

Explaining reactivity

A halogen atom has seven electrons in its outer shell. When a halogen reacts with a metal or hydrogen, each halogen atom gains one electron to complete its outer shell. The less easily a halogen atom gains an electron, the less reactive the halogen is.

Going down group 7:

- the outer shell gets further from the nucleus
- there is more shielding by inner electrons
- the force of attraction between the nucleus and outer shell electrons gets weaker
- electrons are gained less easily
- the elements become less reactive.

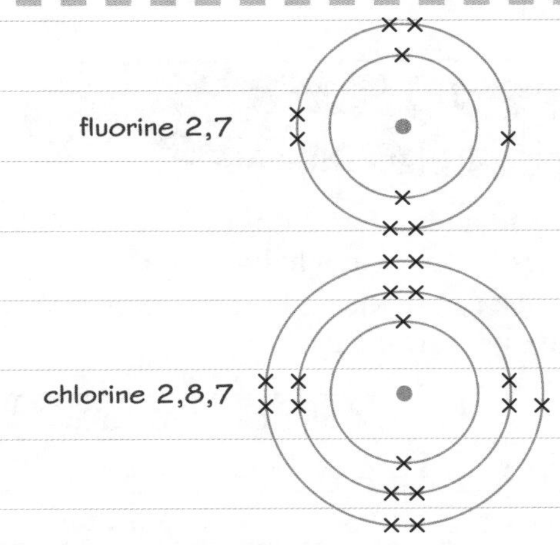

Fluorine is more reactive than chlorine – its atoms gain electrons more easily.

Worked example

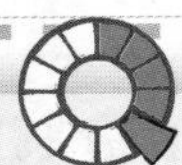

Hydrogen reacts vigorously with bromine vapour to produce hydrogen bromide gas:

$H_2(g) + Br_2(g) \rightarrow 2HBr(g)$

What happens if this gas is added to water? **(1 mark)**

- ☐ **A** It does not dissolve.
- ☐ **B** It dissolves to form a neutral solution.
- ☐ **C** It dissolves to form an alkaline solution.
- ☒ **D** It dissolves to form an acidic solution.

Hydrogen reacts with the halogens to produce hydrogen halides. The reactions become less vigorous down group 7:

- Fluorine reacts explosively in the dark.
- Chlorine reacts explosively in sunlight.
- Bromine reacts vigorously in a flame.

The hydrogen halides all dissolve in water to produce acidic solutions, e.g.

- hydrogen chloride gas, $HCl(g)$, produces hydrochloric acid, $HCl(aq)$. This fully dissociates to form $H^+(aq)$ and $Cl^-(aq)$ ions.

Now try this

1 Cold iron wool burns in fluorine but hot iron wool reacts slowly with iodine vapour. Explain why iodine is less reactive than fluorine. **(2 marks)**

2 Aluminium powder reacts vigorously with iodine to produce aluminium iodide, AlI_3. Write a balanced equation for the reaction. **(2 marks)**

Had a look ☐ Nearly there ☐ Nailed it! ☐

Halogen displacement reactions

A more reactive halogen can displace a less reactive halogen from its compounds.

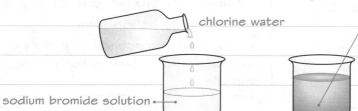

Practical skills: Investigating displacement

You can demonstrate **displacement reactions** by adding a halogen solution to a metal halide solution, then seeing if the mixture darkens. For example:

- Chlorine in 'chlorine water' displaces bromine from aqueous sodium bromide solution.
- balanced equation: $Cl_2(aq) + 2NaBr(aq) \rightarrow 2NaCl(aq) + Br_2(aq)$
- ionic equation (Na⁺ ions are spectator ions – see page 89):

$$Cl_2(aq) + 2Br^-(aq) \rightarrow 2Cl^-(aq) + Br_2(aq)$$

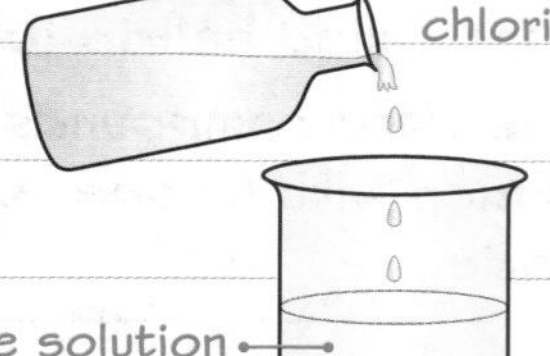
chlorine water
sodium bromide solution

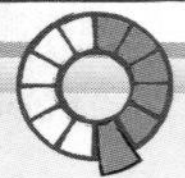
The reddish-brown colour is due to the bromine that has been displaced.
sodium chloride solution + bromine

Worked example

A student adds a few drops of halogen solutions to small volumes of potassium halide solutions on a spotting tile. Put a tick (✓) in the table below to show where she sees the mixture turning darker. **(2 marks)**

	Chloride	Bromide	Iodide
Chlorine	not done	✓	✓
Bromine		not done	✓
Iodine			not done

The halogens are toxic. It is safer to use small volumes of dilute halogen solutions.

Potassium salts form colourless solutions. They darken if bromine or iodine is produced. A halogen cannot displace itself so, for example, chlorine water is not added to potassium chloride solution in the investigation.

Redox reactions

Halogen displacement reactions are **redox** reactions, e.g. when chlorine displaces bromine from bromide ions in solution:

1. Chlorine atoms gain electrons and are reduced to chloride ions:

$$Cl_2(aq) + 2e^- \rightarrow 2Cl^-(aq)$$

2. Bromide ions lose electrons and are oxidised to bromine:

$$2Br^-(aq) \rightarrow Br_2(aq) + 2e^-$$

OIL RIG

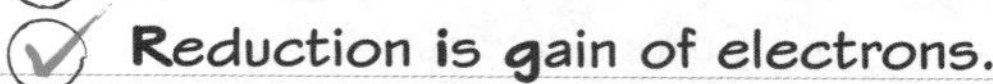

- **O**xidation **i**s **l**oss of electrons.
- **R**eduction **i**s **g**ain of electrons.

These processes happen together, at the same time, in **redox** reactions.

Metal displacement reactions are also redox reactions. You can revise these reactions on page 133.

Now try this

Astatine, At, is at the bottom of group 7.

Aqueous bromine solution reacts with aqueous potassium iodide solution.

(a) For this reaction, write the balanced equation and the ionic equation. **(4 marks)**

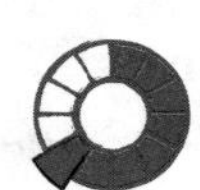

(b) Explain, with the help of half-equations, why this reaction is a redox reaction. **(4 marks)**

(c) State and explain whether astatine will react with potassium iodide solution. **(2 marks)**

The noble gases

The elements in group 0 of the periodic table are known as the **noble gases**.

Chemical properties

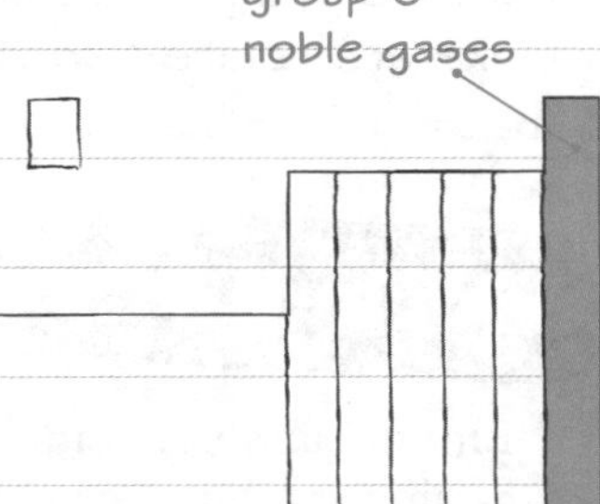

The noble gases are chemically **inert**. Their lack of reactivity is because:

- their atoms have full outer shells of electrons, so
- they have no tendency to lose, gain or share electrons.

Element	He	Ne	Ar
Electronic configuration	2	2.8	2.8.8

Uses of the noble gases

You need to be able to explain how use of a noble gas depends on a particular property. The table shows some typical uses of noble gases.

Noble gas	Use	Property needed		Reason for use
		Low density	Inertness	
helium	**lifting gas** in party balloons and airships	✓	✓	• Helium is less dense than air so the balloons and airships rise. • It is non-flammable so does not ignite.
argon krypton xenon	**filling gas** in filament lamps		✓	• The metal filament becomes hot enough to glow. • The inert gases stop it burning away.
argon	**shield gas** during welding		✓	• Argon is denser than air so it keeps air away from the metal. • It is inert so the metal does not oxidise.

Worked example

The table shows the densities and boiling points of some of the noble gases. Describe the trends in these properties and use them to predict the missing numbers. **(4 marks)**

Element	Density (kg/m^3)	Boiling point (°C)
helium, He	0.15	−269
neon, Ne	1.20	
argon, Ar		−186
krypton, Kr	2.15	−152

The densities increase as you go down the group. The boiling points get higher as you go down the group.

The density of argon could be about 1.7 or 1.8 kg/m^3.

The boiling point of neon could be about −230°C.

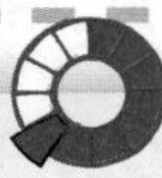

These numbers are about halfway between the numbers above and below in the table.

You need to be able to answer questions like this one that ask you to identify a trend, and use the trend to predict missing values.

Now try this

1 The density of air is 1.225 kg/m^3.
Explain why helium and neon can be used as lifting gases, but argon and krypton cannot. **(2 marks)**

Look at the data in the Worked example.

2 Give **two** reasons why argon may be used in some fire-extinguishing systems. **(2 marks)**

3 Explain, in terms of its electronic configuration, why helium is inert. **(3 marks)**

Extended response – Groups

There will be at least one 6-mark question on your exam paper. For these questions, you will need to think scientifically and structure your answer logically, showing how the points you make are related to each other.

You can revise the topics for this question, which is about **groups in the periodic table**, on pages 143–147.

Worked example

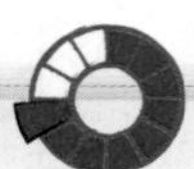

Lithium and potassium react with cold water. Similar products form in these reactions.
Compare and contrast the way in which these two elements react with water. Explain your answers in terms of electronic configurations, reactions and products. **(6 marks)**

Both elements are in group 1 of the periodic table, the alkali metals. Their atoms have one electron in their outer shells, which is why they react in a similar way (lithium is 2.1 and potassium is 2.8.8.1).

Both metals float on water and both produce hydrogen. They also produce soluble metal hydroxides when they react with water, forming alkaline solutions.

However, potassium is more reactive than lithium, so its reaction with water is more vigorous. This is because potassium atoms lose their outer electron more easily. Potassium atoms have more filled electron shells, so their outer electron is further from the nucleus and more shielded, so has a weaker attraction for it.

Lithium fizzes slowly when it reacts with water, gradually becoming smaller until it disappears. Potassium reacts more vigorously. It immediately ignites with a lilac flame when it reacts with water, and disappears with sparks.

Command words: Compare and contrast

The question is looking for the similarities and differences between two or more things, without needing a conclusion. The answer must relate to all things mentioned in the question. It must include at least one similarity and one difference.

You should mention which group the elements are placed in. Their electronic configurations are important in explaining why their reactions are similar.

You should be clear about what is similar in the reactions of lithium and potassium with water.

Your answer should compare the reactivity of the two metals, and explain this in terms of how easily the outer electrons are lost in reactions.

You need to be able to describe the reactions of the alkali metals with water. The key differences in the reactions of lithium and potassium are mentioned here.

Your answer should show comprehensive knowledge and understanding of the topic covered. Take care to cover both similarities and differences. Make sure that you use any information given to you to support your explanations. Organise your explanations in a structured way with clear lines of reasoning.

Now try this

The command word **devise** means that you need to plan or invent a procedure from existing principles or ideas.

The elements in group 7 of the periodic table include chlorine, bromine and iodine. Devise an experiment to determine the order of reactivity of these elements using displacement reactions.
In your answer, assume that you are given dropping pipettes, test tubes and these solutions:

- chlorine solution, bromine solution, iodine solution
- potassium chloride solution, potassium bromide solution, potassium iodide solution.

Describe suitable safety precautions. Include the expected results and explain how you would use them to show the order of reactivity. You may include balanced equations to support your answer. **(6 marks)**

Rates of reaction

The greater the frequency of successful collisions, the greater the rate of reaction.

Colliding particles

For a reaction to happen:
- reactant particles must collide with each other **and**
- the collisions must have enough energy.

The **activation energy** of a reaction is the minimum energy needed by reactant particles for a reaction to happen. A **successful collision** has the activation energy or more.

Rate and time

The greater the **rate of reaction**, the lower the **reaction time**:

✓ A fast reaction happens in a short time, such as combustion.

✓ A slow reaction happens over a long time, such as rusting.

Take care not to confuse the two quantities.

Concentration and pressure

The rate of reaction increases if the **concentration** of a dissolved reactant increases, or if the **pressure** of a reacting gas increases:
- there are more particles in the same volume
- the frequency of successful collisions increases.

Temperature

The rate of reaction increases when the **temperature** increases because the particles gain energy, and:
- the particles move faster
- the frequency of collisions increases **and**
- the energy of collisions increases, so a greater proportion of collisions are successful.

Surface area : volume ratio

The rate of reaction increases when the **surface area : volume ratio** of a solid reactant increases, e.g. when lumps are made into a powder:
- more particles of reactant are available
- the frequency of successful collisions increases.

larger piece

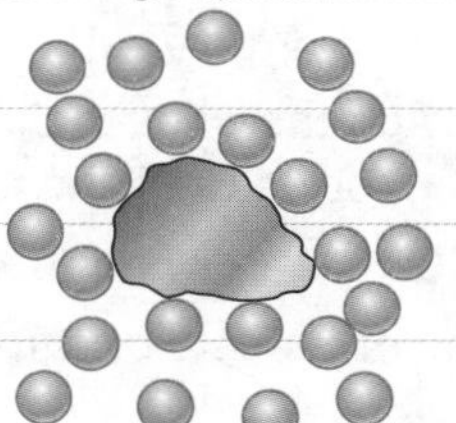

smaller surface area
slower reaction

smaller pieces

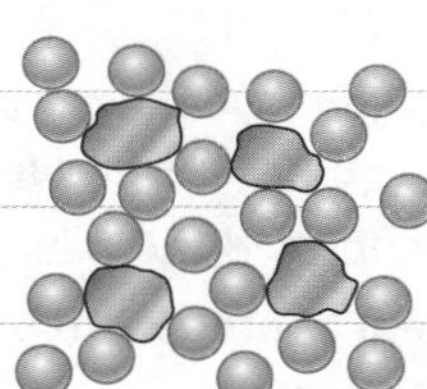

larger surface area
faster reaction

Worked example

(a) Describe three features of a catalyst. **(3 marks)**

Catalysts are substances that speed up the rate of a reaction without altering the products. They are unchanged chemically and in mass at the end of the reaction.

(b) Explain how a catalyst works. **(2 marks)**

Catalysts provide an alternative reaction pathway with a lower activation energy.

Catalysts are often transition metals or their compounds, such as iron in the Haber process.

Enzymes are **biological** catalysts. Enzymes found in yeast catalyse the conversion of glucose to ethanol and carbon dioxide. Enzymes are used in the production of wine and other alcoholic drinks.

With a lower activation energy, a greater proportion of collisions have the necessary activation energy or more. So the frequency of successful collisions is greater at a given temperature.

Now try this

Marble chips react with hydrochloric acid: $CaCO_3(s) + 2HCl(aq) \rightarrow CaCl_2(aq) + H_2O(l) + CO_2(g)$
State and explain the effect of the following on the rate of the reaction:

(a) crushing the chips **(3 marks)**
(b) diluting the acid **(3 marks)**

(c) heating the acid. **(4 marks)**

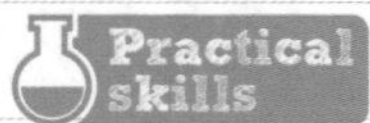

Practical skills

Investigating rates

Core practical

Effect of surface area on rate

Aim

To investigate the effect on the rate of reaction with hydrochloric acid of changing the surface area of calcium carbonate.

You could also investigate the effect of changing the concentration of the hydrochloric acid.

Apparatus

- eye protection
- conical flask
- measuring cylinder
- marble chips (small, medium, large)
- cotton wool
- dilute hydrochloric acid
- ±0.01 g balance
- stop clock.

You could use powdered marble. However, the rate of reaction may be so great that it becomes difficult to measure the mass at a given time.

Instead of recording the change in mass, you could use a gas syringe to measure the volume of carbon dioxide produced.

Method

1 Prepare three sets of marble chips on folded paper, one for each size of chip. Adjust the numbers so that each set has the same mass.
2 Add a measured volume of dilute acid to the conical flask. Plug the flask with cotton wool.
3 Place the flask and a set of marble chips on the balance. Record the reading.
4 Remove the cotton wool. Add the chips to the acid. Start the stop clock, and replace the cotton wool and folded paper.
5 Record the mass each 30 s for a few minutes.

Repeat steps 2 to 5 with the other sets of chips.

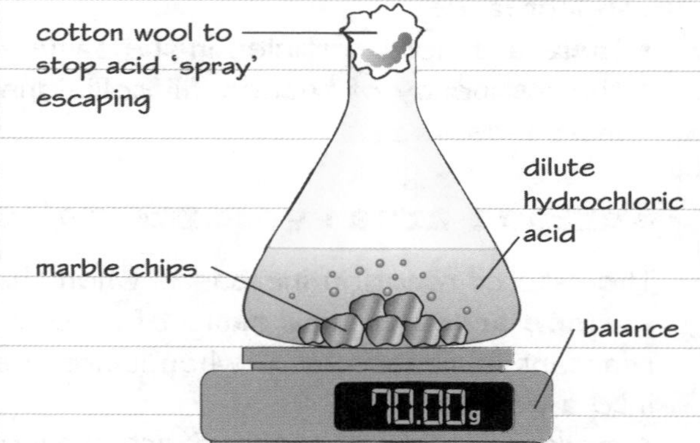

Results

Time (s)	Mass (g)	Change in mass (g)
0	70.00	0.00
30	69.90	0.10
60	69.83	0.17

You should record your results for each set of chips in a suitable table. Include a column for the change in mass at each time. Make sure that you have enough rows for all your readings.

Analysis

Calculate the change in mass at each time, *t*:

- change in mass = mass at start − mass at *t*

For each set of chips, calculate the mean rate of reaction over the same amount of time:

$$\text{mean rate of reaction} = \frac{\text{change in mass}}{\text{chosen time}}$$

You could also plot line graphs:

- change in mass in g on the vertical axis
- time in s on the horizontal axis
- smooth line of best fit, ignoring anomalous results.

You can revise how to interpret the results from experiments similar to this on page 151.

Now try this

A student investigates the effect of acid concentration on the rate of this reaction:

$$Na_2S_2O_3(aq) + 2HCl(aq) \rightarrow 2NaCl(aq) + H_2O(l) + SO_2(g) + S(s)$$

The reaction mixture becomes cloudy as sulfur forms. The student measures the time taken for it to become too cloudy to see a cross drawn on a piece of paper underneath the flask. In each experiment, he uses the same volume and concentration of sodium thiosulfate solution. He uses the same volume of diluted acid each time.

Explain **three** steps that the student takes to obtain results that allow a fair comparison. **(4 marks)**

Exam skills – Rates of reaction

There may be opportunities to draw or complete graphs and to interpret them.

You can revise the topics for this question, which is about **rates of reaction**, on pages 149 and 150.

Worked example

Hydrogen peroxide solution decomposes to form water and oxygen: $2H_2O_2(aq) \rightarrow 2H_2O(l) + O_2(g)$
Manganese dioxide powder, MnO_2, is a catalyst for this reaction. A student investigates the rate of the catalysed reaction. He measures the volume of oxygen produced over 2 minutes at 20 °C.

Time (s)	0	15	30	45	60	75	90	105	120
Volume of oxygen (cm^3)	0	20	34	43	47	49	50	50	50

(a) Plot these results on the grid. **(3 marks)**

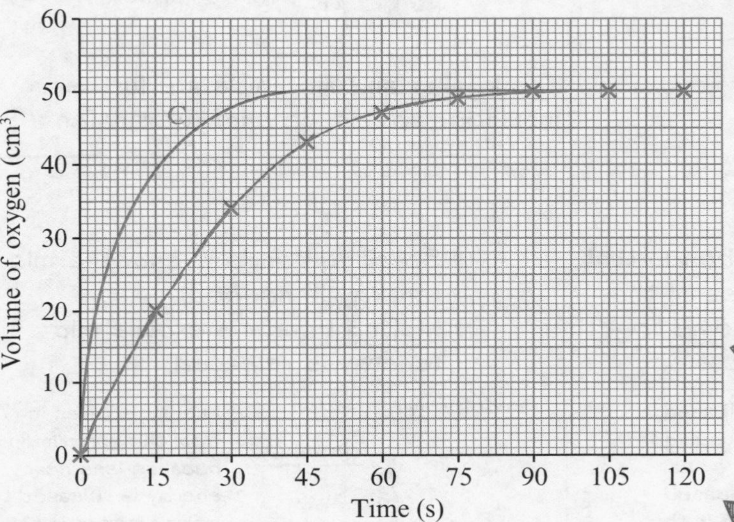

(b) Calculate the mean rate of reaction, giving your answer to two significant figures. **(2 marks)**

volume of oxygen produced = 50 cm^3

reaction time = 90 s

mean rate = $\frac{\text{volume}}{\text{time}} = \frac{50\,cm^3}{90\,s} = 0.56\,cm^3/s$

(c) On the same axes, sketch the curve you would expect if the student used the same volume of hydrogen peroxide solution at 40 °C.
Label this curve **C**. **(2 marks)**

Command word: Plot

You need to produce a graph by marking points accurately on a grid using given data. Then draw a line of best fit through these points. You must include a suitable scale and labelled axes if these are not given to you.

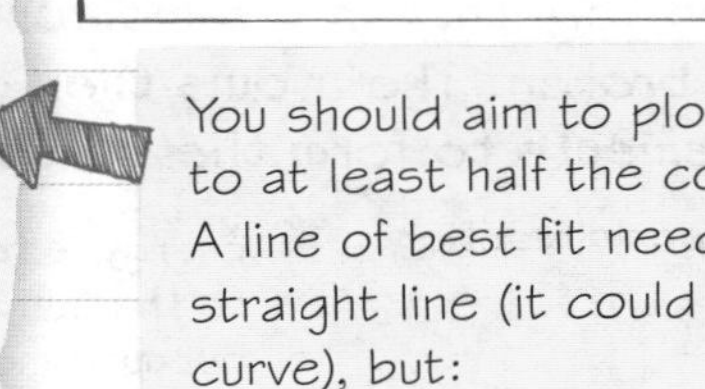

You should aim to plot each point to at least half the correct square. A line of best fit need not be a straight line (it could be a smooth curve), but:

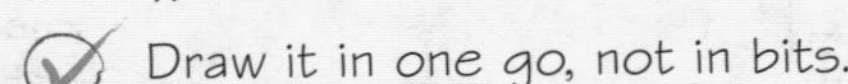

✓ Draw it in one go, not in bits.

Line C, the answer to (**c**), lies to the left of the first line and ends at the same volume.

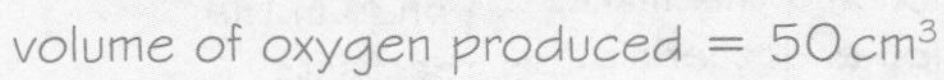

No more oxygen is produced after 90 s, so the reaction time is 90 s, not the duration of the investigation.

Command word: Sketch

You need to do a freehand drawing or graph. You may need labelled axes (not to scale), a line and key features identified.

Now try this

(a) Use the student's results to calculate the mean rate of reaction between 0 s and 45 s, and between 45 s and 90 s. **(4 marks)**

(b) Explain, in terms of particles and their collisions, why the mean rate of reaction differs in each of your answers to (**a**). **(4 marks)**

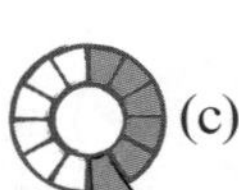

(c) Describe how the student could show that the mass of catalyst is unchanged in the reaction. **(4 marks)**

Had a look ☐ Nearly there ☐ Nailed it! ☐

Heat energy changes

Temperature changes

Most reactions involve temperature change:

- In **exothermic reactions**, heat energy is given **out**, and the reaction mixture or the surroundings increase in temperature.
- In **endothermic reactions**, heat energy is taken **in**, and the reaction mixture or the surroundings decrease in temperature.

Reaction	Exothermic	Endothermic
neutralisation	✓	✓
displacement	✓	
precipitation	✓	✓
dissolving	✓	✓

Temperature can go up or down depending on the precipitate or salt.

Practical skills Measuring temperature changes

You can use this apparatus to investigate the temperature changes in reactions.

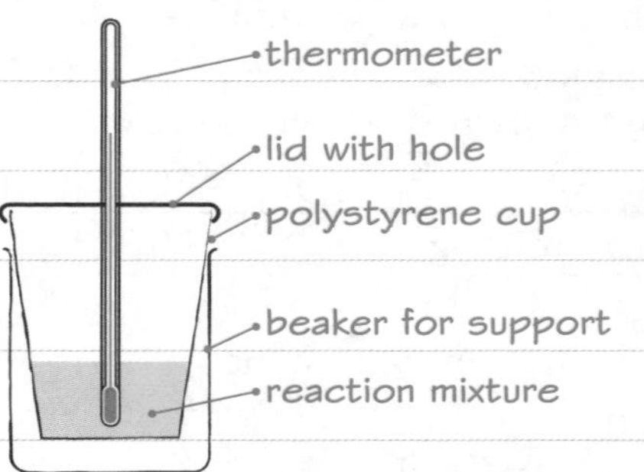

The beaker and air inside are further insulation to reduce energy transfer.

Bonds and energy

When a chemical reaction happens, the bonds that hold the atoms together in the molecules of the reactants are broken. The atoms then come together in new arrangements to form the products.

- Breaking bonds is endothermic (energy is needed).
- Making bonds is exothermic (energy is released).

exothermic reaction

Energy

reactants

products

Overall energy is released to surroundings and this makes the reaction exothermic.

Energy is released to the surroundings because more heat energy is released making bonds in the products than is needed to break bonds in the reactants.

endothermic reaction

Energy

products

reactants

Overall energy is taken in to the reaction and this makes the reaction endothermic.

Energy is taken in from the surroundings because less heat energy is released making bonds in the products than is needed to break bonds in the reactants.

Worked example

Calcium chloride is added to water and stirred. Explain why the mixture warms up. **(3 marks)**

An exothermic change happens. More heat energy is released in forming bonds in the products than is needed to break bonds in the reactants. So, overall, heat energy is given out.

The temperature does not always go up when solutions form. For example, the temperature goes down when ammonium nitrate dissolves.

The temperature increases so this must be an exothermic change. Remember:

- exothermic = energy out or 'energy exits'
- endothermic = energy in or 'energy enters'

Now try this

Copy and complete the table by putting a tick (✓) in each correct box. **(4 marks)**

	Breaking bonds	Making bonds	Temperature of reaction mixture	
			Increases	Decreases
Exothermic process				
Endothermic process				

Reaction profiles

Energy profile diagrams model the energy changes that happen during reactions.

Exothermic reactions

The diagram shows a typical **reaction profile** for an exothermic reaction.

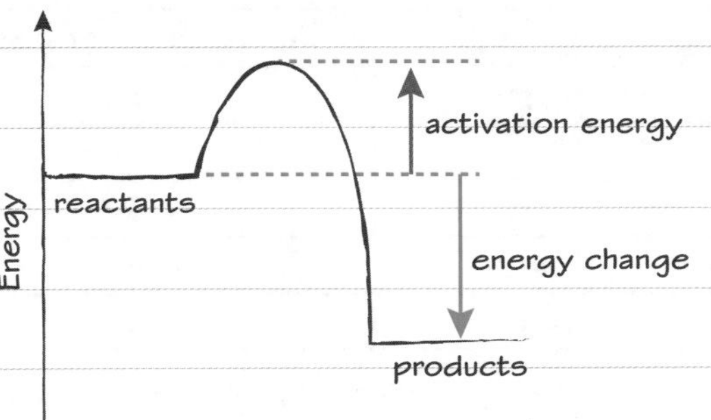

You should see that:

- the **energy level** of the reactants is greater than the energy level of the products
- the **energy change** of the reaction is negative (energy is transferred to the surroundings).

The **activation energy** is the minimum energy needed to start a reaction. It may be supplied by, for example:

- heating the reaction mixture
- applying a flame or a spark.

Endothermic reactions

The diagram shows a typical reaction profile for an endothermic reaction.

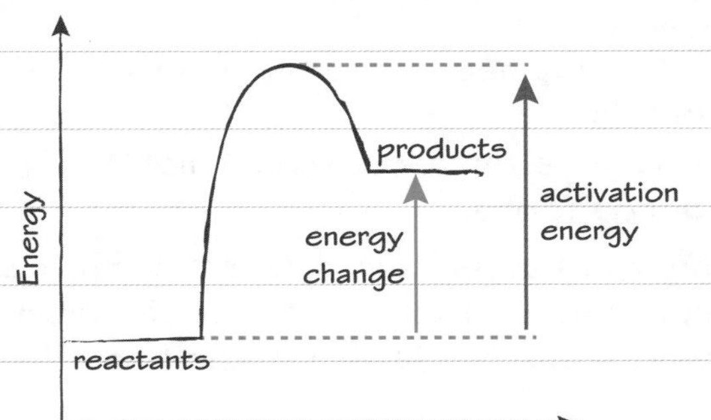

You should see that:

- the energy level of the reactants is lower than the energy level of the products
- the energy change of the reaction is positive (energy is transferred from the surroundings).

The activation energy may be supplied by, for example:

- continually heating the reaction mixture
- passing an electric current through an electrolyte (as in electrolysis).

Worked example

The reaction profile represents an exothermic reaction.

(a) Explain what the energy change shown as A represents. **(2 marks)**

the activation energy, which means that this is the energy needed to start the reaction

(b) Draw a line to represent the change in energy level when a catalyst is added. **(1 mark)**

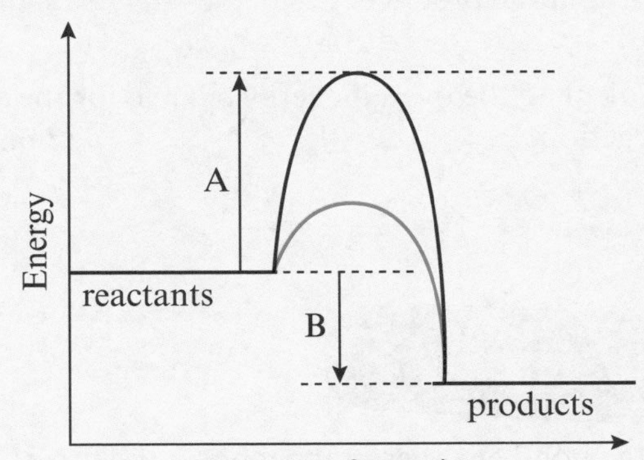

In this diagram, B represents the overall energy change. A catalyst provides an alternative pathway with a lower activation energy for the chemical reaction.

Now try this

1 Explain the term **activation energy**. **(2 marks)**

2 Sketch reaction profiles for the following reactions:

(a) $CH_4 + 2O_2 \rightarrow CO_2 + 2H_2O$ (exothermic) **(4 marks)**

(b) $CaCO_3 \rightarrow CaO + CO_2$ (endothermic) **(4 marks)**

Had a look ☐ Nearly there ☐ Nailed it! ☐

Calculating energy changes

You can calculate the energy change in a reaction using bond energies.

Breaking and making bonds

A **bond energy** is the energy needed to break 1 mol of a particular covalent bond, for example:

- 413 kJ is needed to break 1 mol of C–H bonds
- 413 kJ is released when 1 mol of C–H bonds forms.

Different bonds have different bond energies, depending on factors such as the elements involved and the length of the bond.

Some bond energies

The table shows some examples of bond energies, which are all positive.

Bond	Bond energy (kJ mol^{-1})
C–H	413
H–H	436
O–H	464
O=O	498
C=O	805

Do not try to learn any of these values – you will be given any that you need for calculations.

Worked example

Hydrogen burns in oxygen to form water. The equation shows the structures of the substances involved in this balanced reaction.

$$2 \times (H—H) + O{=}O \longrightarrow 2 \times$$ 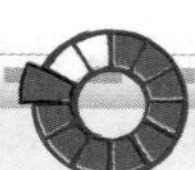

You need to use the displayed formula of each substance to work out how many bonds of each type it contains.

(a) Calculate the energy needed to break all the bonds in the reactants. **(3 marks)**

$2 \times (H–H) = 2 \times 436 = 872\ kJ\ mol^{-1}$

$1 \times (O{=}O) = 1 \times 498 = 498\ kJ\ mol^{-1}$

total $= 872 + 498 = 1370\ kJ\ mol^{-1}$

Multiply the bond energy for each type of bond by the number of those bonds present. Take care to use the correct values.

Notice that O=O is the energy needed to break the double bond.

(b) Calculate the energy released when new bonds form in the products. **(2 marks)**

$4 \times (O–H) = 4 \times 464 = 1856\ kJ\ mol^{-1}$

Each water molecule has two O–H bonds. The total number of O–H bonds in two water molecules is $(2 \times 2) = 4$.

(c) Calculate the overall energy change for the reaction. **(2 marks)**

energy change $= 1370 - 1856 = -486\ kJ\ mol^{-1}$

energy change = total energy in − total energy out

The negative sign in the final answer shows that the reaction is exothermic.

The energy change for an endothermic reaction is positive.

Now try this

Use the bond energies in the table at the top of the page to help you answer these questions.

1 During electrolysis, water decomposes to form hydrogen and oxygen.

(a) Calculate the overall energy change in the reaction. **(4 marks)**

(b) Explain whether the process is exothermic or endothermic. **(2 marks)**

$$2 \times \left(\begin{smallmatrix} & O & \\ H & & H \end{smallmatrix}\right) \longrightarrow 2 \times (H—H) + O{=}O$$

2 Methane burns in oxygen to form carbon dioxide and water.
Calculate the overall energy change in the reaction. **(5 marks)**

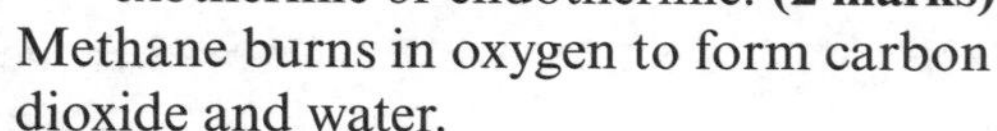

$$H—\underset{H}{\overset{H}{C}}—H + 2 \times (O{=}O) \longrightarrow 2 \times \left(\begin{smallmatrix} & O & \\ H & & H \end{smallmatrix}\right) + O{=}C{=}O$$

Had a look ☐ Nearly there ☐ Nailed it! ☐

Crude oil

Crude oil is a fossil fuel, formed over millions of years from ancient remains of marine organisms.

Hydrocarbons

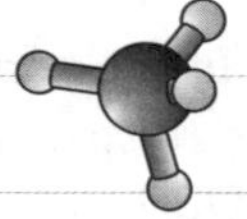

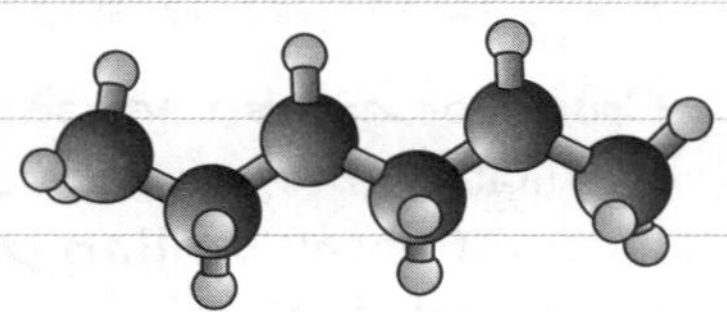

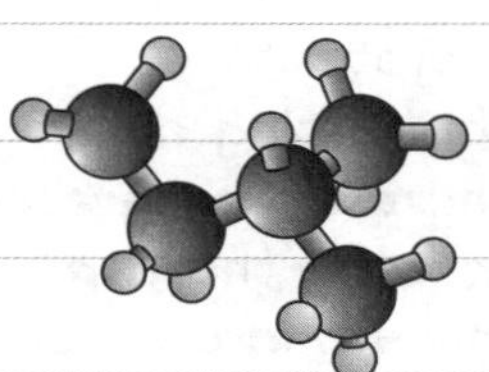

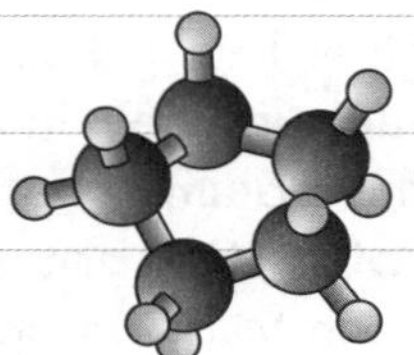

Hydrocarbons are compounds of carbon and hydrogen atoms **only**. Carbon atoms can form four covalent bonds. In a hydrocarbon molecule, these bonds can be:

- carbon–carbon bonds
- carbon–hydrogen bonds.

Hydrocarbon molecules can consist of:

- chains (with or without branches) or rings of carbon atoms.

Hydrocarbons in crude oil

Crude oil is:

- a complex mixture of hydrocarbons, with their carbon atoms in chains or rings
- an important source of useful substances
- a finite resource.

Finite resources

Finite resources:

✓ are no longer being made, or

✓ are being made extremely slowly.

Crude oil takes millions of years to form.

Worked example

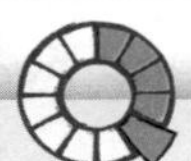

(a) Hydrocarbons are found in crude oil. Which of the following formulae represents a hydrocarbon? **(1 mark)**

☐ **A** C_2H_5OH

☒ **B** C_2H_6

☐ **C** $C_6H_{12}O_6$

☐ **D** CCl_4

(b) Describe two ways in which crude oil is an important source of useful substances. **(2 marks)**

Hydrocarbons from crude oil are useful as fuels and as feedstock for the petrochemical industry.

The chemical symbols you see in the formulae for hydrocarbon molecules are:

- C for carbon
- H for hydrogen.

In **ball-and-stick models** (similar to the ones above), atoms are usually modelled as:

- black for carbon atoms
- white for hydrogen atoms.

Familiar fuels such as petrol and diesel oil (used in cars) come from crude oil.

A **feedstock** is a starting material for an industrial chemical process.

The **petrochemical** industry involves the use and manufacture of substances from crude oil.

Now try this

1 (a) Explain the meaning of the term **hydrocarbon**. **(2 marks)**

(b) State the type of chemical bond found in a hydrocarbon molecule. **(1 mark)**

2 Explain why crude oil is described as a **finite** resource. **(2 marks)**

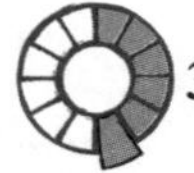

3 Describe the possible arrangements of carbon atoms in hydrocarbon molecules. **(2 marks)**

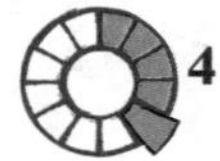

4 (a) Give an example of a fuel obtained from crude oil. **(1 mark)**

(b) Give an example of a substance manufactured using crude oil as a feedstock. **(1 mark)**

Had a look ☐ Nearly there ☐ Nailed it! ☐

Fractional distillation

Fractional distillation is used to separate crude oil into simpler, more useful mixtures.

Fractional distillation

Crude oil can be separated by fractional distillation because its different hydrocarbons have different **boiling points**. During fractional distillation:

- oil is heated to evaporate it
- vapours rise in a fractionating column
- the column has **a temperature gradient** – hot at the bottom, cool at the top
- each fraction condenses where it becomes cool enough, and is piped out of the column.

The **gases** fraction does not condense and leaves at the top.

The **bitumen** fraction does not evaporate and leaves at the bottom.

The other fractions are liquid at room temperature and are useful as fuels.

Fractions

A **fraction** is a mixture of hydrocarbons with **similar** boiling points and numbers of carbon atoms. There are trends in the properties of the different fractions from crude oil:

Number of C and H atoms	Boiling point	Ease of ignition	Viscosity
smallest	lowest	most flammable	least viscous
↑	↑	↑	↑
largest	highest	least flammable	most viscous

Substances with a low viscosity are runny, and those with a high viscosity are 'thick'.

Alkanes

Most of the hydrocarbons from crude oil are **alkanes**, a **homologous series** of compounds.

You can revise alkanes and homologous series on page 157.

Boiling point and molecule size

As the number of carbon and hydrogen atoms in a hydrocarbon molecule increases:

- ✓ the strength of the **intermolecular forces** increases
- ✓ more energy must be transferred to overcome these forces
- ✓ the boiling point increases.

Worked example

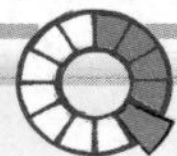

Draw one straight line from the name of each fraction to its correct typical use. **(5 marks)**

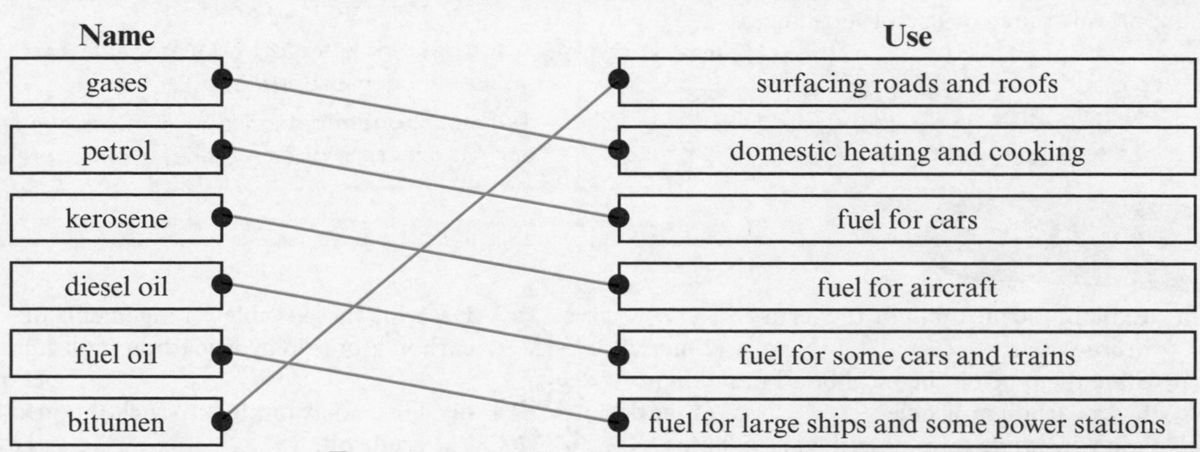

Now try this

1 Suggest **two** reasons to explain why bitumen is not suitable for use as a fuel. **(2 marks)**

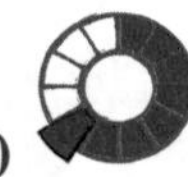

2 Describe how crude oil is separated into fractions by fractional distillation. **(3 marks)**

Alkanes

The **alkanes** are a homologous series of hydrocarbons.

Features of a homologous series

A **homologous series** is a series of compounds:

- in which **molecular formulae** of neighbouring members differ by CH_2
- that show a gradual variation in physical properties, such as boiling points
- that have similar chemical properties.

Chemical properties of alkanes

The alkanes undergo **complete combustion**. When they react completely with oxygen, carbon dioxide and water vapour form.

For example, methane (from natural gas):

methane + oxygen → carbon dioxide + water

$$CH_4(g) + 2O_2(g) \rightarrow CO_2(g) + 2H_2O(g)$$

Alkanes

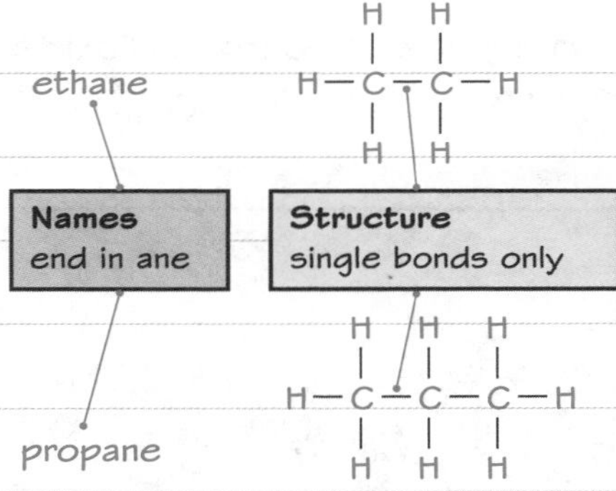

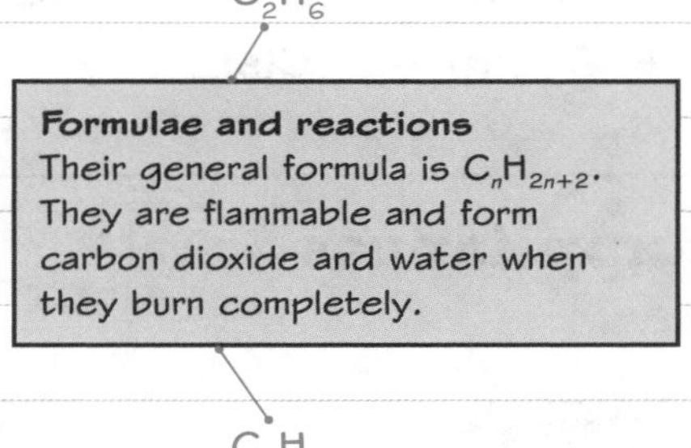

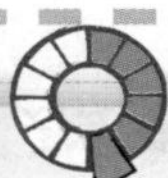

Worked example

The molecular formula shows the number of carbon atoms (and hydrogen atoms) in the molecule.

Going from one alkane to the next, the formula changes by CH_2.

The table shows the boiling points of some alkanes.

Molecular formula of alkane	CH_4	C_2H_6	C_4H_{10}	C_6H_{14}	C_8H_{18}	$C_{10}H_{22}$	$C_{16}H_{34}$
Boiling point of alkane (°C)	–162	–89	0	69	126	174	287

Complete the graph below, and draw a line of best fit. **(3 marks)**

The boiling point increases as the number of carbon atoms in the alkane molecule increases.

This is why fractional distillation produces fractions with similar boiling points and number of carbon atoms in their molecules.

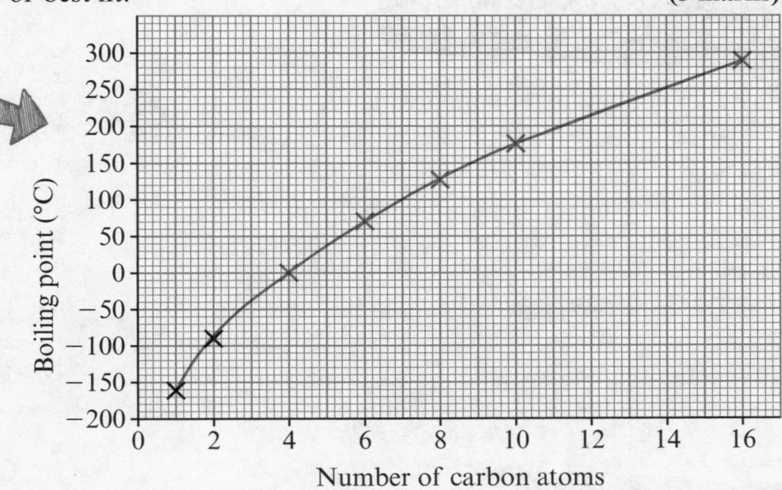

Now try this

1 State **three** features of a homologous series. **(3 marks)**

2 The molecular formula for icosane is $C_{20}H_{42}$. Predict the molecular formula for heneicosane, the next member of the alkane homologous series. **(1 mark)**

3 Use the Worked example graph to predict the boiling point of propane, C_3H_8. **(1 mark)**

Incomplete combustion

Incomplete combustion happens when the supply of oxygen to a burning fuel is limited.

Complete versus incomplete combustion

During **complete combustion** of a hydrocarbon fuel, such as petrol, kerosene or diesel oil:

- hydrogen is oxidised to water vapour, H_2O
- carbon is oxidised to carbon dioxide, CO_2
- energy is given out (transferred to the surroundings by radiation as heat and light).

During **incomplete combustion**, hydrogen is still oxidised to water vapour, but:

- carbon may be partially oxidised to carbon monoxide, CO
- carbon may be released as carbon particles or **soot**
- less energy is given out.

During incomplete combustion, different amounts of the different carbon products form, depending on how much oxygen is available for oxidation. For example:

methane + oxygen → carbon + carbon monoxide + carbon dioxide + water

$$4CH_4(g) + 6O_2(g) \rightarrow C(s) + 2CO(s) + CO_2(g) + 8H_2O(g)$$

Cars and appliances

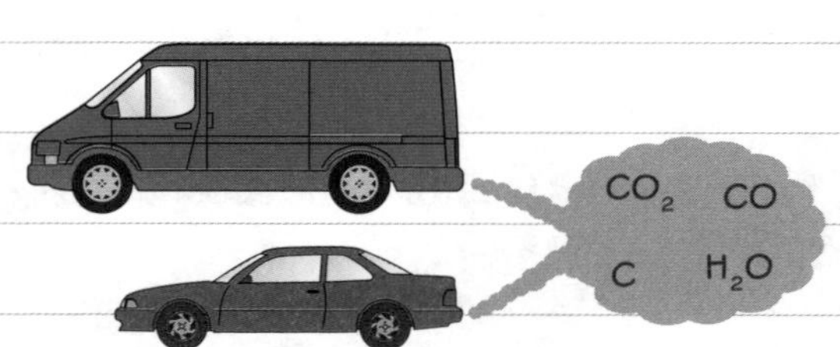

There is always incomplete combustion in vehicle engines.

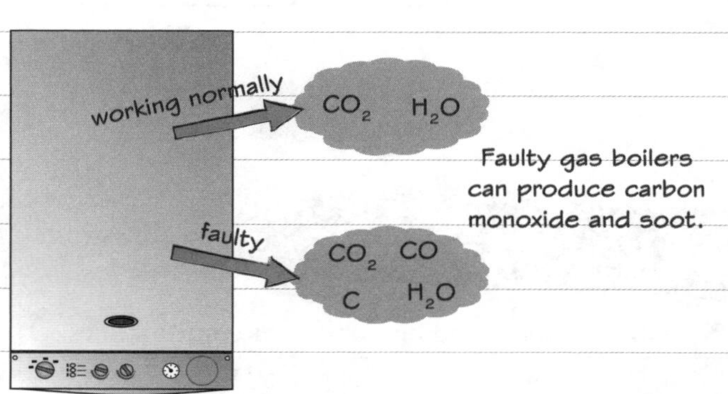

Faulty gas boilers can produce carbon monoxide and soot.

Worked example

(a) Explain how carbon monoxide behaves as a toxic gas. **(2 marks)**

Carbon monoxide attaches to haemoglobin in red blood cells, preventing oxygen attaching instead. This reduces the amount of oxygen carried around the body by the bloodstream.

Carbon monoxide can cause unconsciousness and even death. It is:

- colourless
- odourless (has no smell).

Electronic carbon monoxide detectors are used to warn us when the gas is present.

(b) Describe **two** problems caused by soot. **(2 marks)**

Soot can build up in chimneys where it may eventually cause fires; it also blackens buildings.

Tiny soot particles can be breathed in. This may cause lung diseases such as bronchitis, or make existing lung disease worse.

Now try this

When are soot and carbon monoxide produced, and are they always formed together?

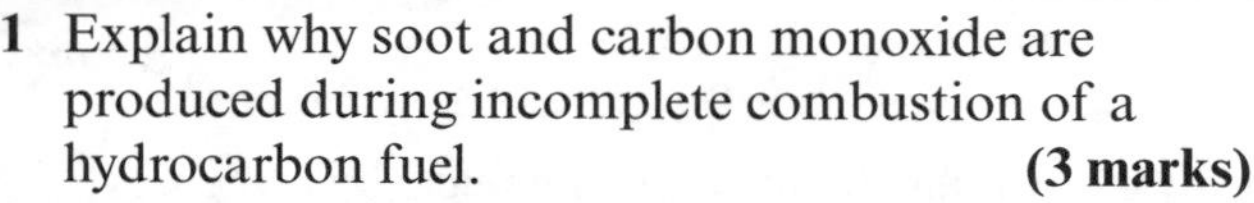

1 Explain why soot and carbon monoxide are produced during incomplete combustion of a hydrocarbon fuel. **(3 marks)**

2 Explain why carbon monoxide is difficult for our bodies to detect. **(2 marks)**

3 A householder sees soot marks around his gas boiler. He thinks that this shows that carbon monoxide is being produced. Comment on the accuracy of this thought. **(2 marks)**

Acid rain

Rainwater is naturally acidic, but **acid rain** is more acidic than normal.

Sulfur dioxide

Hydrocarbon fuels may contain impurities such as sulfur compounds. When the fuel burns, the sulfur in these impurities is oxidised to form sulfur dioxide:

sulfur + oxygen → sulfur dioxide

$S(s) + O_2(g) \rightarrow SO_2(g)$

Sulfur dioxide is a non-metal oxide:

- It dissolves in rainwater to form an acidic solution.

Reducing environmental damage

The problems caused by acid rain can be reduced in several ways, including:

- ✓ removing sulfur from petrol, diesel oil and fuel oil at the oil refinery before selling it
- ✓ preventing sulfur dioxide leaving power station chimneys – 'flue gas desulfurisation'
- ✓ adding calcium carbonate or calcium hydroxide to fields and lakes to neutralise excess acid from acid rain.

The effects of acid rain

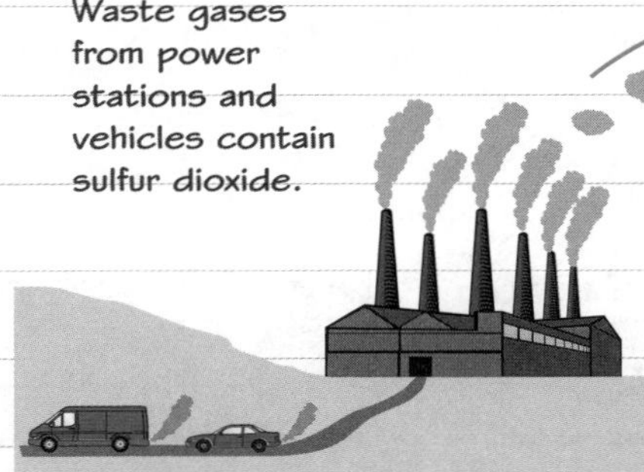
Waste gases from power stations and vehicles contain sulfur dioxide.
Sulfur dioxide dissolves in water in the air.
Acid rain speeds up the weathering of buildings and statues.
Trees are damaged.
Rain is more acidic than normal.
Rivers, lakes and soils are more acidic, which harms organisms living in them.

Worked example

Oxides of nitrogen, NO_x, are atmospheric pollutants. They may contribute to acid rain.

(a) Explain why oxides of nitrogen are produced when hydrocarbon fuels are used in vehicle and aircraft engines. **(2 marks)**

Air goes into the engine so that the fuel can burn. Nitrogen and oxygen from the air react together at the high temperatures in the engine to produce oxides of nitrogen.

(b) Balance the equation below to represent the production of nitrogen dioxide. **(1 mark)**

$N_2(g) + \underline{2}\,O_2(g) \rightarrow \underline{2}\,NO_2(g)$

Unlike sulfur dioxide, oxides of nitrogen are **not** produced from impurities in the hydrocarbon fuel.

NO_2 is a non-metal oxide that dissolves in rainwater to form an acidic solution.

Nitrogen dioxide is a toxic, orange-brown gas. Unlike carbon monoxide (produced during incomplete combustion), nitrogen dioxide has a sharp smell.

You can revise incomplete combustion on page 158.

Now try this

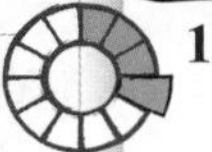

1 State **two** environmental problems caused by acid rain. **(2 marks)**

Look at the diagram to help you.

2 Explain why, when hydrocarbon fuels are used:
(a) sulfur dioxide is produced **(2 marks)**
(b) oxides of nitrogen are produced. **(2 marks)**

Had a look ☐ Nearly there ☐ Nailed it! ☐

Choosing fuels

Most cars use petrol or diesel oil, but hydrogen may also be used as a fuel for cars.

Fossil fuels

These **fossil fuels** are obtained from crude oil:

- petrol (for cars)
- diesel oil (for some cars and trains)
- fuel oil (for large ships and some power stations).

This fossil fuel is obtained from natural gas:

- methane (for domestic cooking and heating).

Non-renewable resources

Non-renewable resources are used up faster than they are formed.

Crude oil and natural gas take millions of years to form (see page 155 for more about this). The fossil fuels obtained from these resources are being used up very quickly, so they are non-renewable fuels.

Hydrogen

The combustion of hydrogen produces only water vapour:

hydrogen + oxygen → water

$$2H_2(g) + O_2(g) \rightarrow 2H_2O(g)$$

Hydrogen is manufactured in several ways, including:

- electrolysis of water (the reverse of the process above)
- cracking of oil fractions (for more about cracking, see page 161)
- reaction of natural gas with steam:

$$CH_4(g) + 2H_2O(g) \rightarrow CO_2(g) + 4H_2(g)$$

Worked example

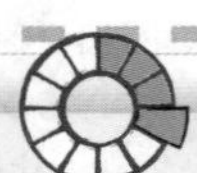

Describe three features of a good fuel. **(3 marks)**

A good fuel should:

- burn easily – it should be easy to ignite and stay alight
- not produce soot, smoke or ash
- release a lot of energy when it burns.

A good fuel should also be easy to store and transport safely.

Petrol versus hydrogen

Petrol	Hydrogen
👍 burns easily	👍 burns easily
👍 does not produce ash	👍 does not produce ash or smoke
👎 produces carbon dioxide and carbon monoxide as well as water when it burns	👍 only produces water when it burns
👍 releases more energy per kg when it burns than fuels such as coal or wood	👍 releases nearly three times as much energy per kg as petrol
👍 is a liquid, so it is easy to store and transport	👎 is a gas, so it has to be stored at high pressure 👎 filling stations would need to be adapted for hydrogen to be used in cars

Now try this

1 Explain why the combustion of petrol produces carbon dioxide, but the combustion of hydrogen does not. **(3 marks)**

2 (a) Suggest reasons to explain why hydrogen produced from crude oil or natural gas may not be renewable. **(2 marks)**

Look at the ways hydrogen can be manufactured.

(b) Carbon dioxide is a greenhouse gas linked to global warming and climate change. Hydrogen is often regarded as a 'carbon-neutral fuel', a fuel with production, transport and use that have no overall emission of carbon dioxide. Suggest reasons to explain why hydrogen may not actually be carbon neutral. **(2 marks)**

Cracking

Cracking is a process carried out on fractions in oil refineries after fractional distillation.

Making fractions more useful

Cracking involves breaking down larger alkanes into smaller, more useful alkanes and **alkenes**.

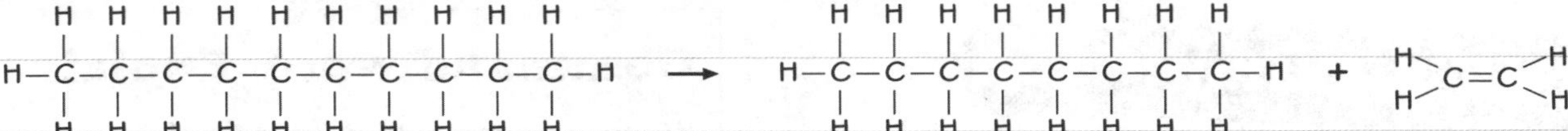

The long molecules are not very useful.

Cracking breaks down molecules by heating them.

Cracking produces shorter-chain alkanes, which are useful as fuels.

Cracking also produces **alkenes**, which are used to make **polymers**.

Cracking in the lab

Paraffin is an alkane. Liquid paraffin can be cracked in the laboratory using this apparatus.

- The porous pot catalyst is heated strongly.
- The liquid paraffin is heated and evaporates.
- The paraffin vapour passes over the hot porous pot and the hydrocarbon molecules break down.
- One of the products is ethene, which is a gas and collects in the other tube.

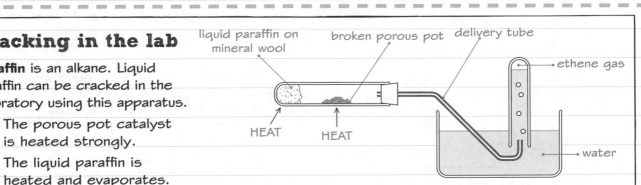

liquid paraffin on mineral wool

broken porous pot

delivery tube

ethene gas

HEAT

HEAT

water

Worked example

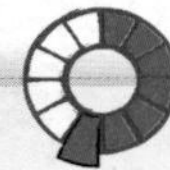

The bar chart shows the supply and demand of different fractions from crude oil.

Explain how cracking helps to balance supply with demand. **(2 marks)**

Some larger alkanes, such as bitumen, are in greater supply than their demand. Cracking converts these alkanes into smaller hydrocarbons, such as petrol, which are in greater demand than can be supplied by fractional distillation alone.

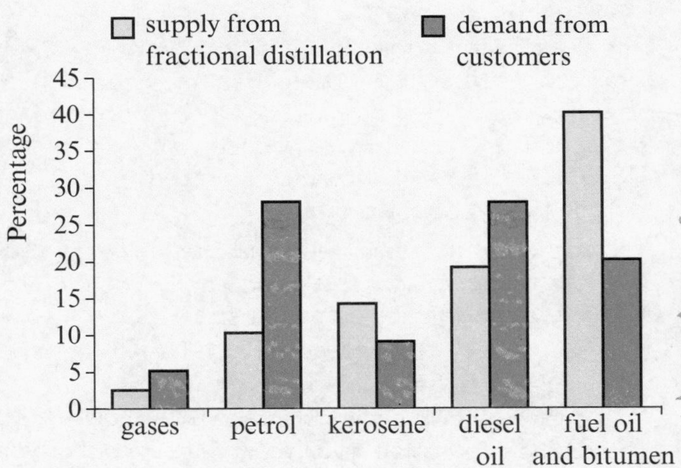

Now try this

(a) Give the formula of the substance needed to balance this equation (which represents a cracking reaction):

$C_8H_{18} \rightarrow C_6H_{14}$ + ... **(1 mark)**

(b) State a use for the hydrocarbon in your answer to **(a)**, other than as a fuel. **(1 mark)**

Study the diagram at the top of the page.

Had a look ☐ Nearly there ☐ Nailed it! ☐

Extended response – Fuels

There will be at least one 6-mark question on your exam paper. For these questions, you will need to think scientifically and structure your answer logically, showing how the points you make are related to each other.

You can revise the topics for this question, which is about **fuels**, on pages 155–160.

Worked example

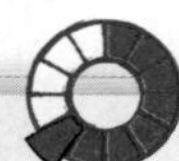

Crude oil is a complex mixture of hydrocarbons. Most belong to the alkane homologous series. Diesel oil is a fuel produced from crude oil.

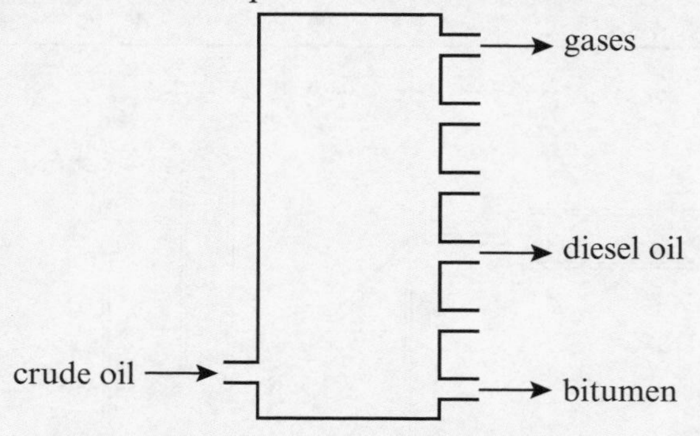

Explain how diesel oil is separated from crude oil. Use your knowledge and understanding, and the diagram above, in your answer. **(6 marks)**

Crude oil is separated into different fractions by fractionation distillation. The crude oil is heated to around 350°C to evaporate it. Some of the hydrocarbons stay in the liquid state and leave at the bottom of the column. The other hydrocarbons rise inside the fractionating column in the gas state.

The fractionating column has a temperature gradient inside it. The column becomes cooler towards the top, so the hydrocarbons cool down as they rise. They condense to the liquid state when they reach a part of the column that is cool enough.

The gases do not condense and they leave at the top. Diesel oil leaves through a pipe as a liquid fraction when it condenses.

Command word: Explain

When you are asked to **explain** something, it is not enough just to state or describe it.

Your answer must contain some reasoning or justification of the points you make.

The labelled diagram in the stem of the question helps you in two ways:

- It may remind you that crude oil is separated by fractionation distillation (not, for example, by cracking).
- It shows approximately where diesel oil leaves the fractionating column.

You should state clearly which separation method is involved, and outline what happens to the crude oil immediately after heating it.

The temperature gradient inside a fractionating column is important to the process. You could also mention that the hydrocarbons condense when they cool to their boiling point, because this is an important feature too.

Your answer could also explain that smaller hydrocarbons have lower boiling points because they have weaker intermolecular forces.

Your answer should show comprehensive knowledge and understanding of the topic covered. Take care to use the information given to you in support of your explanations. Organise your answer in a structured way with clear lines of reasoning.

Now try this

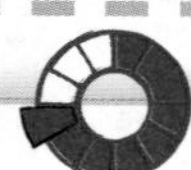

Diesel oil is used as a fuel for some cars and trains. Scientists and engineers are researching ways to use hydrogen as a fuel for cars instead of diesel oil. Evaluate the use of hydrogen as a fuel for cars, rather than diesel oil. **(6 marks)**

You must be able to evaluate the use of hydrogen rather than petrol as a fuel for cars. Therefore, here you can discuss its advantages and disadvantages compared with a different fossil fuel.

The early atmosphere

Gases produced by volcanic activity formed the Earth's early atmosphere.

Gases in the early atmosphere

Scientists believe that the Earth's early atmosphere, billions of years ago, contained:

- little or no oxygen
- a large amount of carbon dioxide
- water vapour
- small amounts of other gases.

Evidence for this includes:

- the mixture of gases released by volcanoes
- the atmospheres of other planets in our solar system today, which have not been changed by living organisms.

Today's atmosphere

Nitrogen and oxygen are the main gases in the modern atmosphere.

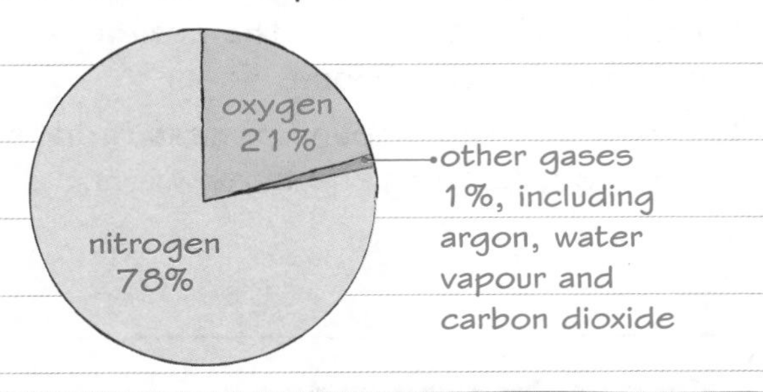

Decreasing carbon dioxide and increasing oxygen

The Earth was very hot to start with. As it cooled, water vapour condensed and fell as rain to form the oceans.

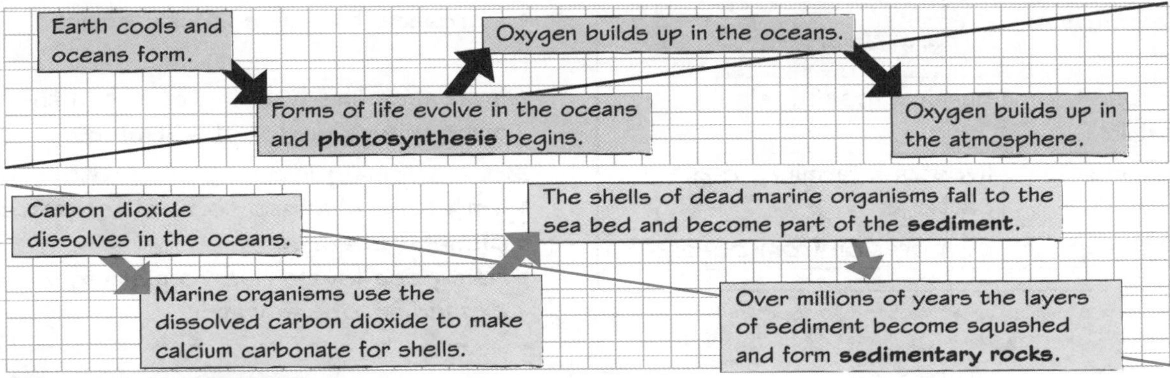

Earth cools and oceans form.
Forms of life evolve in the oceans and photosynthesis begins.
Oxygen builds up in the oceans.
Oxygen builds up in the atmosphere.
Carbon dioxide dissolves in the oceans.
Marine organisms use the dissolved carbon dioxide to make calcium carbonate for shells.
The shells of dead marine organisms fall to the sea bed and become part of the sediment.
Over millions of years the layers of sediment become squashed and form sedimentary rocks.

Worked example

Some rocks contain iron. Very old rocks do not contain iron oxides but later ones do.

(a) Explain how these rocks provide evidence for changes to the Earth's atmosphere. **(3 marks)**

The very old rocks are evidence that the early atmosphere contained little or no oxygen, because they do not contain iron oxides. Later rocks do contain iron oxides, which is evidence that oxygen was released into the atmosphere and reacted with iron in the rocks.

(b) Describe a simple test for oxygen. **(2 marks)**

A glowing wooden splint relights in oxygen.

When oxygen was first produced, it reacted with iron in rocks to produce iron oxides. This meant that, even though primitive plants were photosynthesising, oxygen levels in the atmosphere did not begin to rise straight away.

When you describe a laboratory test, say what you would do, and what you would observe.

Remember:

- A lighted wooden splint ignites hydrogen with a pop.
- Limewater turns milky in the presence of carbon dioxide.

Now try this

1 (a) State how the formation of oceans was a cause of decreasing carbon dioxide levels. **(1 mark)**

(b) Describe why the levels of oxygen in the atmosphere increased. **(2 marks)**

2 Suggest a reason that explains why scientists cannot be certain about the Earth's early atmosphere. **(1 mark)**

Had a look ☐ Nearly there ☐ Nailed it! ☐

Greenhouse effect

The greenhouse effect helps to keep the Earth warm enough for living organisms to exist.

Reducing radiation to space

In the **greenhouse effect**:

- **Greenhouse gases** in the atmosphere absorb heat radiated from the Earth.
- The greenhouse gases then release energy in all directions.

This reduces the amount of heat radiated into space, keeping the Earth warm.

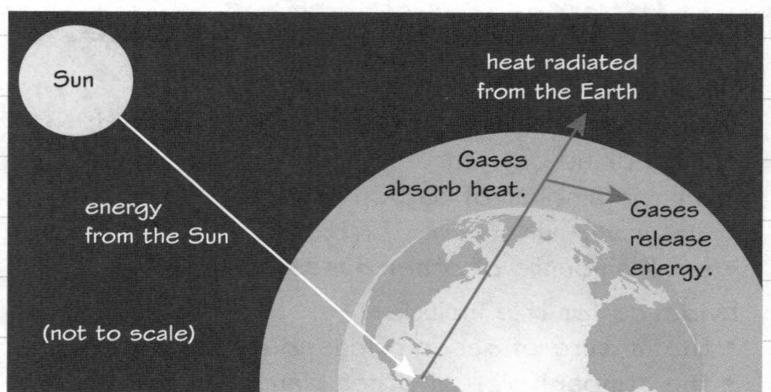

Greenhouse gases

The table shows gases that are particularly good at absorbing and emitting energy by radiation.

Greenhouse gas	Typical source
carbon dioxide	burning fossil fuels
methane	livestock farming, e.g. cattle
water vapour	evaporation from oceans

This is not seen as a problem because excess water vapour leaves the atmosphere as rain and snow.

Global warming and climate change

The accumulation of greenhouse gases in the atmosphere:

- increases the greenhouse effect.

This:

- increases the warming effect in the atmosphere – **global warming**.

Global warming is associated with:

- long-term changes to weather patterns – **climate change**
- rising sea levels due to melting ice and expanding ocean water.

Worked example

Evaluate how far the graph provides evidence for global warming. **(2 marks)**

The graph shows that both temperature and carbon dioxide concentration have been rising since 1950. There is a correlation but this does not prove that rising carbon dioxide levels are causing the warming.

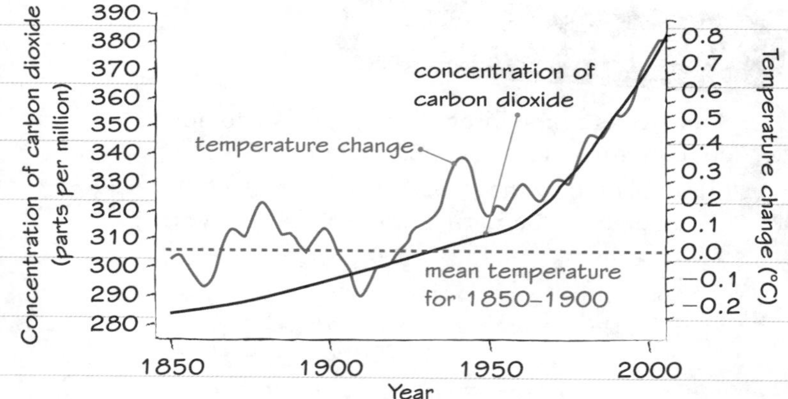

Many scientists believe that carbon dioxide is the main cause of global warming, but this opinion is based on computer modelling and other information, not just on graphs like this.

Now try this

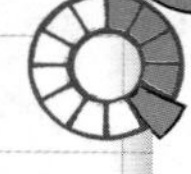

1 (a) Name **two** greenhouse gases, other than water vapour. **(2 marks)**
(b) State **one** typical source for each gas named in your answer to (**a**). **(2 marks)**

2 Describe how the greenhouse effect keeps the Earth warm. **(2 marks)**

3 Explain how the increased use of fossil fuels, which contain carbon and carbon compounds, may lead to global warming. **(4 marks)**

Extended response – Atmospheric science

There will be at least one 6-mark question on your exam paper. For these questions, you will need to think scientifically and structure your answer logically, showing how the points you make are related to each other.

You can revise the topics for this question, which is about **the atmosphere**, on pages 163 and 164.

Worked example

The graph shows the mean annual release of carbon dioxide from burning fossil fuels during each decade from the 1900s to the 2000s. It also shows the mean change in global temperature during that time.

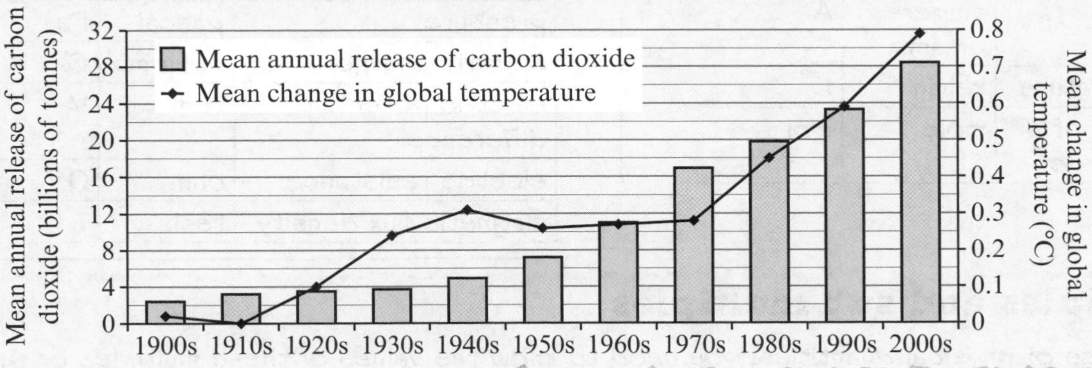

Evaluate this evidence of human activity as a cause of global warming. **(6 marks)**

The mass of carbon dioxide released increased each decade from the 1900s to the 2000s. For example, in the 1900s about 2.5 billion tonnes was released on average each year, but by the 2000s it was over 28 billion tonnes, more than 10 times greater.

Over the same period, the mean global temperature increased by nearly 0.8 °C. As the carbon dioxide was released by burning fossil fuels, this shows that there is a positive correlation between the consumption of fossil fuels and temperature change.

However, the temperature increased more rapidly in the 1930s and 1940s than the release of carbon dioxide. It also went down in the 1950s and barely changed until the 1980s, whereas the release of carbon dioxide increased greatly. By itself, the evidence given does not show that human activity caused the global warming observed, but it strongly suggests that it did.

Command word: Evaluate

When you are asked to **evaluate** something, you need to review information given to you, and then bring it together to form a conclusion. You should include evidence such as strengths and weaknesses, and arrive at a judgement.

You should use the information in your answer, e.g. by interpreting the bar chart.

A **correlation** between two sets of data means that they are connected in some way. In this case, there is a **positive correlation** – as one variable increases, the other one does too.

You should point out evidence for weaknesses as well as strengths for the main argument.

A **causal** relationship means that a change in one variable causes a change in another variable.

Now try this

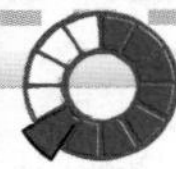

The concentration of carbon dioxide in the atmosphere increased by about 35% between 1900 and 2015.

Describe the processes that remove carbon dioxide from the atmosphere, and suggest reasons to explain the observed increase in its concentration. **(6 marks)**

Had a look ☐ Nearly there ☐ Nailed it! ☐

Key concepts

You need to know the SI units of physical quantities, as well as their multiples and sub-multiples, how to convert between units and how to use significant figures and standard form.

Base SI units

There are six main **base SI units** you need to know.

Name	Unit	Abbreviation
length	metre	m
mass	kilogram	kg
time	second	s
current	ampere or amp	A
temperature	kelvin	K
amount of a substance	mole	mol

Derived SI units

There are nine derived SI units you need to know.

Name	Unit	Abbreviation
frequency	hertz	Hz
force	newton	N
energy	joule	J
power	watt	W
pressure	pascal	Pa
electric charge	coulomb	C
electric potential difference	volt	V
electric resistance	ohm	Ω
magnetic flux density	tesla	T

Multiples and sub-multiples

For values of physical quantities, you need to know the values of these multiples or **prefixes**.

10^9 10^8 10^7 10^6 10^5 10^4 10^3 10^2 10^1 10^0 10^{-1} 10^{-2} 10^{-3} 10^{-4} 10^{-5} 10^{-6} 10^{-7} 10^{-8} 10^{-9}

giga-	**mega-**	**kilo-**	**centi-**	**milli-**	**micro-**	**nano-**
prefix **G**	prefix **M**	prefix **k**	prefix **c**	prefix **m**	prefix **μ**	prefix **n**
× 1 000 000 000	× 1 000 000	× 1000	÷ 100	÷ 1000	÷ 1 000 000	÷ 1 000 000 000
e.g. 5 GHz	e.g. 5 MW	e.g. 2 km	e.g. 8 cm	e.g. 6 mA	e.g. 7 μs	e.g. 2 nm

Significant figures

Significant figures are either non-zero digits (i.e. the digits 1 to 9) or zeros between non-zero digits.

- The number **543 563** has 6 significant figures.
- The number 0.**23** has two significant figures.
- The number 0.000 000 000 **254** has 3 significant figures. The zeros before the '2' are not significant figures.
- The number **30 340 0**00 has 4 significant figures.

The number 30 339 900 rounded to 5 significant figures is 30 340 000.

Standard form

It is useful to write very large and very small numbers in **standard form**.

$$A \times 10^B$$

This is a number between 1 and 10. (A)

This is an integer – a whole number that can be positive or negative. (B)

You also need to be able to convert between units such as kilometres to metres, or hours to seconds.

Now try this

1 State the number of significant figures in each number.
(a) 4.56 **(1 mark)**
(b) 0.000 564 5 **(1 mark)**
(c) 3.0046 **(1 mark)**

2 Convert:
(a) 12 hours to seconds **(2 marks)**
(b) 64 km/h to m/s. **(2 marks)**

3 Convert 58.3 MW to watts. Give your answer in standard form. **(2 marks)**

4 The speed of light is 186 000 miles per second. Convert this to metres per second and write it in standard form. **(3 marks)**

1 mile is equal to 1609 m

Scalars and vectors

All physical quantities can be described as either a scalar or a vector quantity.

Scalar quantities

Scalar quantities have a size or a magnitude but no specific direction.

Examples include:

- mass
- speed
- distance
- energy
- temperature.

Vector quantities

Vector quantities have a size or magnitude and a specific direction.

Examples include:

- force or weight
- velocity
- displacement
- acceleration
- momentum.

Speed and velocity

The girl is running to the right so she has a velocity of 5 m/s to the right.

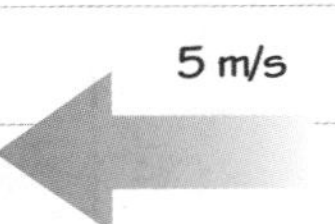

5 m/s

Both the girl and the boy are running at 5 m/s. They have the same speed.

5 m/s

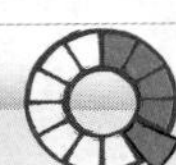

If we take 'to the right' as the positive direction, then the girl has a velocity of +5 m/s and the boy has a velocity of −5 m/s.

The boy is running to the left so he has a velocity of 5 m/s to the left.

Speed has a size but velocity has a size AND a direction.

Velocity is speed in a stated direction.

Worked example

(a) Describe the difference between a scalar and a vector. **(2 marks)**

A scalar has a size, whereas a vector has a size and a specific direction.

(b) Describe an example which shows the difference between a scalar and a vector. **(2 marks)**

Speed is a scalar quantity and may have a value of 5 m/s. Velocity may also have a value of 5 m/s but a direction of north.

All vector quantities can be given positive and negative values to show their direction. Examples include:

1. A force of +4 N may be balanced by a force acting in the opposite direction of −4 N.
2. A car that accelerates at a rate of $+2\ m/s^2$ could decelerate at $-2\ m/s^2$.
3. If a distance walked to the right of 20 m is a displacement of +20 m, then the same distance walked to the left from the same starting point is a displacement of −20 m.

Now try this

1 (a) Which of these quantities is a scalar? **(1 mark)**

☐ **A** velocity ☐ **B** acceleration ☐ **C** mass ☐ **D** weight

(b) Which of these quantities is a vector? **(1 mark)**

☐ **A** temperature ☐ **B** energy ☐ **C** speed ☐ **D** electric field

2 A boy has a speed of 4 m/s when running. State his (a) speed and (b) velocity when he is running in the opposite direction. **(2 marks)**

3 Explain why a satellite can be said to be moving at a constant speed but not at a constant velocity. **(3 marks)**

Speed, distance and time

Speed is the distance that a moving object covers each second.

Calculating speed

When a body covers the same distance per second throughout its journey, you can use this equation to calculate its speed:

$$\text{speed (m/s)} = \frac{\text{change in distance (m)}}{\text{time taken (s)}}$$

The greater the change in distance per second, the faster the object is moving.

When the change in distance over a period of time is zero, the speed is zero and the object is stationary.

Average speed

LEARN IT! IT'S NOT ON THE EQUATIONS LIST

Objects often change speed during a journey, so it is better to use average speed:

$$\text{average speed (m/s)} = \frac{\text{total distance travelled (m)}}{\text{total time taken (s)}}$$

You can also write this as

$$\text{distance travelled (m)} = \text{average speed (m/s)} \times \text{time taken (s)}$$

Distance/time graphs

Distance/time graphs have distance on the *y*-axis and time on the *x*-axis. The gradient or slope of the graph tells us about the motion of the vehicle.

A
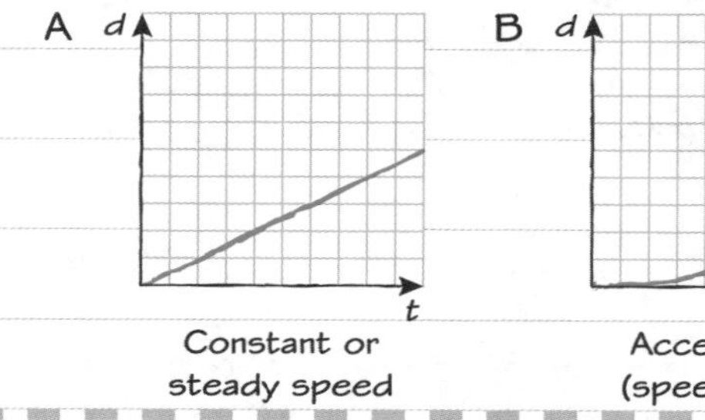
d
t
Constant or steady speed

B
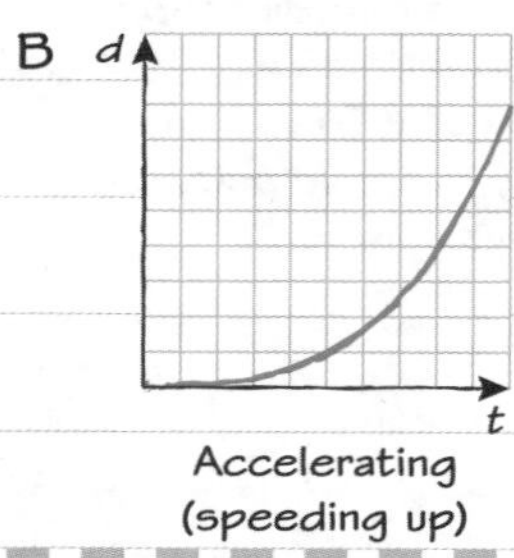
d
t
Accelerating (speeding up)

C
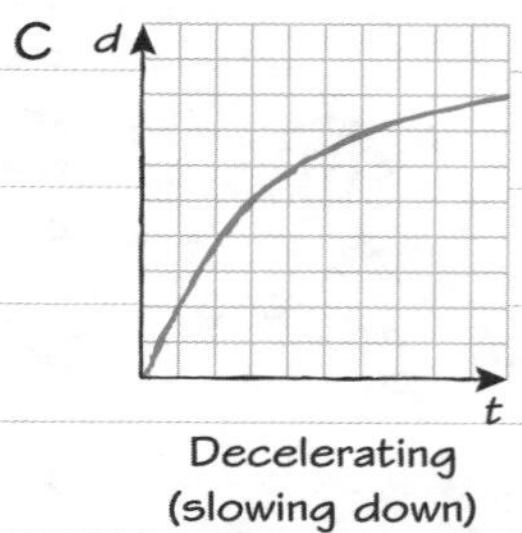
d
t
Decelerating (slowing down)

D
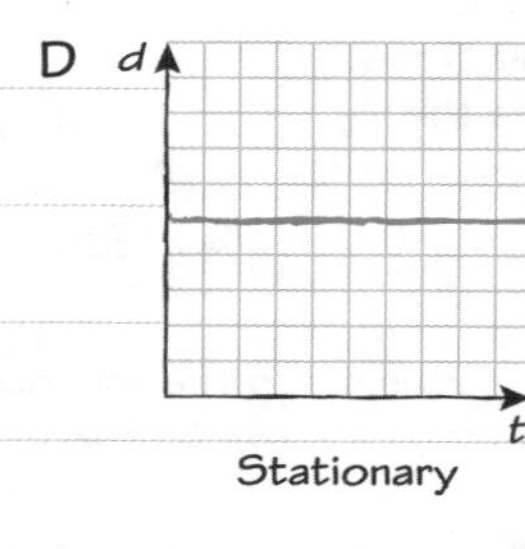
d
t
Stationary

Journeys and distance/time graphs

This distance/time graph tells you about a student's journey.

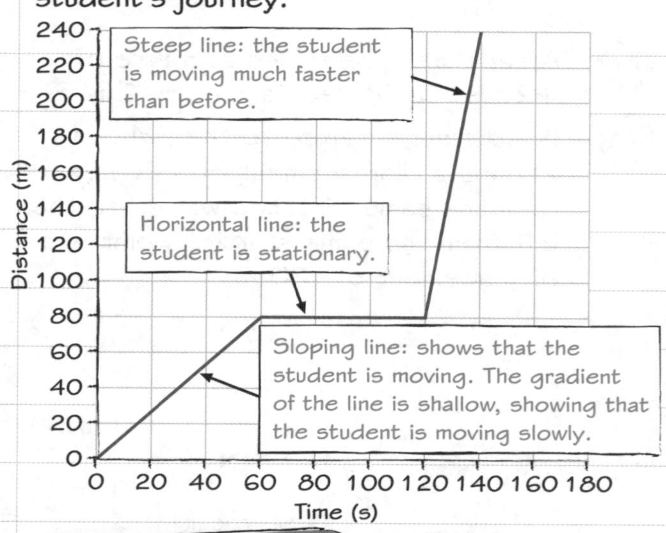
Distance (m)
Time (s)
0
20
40
60
80
100
120
140
160
180
200
220
240
0 20 40 60 80 100 120 140 160 180
Steep line: the student is moving much faster than before.
Horizontal line: the student is stationary.
Sloping line: shows that the student is moving. The gradient of the line is shallow, showing that the student is moving slowly.

Worked example

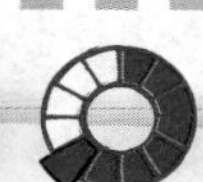

(a) Calculate the speed for each part of the journey shown in the graph on the left. **(3 marks)**

speed between 0 and 60 s = 80 m ÷ 60 s = 1.33 m/s

speed between 60 s and 120 s = 0 m ÷ 60 s = 0 m/s

speed between 120 s and 140 s = 160 m ÷ 20 s = 8 m/s

(b) Calculate the average speed for the whole journey. **(2 marks)**

average speed = 240 m ÷ 140 s = 1.7 m/s

Now try this

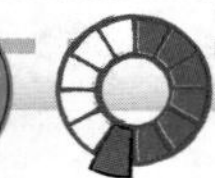

1 Calculate the average speed of a motorbike that travels 120 m in 8 s. **(3 marks)**

2 A car accelerates from 0 m/s to 12 m/s in 5 s. Sketch a distance/time graph to show this. **(3 marks)**

3 The speed limit for a car on a motorway in the UK is 112.7 km/h. Calculate this speed in m/s. **(3 marks)**

Equations of motion

You can use equations to work out the velocity and acceleration of moving bodies.

Acceleration

Acceleration is a change in velocity per second. Acceleration is a vector quantity.

$$\text{acceleration (m/s}^2) = \frac{\text{change in velocity (m/s)}}{\text{time taken (s)}}$$

$$a = \frac{(v - u)}{t}$$

LEARN IT!
IT'S NOT ON THE EQUATIONS LIST

a is the acceleration

v is the final velocity

u is the initial velocity

t is the time taken

Velocity

Velocity is the change in distance per second.

Velocity is a vector quantity.

(final velocity)^2 – (initial velocity)^2 = 2 × acceleration × distance

$$v^2 - u^2 = 2 \times a \times x$$

x is the distance travelled.

You can also write this as:

$$v^2 = u^2 + 2ax$$

Worked example

A cat changes its speed from 2.5 m/s to 10.0 m/s over a period of 3.0 s. Calculate the cat's acceleration. **(3 marks)**

$v = 10\text{ m/s},\ u = 2.5\text{ m/s},\ t = 3\text{ s},\ a = ?$

$$a = \frac{(v - u)}{t}$$

$$a = \frac{(10.0\text{ m/s} - 2.5\text{ m/s})}{3.0\text{ s}}$$

$$a = \frac{(7.5\text{ m/s})}{3.0\text{ s}}$$

$$a = 2.5\text{ m/s}^2$$

Write down all the quantities you know and the one you need to work out. Then decide which equation you need to use.

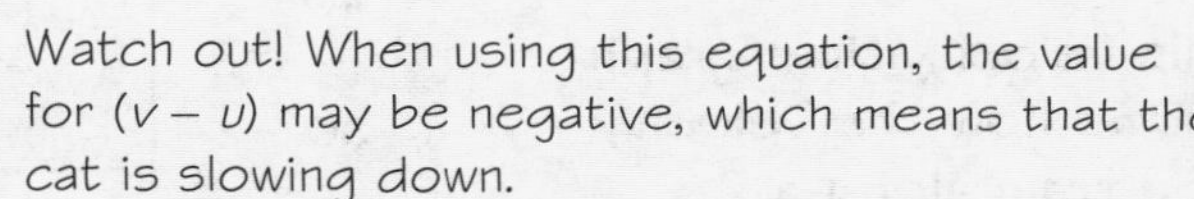

Watch out! When using this equation, the value for $(v - u)$ may be negative, which means that the cat is slowing down.

Worked example

A motorcyclist passes through green traffic lights with an initial velocity of 4 m/s and then accelerates at a rate of 2.4 m/s^2, covering a total distance of 200 m. Calculate the final velocity of the motorcycle. **(4 marks)**

$u = 4\text{ m/s},\ a = 2.4\text{ m/s}^2,\ x = 200\text{ m},\ v = ?$

$$v^2 = u^2 + 2ax$$

$$v^2 = (4)^2 + 2 \times 2.4 \times 200$$

$$v^2 = 16 + 960 = 976$$

$$v = \sqrt{976} = 31.2\text{ m/s}$$

Maths skills When working out x using this equation, you will need to rearrange it. Subtract u^2 from both sides and then divide both sides by $2a$, to give $x = \frac{v^2 - u^2}{2a}$. Substitute the other values and work out the value of x.

Watch out! When you substitute the values for u, a and x, you get a value for v^2. So you need to find the square root of this value to get the value for the final velocity, v.

Now try this

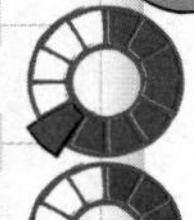

1 A dog changes its speed from 2 m/s to 8 m/s in 5 s. Calculate the acceleration of the dog. **(3 marks)**

2 An aeroplane starts at rest and accelerates at 1.6 m/s^2 down a runway. After 1.8 km it takes off. Calculate its speed at take-off. **(3 marks)**

3 A car passes through traffic lights at a speed of 5 m/s and then accelerates at 1.2 m/s^2 until it has reached a final speed of 18 m/s. Calculate the distance it has travelled from the traffic lights. **(4 marks)**

If a vehicle is said to be 'starting from rest' then its initial velocity, u, will be zero, so the equation simplifies to $v^2 = 2ax$.

Had a look ☐ Nearly there ☐ Nailed it! ☐

Velocity/time graphs

Velocity/time graphs show how the velocity of a vehicle changes with time. You can also work out acceleration and distance travelled from the graph.

Interpreting velocity/time graphs

Velocity/time graphs have velocity plotted on the *y*-axis and time plotted on the *x*-axis. The graph shows you how the velocity changes with time.

- The **slope** or **gradient** of the graph tells us the acceleration of the vehicle.
- The **area under the graph** tells us the **distance** travelled.

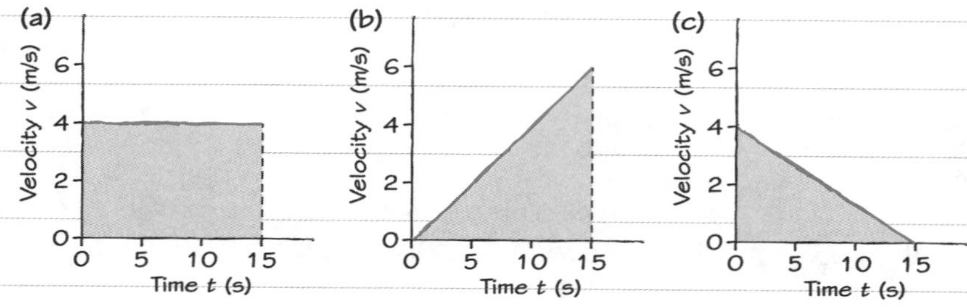
(a)
(b)
(c)
Velocity v (m/s)
0
2
4
6
Time t (s)
0
5
10
15

Maths skills Acceleration (gradient) = change in velocity/change in time. For distance travelled (area) use the formulae for the area of a rectangle (base × height, $b \times h$) and area of a triangle ($\frac{1}{2} \times b \times h$). The area will have units of metres, as m/s × s = m.

(a) Acceleration is zero, distance travelled is 60 m.

(b) Acceleration is 0.4 m/s^2, distance travelled is 45 m.

(c) Acceleration is –0.27 m/s^2, distance travelled is 30 m.

Velocity/time graphs

This velocity/time graph shows how the velocity of a train along a straight track changes with time.

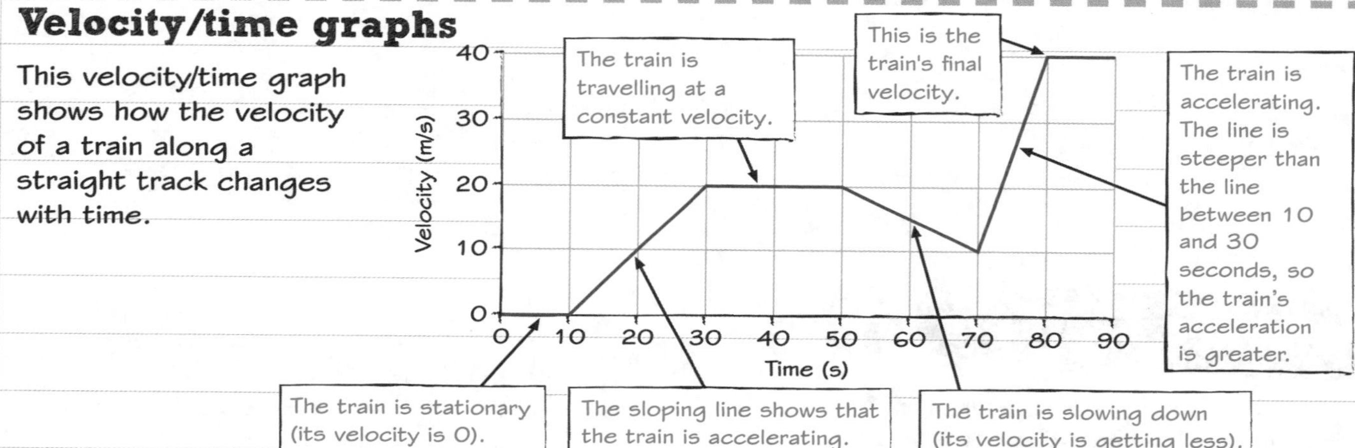
Velocity (m/s)
0
10
20
30
40
Time (s)
0
10
20
30
40
50
60
70
80
90
The train is travelling at a constant velocity.
This is the train's final velocity.
The train is accelerating. The line is steeper than the line between 10 and 30 seconds, so the train's acceleration is greater.
The train is stationary (its velocity is 0).
The sloping line shows that the train is accelerating.
The train is slowing down (its velocity is getting less).

Worked example

Use the velocity/time graph of the train to calculate:

(a) the distance travelled after 50 s **(3 marks)**

area beneath the graph = $\left[\frac{1}{2} \times 20 \times (30 - 10)\right] + [20 \times (50 - 30)]$
= 200 m + 400 m = 600 m

(b) the acceleration of the train between 50 s and 70 s. **(3 marks)**

change in velocity = 10 m/s – 20 m/s = –10 m/s
change in time = 70 s – 50 s = 20 s
acceleration = $\frac{-10\text{ m/s}}{20\text{ s}}$ = –0.5 m/s^2

Now try this

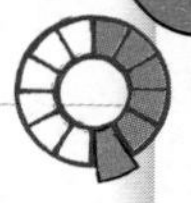

1 State what can be determined from the (a) gradient and (b) area beneath a velocity/time graph. **(2 marks)**

2 A car travels at a constant speed of 8 m/s for 12 s before accelerating at 1.5 m/s^2 for the next 6 s.

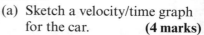

(a) Sketch a velocity/time graph for the car. **(4 marks)**

(b) Calculate the total distance travelled by the car. **(4 marks)**

Determining speed

Speed can be determined in the laboratory using light gates and other equipment.

You can measure speeds of objects in the lab by using light gates connected to a computer or data logger.

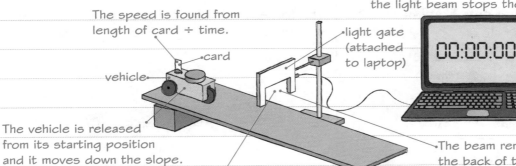

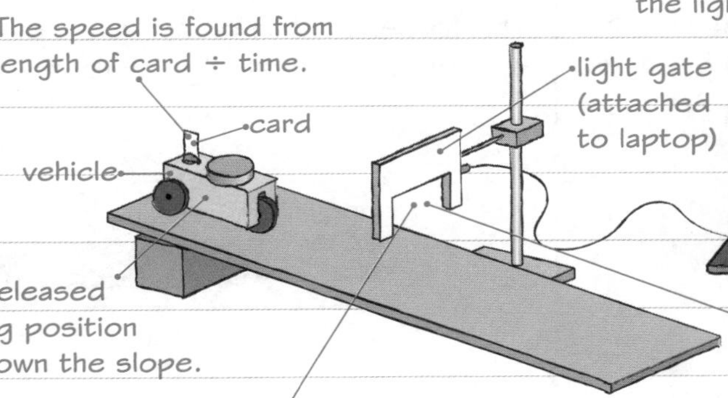

When the card leaves the light gate, the light beam stops the timer.
The speed is found from length of card ÷ time.
card
vehicle
light gate (attached to laptop)
00:00:00
The vehicle is released from its starting position and it moves down the slope.
The beam remains cut until the back of the card passes out of the light gate.
The light beam is cut by the card as it enters the light gate and the timer starts.

Falling objects

You can measure the speed of falling objects using light gates.

The acceleration due to gravity on Earth is close to $10\,m/s^2$. This means that a falling object will increase its speed by 10 m/s every second when falling in the absence of frictional or resistive forces. However, for a piece of falling paper this terminal velocity value will decrease considerably because of resistive forces.

Typical speeds

Different activities occur at different speeds.

Activity	Typical speed
walking	1.5 m/s
running	3.0 m/s
cycling	6.0 m/s
driving	14 m/s
speed of sound in air	330 m/s
airliner	250 m/s
commuter train	55 m/s
gale-force wind	16 m/s

Worked example

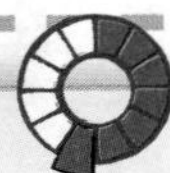

Describe how you can determine the average speed of a moving trolley as it moves down a slope. **(3 marks)**

When you know the distance that a body moves, and the time taken for the body to move that distance, you can use the equation speed = distance ÷ time to determine its speed. This can be done by measuring the distance between two points that the trolley moves past and dividing it by the time taken to travel between those points, or by using light gates to measure the time for which a known length of card breaks a light beam.

Now try this

1 State what you need to measure to determine the speed of an object. **(2 marks)**

2 Explain how two light gates can be used to find the acceleration of a ball rolling down a slope. **(4 marks)**

3 Describe how a falling ball can be used to find an accurate value for the acceleration due to gravity, g. **(5 marks)**

Had a look ☐ Nearly there ☐ Nailed it! ☐

Newton's first law

A body will remain at rest or continue in a straight line at a constant speed as long as the forces acting on it are balanced.

Stationary bodies

The forces acting on a stationary body are balanced.

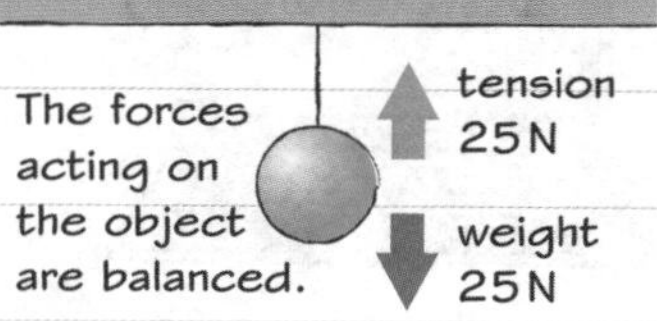

Bodies moving at a constant speed

The forces acting on a body moving at a constant speed, and in a straight line, are balanced.

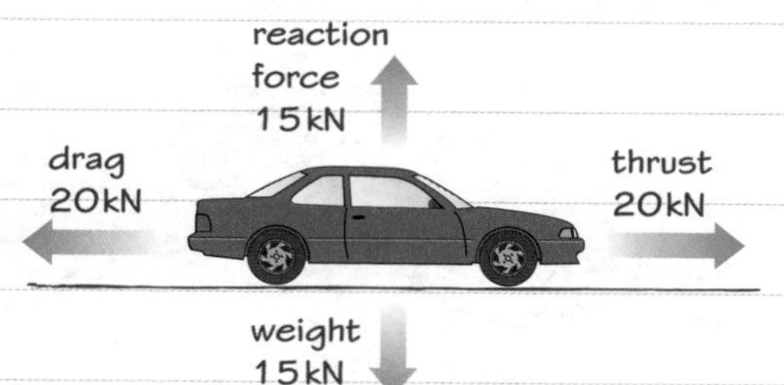

The forces on the car are balanced. The car will continue to move at a constant speed in a straight line until another external force is applied.

A common mistake is to think that when the resultant force on an object is zero, the object is stationary. The object may also be travelling at a constant speed.

Unbalanced forces

This body will accelerate to the right, since there is a resultant force of 5 N acting to the right.

Worked example

A resultant force acting in the opposite direction to the movement of a body will slow it down. It can also reverse the direction of motion.

Explain the effect that each of these forces will have on a car.

(a) 300 N forward force from the engine, 200 N drag. **(3 marks)**

resultant force = 300 N − 200 N = 100 N
The car will accelerate in the direction of the resultant force.
Its velocity will increase.

(b) 200 N forward force from the engine, 400 N friction from brakes. **(3 marks)**

resultant force = 200 N − 400 N = −200 N (200 N acting backwards)
The car will accelerate in the direction of the resultant force. This is in the opposite direction to its velocity, so the car will slow down.

(c) 300 N forward force, 300 N drag. **(3 marks)**

resultant force = 0 N
The car will continue to move at the same velocity.

Now try this

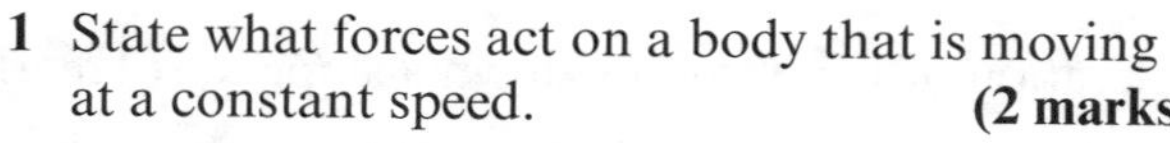

1 State what forces act on a body that is moving at a constant speed. **(2 marks)**

2 Explain how the direction of a moving body can be made to change. **(2 marks)**

3 Draw diagrams to show how two forces of 100 N acting on a mass can make it:
(a) stationary **(2 marks)**
(b) move at a constant speed **(2 marks)**
(c) accelerate to the left. **(2 marks)**

Newton's second law

When a resultant force acts on a mass then there will be a change in its velocity. The resultant force determines the size and direction of the subsequent acceleration of the mass.

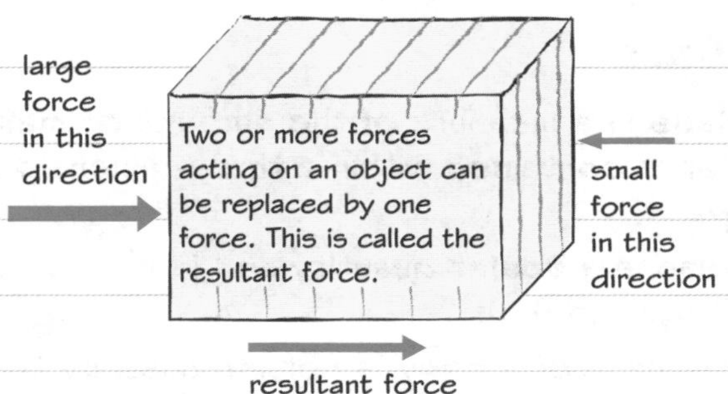

Forces have direction, so a force of −1 N is in the opposite direction to a force of 1 N.

When two or more forces act on the same straight line or are parallel, they can be added together to find the **resultant force**.

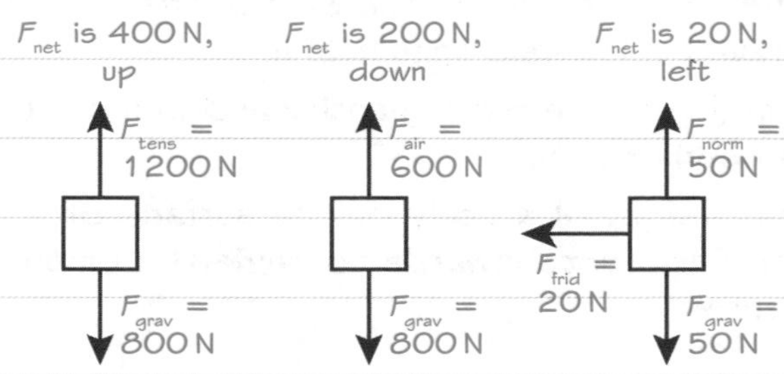

Force, mass and acceleration

You can calculate the acceleration of an object when you know its mass and the resultant force acting on it using the equation:

$$\text{acceleration (m/s}^2) = \frac{\text{force (N)}}{\text{mass (kg)}}$$

$$a = \frac{F}{m}$$

Rearrange the equation to give the force:

$$F = m \times a$$

LEARN IT! IT'S NOT ON THE EQUATIONS LIST

- The acceleration is in the same direction as the force.
- When the resultant force is zero, the acceleration is zero.
- A negative force means that the object is accelerating backwards or is slowing down.

Inertial mass is a measure of how difficult it is to change the velocity of a moving object and is defined as the ratio 'force over acceleration'.

Worked example

The diagram shows the horizontal forces acting on a boat. The boat has a mass of 400 kg.

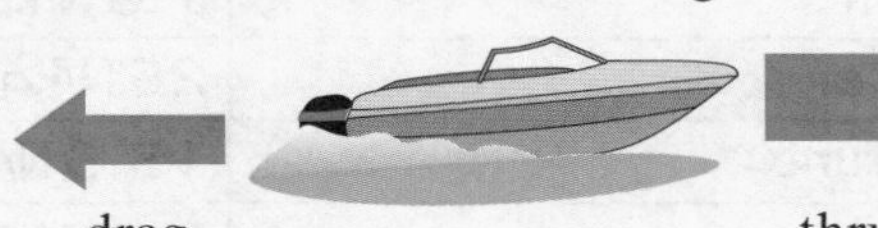

Calculate the acceleration of the boat at the instant shown in the diagram. **(3 marks)**

resultant force on boat = 900 N − 600 N = 300 N forwards

mass = 400 kg

acceleration $= \frac{F}{m} = \frac{300\text{ N}}{400\text{ kg}}$

$= 0.75\text{ m/s}^2$

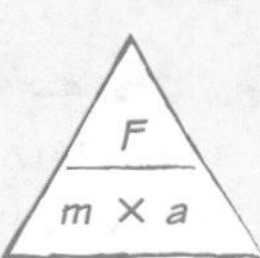

Worked example

A basketball player catches a ball. The force acting on the ball is −1.44 N and its acceleration is -2.4 m/s^2.

(a) Calculate the mass of the ball. **(3 marks)**

$m = \frac{F}{a}$

$= \frac{-1.44\text{ N}}{-2.4\text{ m/s}^2} = 0.6\text{ kg}$

(b) Which of the following describes the effect of the force on the ball? **(1 mark)**

☐ **A** The ball is moving backwards.

☒ **B** The ball is slowing down moving forwards.

☐ **C** The ball is moving faster forwards.

☐ **D** The ball is slowing down moving backwards.

Make sure you include minus signs in your calculations where necessary.

Now try this

1 Calculate the resultant force that causes a 1.2 kg mass to accelerate at 8 m/s^2. **(3 marks)**

2 Calculate the size of the mass that accelerates at 0.8 m/s^2 when the resultant force is 18.8 N. **(3 marks)**

3 Calculate the acceleration of a mass of 80 g when the resultant force is 0.6 kN. **(5 marks)**

Had a look ☐ Nearly there ☐ Nailed it! ☐

Weight and mass

It is important that you understand the difference between weight and mass. These words are often used interchangeably but are actually different things.

Weight

Weight is the **force** that a body experiences due to its mass and the size of the gravitational field that it is in.

Weight is a **vector** quantity and is measured in **newtons** (N).

The weight of a body on the surface of the Earth acts inwards towards the Earth's centre.

Mass

Mass is a measure of the **amount of matter** that is contained within a three-dimensional space.

Mass is a **scalar** quantity and is measured in **kilograms** (kg).

The mass of a body is not affected by the size of the gravitational field that it is in.

Connection between mass and weight

To find the weight of an object, use the equation:

weight (N) = mass (kg) × gravitational field strength (N/kg)

$W = m \times g$

LEARN IT! IT'S NOT ON THE EQUATIONS LIST

The weight of an object is directly proportional to the value of g, so a mass will weigh more on Earth than it does on the Moon.

Worked example

An astronaut has a mass of 58 kg on Earth. State the astronaut's mass and weight:

(a) on the surface of the Moon **(3 marks)**

Mass does not change so it is still 58 kg.

weight = 58 kg × 1.6 N/kg = 92.8 N

(b) on the surface of Jupiter. **(3 marks)**

Mass does not change, so it is still 58 kg.

weight = 58 kg × 26 N/kg = 1508 N

Gravitational field strength

The gravitational field strength of a body, such as a planet, depends on:

1. the **mass** of the body
2. the **radius** of the body.

When a body has a large mass and a small radius, it will have a large gravitational field strength.

The units of gravitational field strength are newtons per kilogram (N/kg), but it can also be given as m/s².

Astronomical body	Value for g
Earth	10 N/kg
Moon	1.6 N/kg
Jupiter	26 N/kg
Neptune	13.3 N/kg
Mercury	3.6 N/kg
Mars	3.75 N/kg
neutron star	10^{12} N/kg

Measuring weight

Weight is measured using a newtonmeter.

Reading the scale tells you the weight of the mass in newtons.

The greater the mass attached, the more weight it will experience due to gravity and the more the spring will stretch.

A 2 kg mass has twice the weight on Earth as a 1 kg mass, so the extension of the spring will be twice as big.

Now try this

1 Calculate the value for g on the surface of a planet where a mass of 18 kg experiences a weight of 54 N. **(3 marks)**

2 The value of g on Earth is 10 N/kg. On Planet X, the value of g is 25 N/kg. Explain what the mass and weight of a 5 kg brick will be on the surface of Planet X. **(3 marks)**

Force and acceleration

You can determine the relationship between force, mass and acceleration by varying the masses added to a trolley and measuring the time it takes to pass between two light gates that are a small distance apart.

Core practical

There is more information about the relationship between force, mass and acceleration on pages 173 and 174.

Investigating force and acceleration

Aim

To investigate the effect of mass on the acceleration of a trolley.

Apparatus

trolley, light gates, data logger, card of known length, slope or ramp, masses

An accelerating mass of greater than a few hundred grams can be dangerous, and may hurt somebody if it hits them at speed. Bear this in mind when designing your investigation.

Method

1 Set up the apparatus as shown.

It is better to use light gates and other electronic equipment to record values as this is more accurate than using a ruler and a stopwatch.

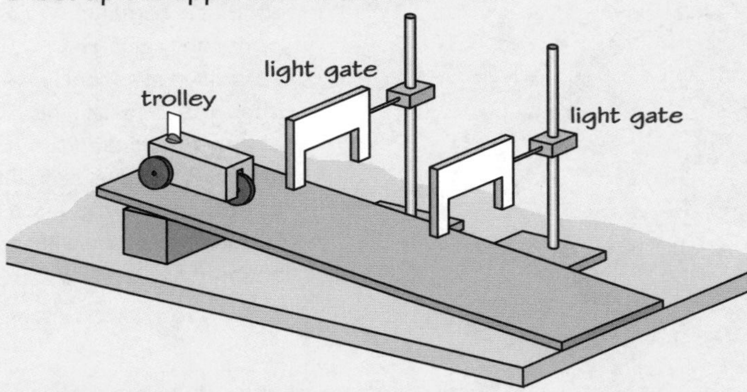

2 Set up the light gates to take the velocity and time readings for you.

3 Record velocity and time for different values of mass on the trolley.

4 Work out acceleration by dividing the difference in the velocity values by the time taken for the card to pass between both gates.

5 If you are changing the mass, the slope or gradient needs to remain the same throughout the investigation.

Results

The results for a slope of 20° to the horizontal and a card width of 10 cm are shown in the table below.

Mass (g)	Δv (m/s)	Δt (s)	a (m/s^2)
200	7.0	2.2	3.2
400	6.7	2.1	3.2
600	6.8	2.2	3.1
800	6.9	2.1	3.3

Conclusion

The acceleration of the trolley does not depend on the mass of the trolley and will remain fairly constant throughout.

Velocity and acceleration

Key points to remember for this investigation are:

Acceleration is change in speed ÷ time taken, so two velocity values are needed, along with the time difference between these readings, to obtain a value for the acceleration of the trolley.

Velocity is the rate of change of displacement and acceleration is the rate of change of velocity. The word rate means 'per unit time':

$$v = \frac{\Delta x}{\Delta t} \quad \text{and} \quad a = \frac{\Delta v}{\Delta t}$$

Now try this

1 Explain why it is better to use light gates and a data logger than a stopwatch to record time values. **(3 marks)**

2 Suggest how the investigation described above could be improved to obtain better results. **(4 marks)**

3 Design a similar experiment to determine how acceleration depends on the gradient of the slope. **(5 marks)**

Circular motion

A body will move in a circular path if its motion is at right angles to a force that acts inwards along the radius of the path.

Speed and velocity

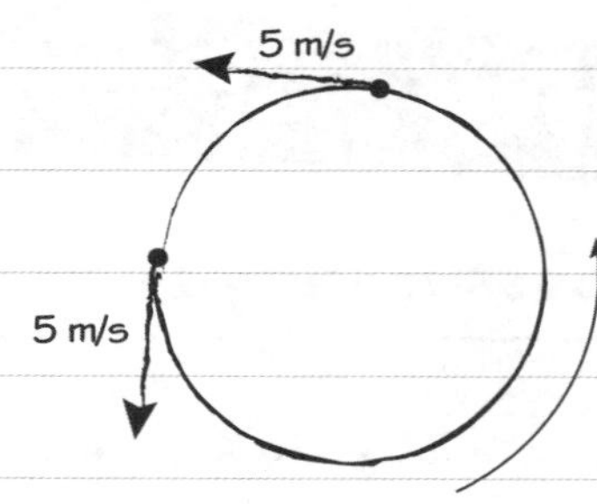

A body moving in a circular path with **constant speed** still has a **changing velocity.**

Speed is a **scalar** quantity and has size but no direction. Velocity is a **vector** quantity, which means it has both size and direction.

The velocity of a body changes when **either** its **speed** or its **direction** of motion changes.

Why circular motion happens

For circular motion to occur there must be a force acting inwards along the radius of the circle. This is called the **centripetal force**. The object moves at **right angles** to this force.

One example of this is the motion of a satellite in orbit around the Earth.

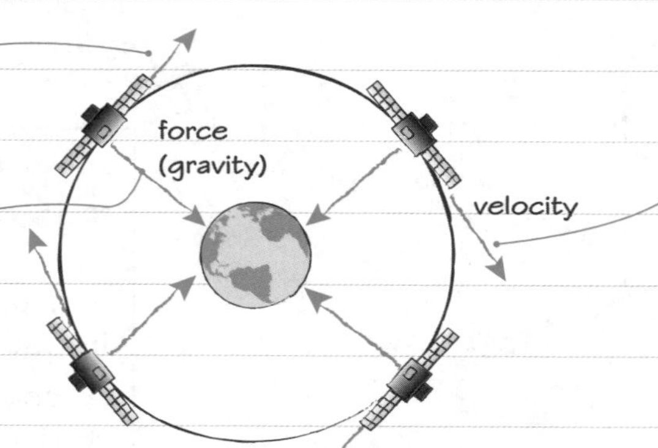

If there were no forces acting on the satellite it would move in a straight line.

The gravitational force, acting towards the Earth, acts on the moving satellite. This is the **centripetal force** and it acts at 90° to the direction of motion of the satellite.

The speed of the satellite remains constant, but the velocity constantly changes due to the change in direction. This means that the satellite is **accelerating** even though it is travelling at a constant speed.

Worked example

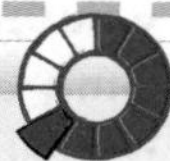

A motorcyclist is travelling around a circular curve in a flat road. What is providing the centripetal force? **(1 mark)**

- ☐ **A** gravity
- ☒ **B** the friction between the tyres and the road
- ☐ **C** the normal reaction of the road on the tyres
- ☐ **D** the thrust of the engine

The centripetal force must act at right angles to the direction of motion and **towards** the **centre of the circle**.

Centripetal forces

There are four main types of centripetal force that will result in circular motion. Here are some examples.

Force	Example
gravitational	orbiting satellite
frictional	car on a roundabout
tension	a hammer thrower
electrostatic	electron orbiting nucleus

Now try this

The figure shows an aircraft doing a circular loop.

(a) Draw arrows to show two forces acting on the aeroplane at point A. Label your arrows with the name of each force. **(2 marks)**

(b) Explain what happens to the velocity of the aircraft between point A and point B. **(2 marks)**

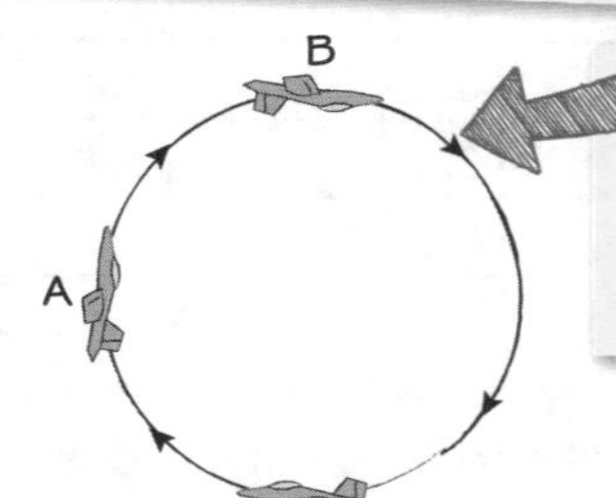

This is the motion of the aircraft in a horizontal circle, as seen from above.

Momentum and force

The momentum and the force of a moving body are closely related and can be understood by the use of Newton's second law of motion.

The **momentum** of a moving object depends on its mass and its velocity. Momentum is a **vector** quantity.

$$\underset{(\text{kg m/s})}{\text{momentum}} = \underset{(\text{kg})}{\text{mass}} \times \underset{(\text{m/s})}{\text{velocity}}$$

$$p = m \times v$$

LEARN IT! IT'S NOT ON THE EQUATIONS LIST

Momentum is mass multiplied by velocity, so the units for momentum are a combination of the units for mass and velocity.

Momentum and force

Force is the rate of change of momentum.

$$\underset{(\text{N})}{\text{force}} = \frac{\text{change in momentum (kg m/s)}}{\text{time taken for the change (s)}}$$

$$F = \frac{mv - mu}{t}$$

Since $a = \frac{v - u}{t}$, then we also get the equation from Newton's second law, $F = ma$:

$$F = \frac{m(v - u)}{t}$$

Deceleration in cars and forces

When the mass or the deceleration of a vehicle are large, the forces exerted on the passengers are also large. This can be very dangerous.

Car safety features such as seat belts and crumple zones reduce the size of the forces on the driver and passengers by increasing the time over which the vehicle comes to rest.

Changing the size of the force

Changing the time changes the force needed to change the momentum.

By increasing the time over which the change in momentum takes place, the force needed is reduced.

$$\downarrow F = \frac{m(v - u)}{t \uparrow}$$

Worked example

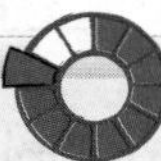

A car of mass 1450 kg, travelling at 18 m/s, is brought to rest in 1.2 s. Calculate the force exerted on the car. **(4 marks)**

$m = 1450\,\text{kg}$, $u = 18\,\text{m/s}$, $v = 0\,\text{m/s}$, $t = 1.2\,\text{s}$

$$F = \frac{m(v - u)}{t}$$

$$F = \frac{1450\,\text{kg}\,(0\,\text{m/s} - 18\,\text{m/s})}{1.2\,\text{s}}$$

$$= -21\,750\,\text{N}$$

Worked example

A golfer wants to hit a golf ball with as much force as possible. Explain how the golfer does this. **(3 marks)**

The change in speed of the club and ball needs to be as high as possible and the contact time between the club and ball needs to be as small as possible. Both of these make the force as large as possible.

Now try this

1 Calculate the momentum of a 25 kg trolley moving at 6 m/s. **(3 marks)**

2 Calculate the force a car experiences when braking from 16 m/s to 7 m/s over a time period of 0.8 s when the car's mass is 1580 kg. **(3 marks)**

3 Explain how the size of the force exerted on passengers in a car crash can be reduced. **(3 marks)**

Newton's third law

Newton's third law relates to bodies in equilibrium and can be applied to collisions when considering the conservation of momentum.

Action and reaction

Newton's third law states that for every action there is an equal and opposite reaction.

The action force and the reaction force **act on different bodies**.

The man and wall are in equilibrium because the forces in the system are all balanced. There is no overall force in any direction.

Newton's third law examples

Watch out! Just because two forces are equal and opposite it does not always mean that they are an example of Newton's third law.

$R = -F = -mg$

$F = mg$

The reaction force of the table pushing up on the book, and the force of gravity acting downwards, are both acting on the same object – the book – so this is **not** an example of Newton's third law.

The gravitational pull of the Earth on the book and the gravitational pull of the book on the Earth **are** an example of Newton's third law, since they act on different objects, and are equal and opposite.

Newton's third law and collisions

When the truck and car collide, they exert equal and opposite forces on each other and are in contact for the same amount of time. Newton's third law is obeyed.

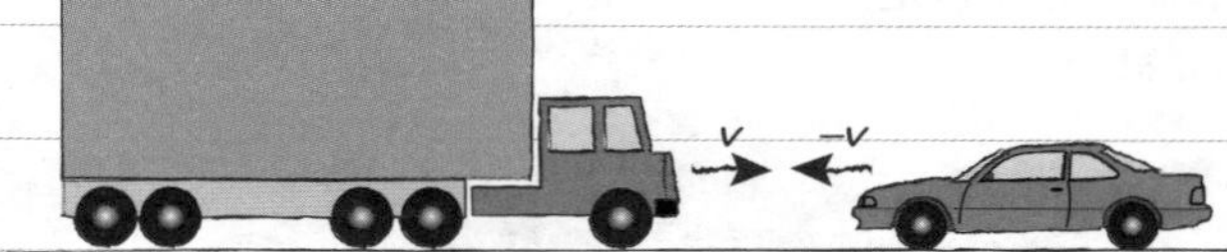

Momentum is conserved: the total momentum before the collision equals the total momentum after the collision.

Worked example

Before collision

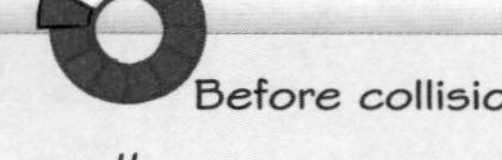

After collision

A model railway wagon (m) of mass 1 kg travelling at 12 m/s runs into a second stationary wagon (M) of mass 3 kg. After the collision the wagons stay linked together and move together. Calculate the velocity of the two wagons after the collision. **(4 marks)**

momentum = mass × velocity:

$1\ \text{kg} \times 12\ \text{m/s} = (1 + 3)\ \text{kg} \times v$

$v = 12\ \text{kg m/s} / 4\ \text{kg}$

$v = 3\ \text{m/s}$

The wagon with larger mass is stationary before the collision so has no momentum. The wagons move together after the collision with a combined mass of $(M + m)$.

Now try this

1 Explain how Newton's third law applies to collisions. **(2 marks)**

2 A truck of mass 12 kg moving at 8 m/s collides with a truck of mass 8 kg that is stationary. They collide and move off together. Calculate their shared velocity after the collision. **(4 marks)**

Human reaction time

Human reaction time is the time between a **stimulus** occurring and a **response.** It is related to how quickly the human brain can process information and react to it.

Human reaction time

It takes a typical person between 0.20 s and 0.25 s to react to a stimulus. Some people, such as international cricketers, 100 m sprinters and fighter pilots, train themselves to have improved reaction times.

For example, a top cricketer has a total time of 0.5 s to play a batting stroke. The first 0.2 s of this is the reaction time, the next 0.2 s is the batsman's preparation to play the shot and the final 0.1 s involves hitting the ball.

Reaction times and driving

Drivers have to react to changes in the traffic when driving. This may involve reacting to traffic lights changing colour, traffic slowing on a motorway or avoiding people or animals that may have walked in front of the vehicle.

The reaction time of humans may be affected by:

- tiredness
- distractions
- alcohol and drugs
- age.

Measuring reaction time

You can determine human reaction time by using the ruler drop test.

The reaction time is determined from the equation:

$$\text{reaction time} = \sqrt{\frac{2 \times \text{distance ruler falls}}{\text{gravitational field strength}}}$$

Repeats can be used to get a mean value for the reaction time.

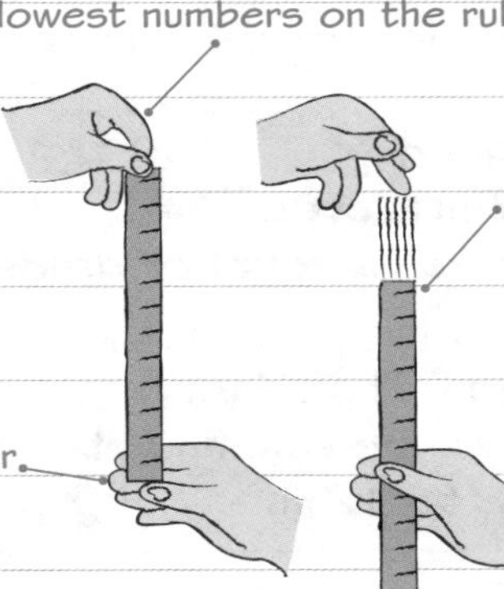

A metre ruler is held, by a partner, so that it is vertical and exactly level with the person's finger and thumb, with the lowest numbers on the ruler at the bottom.
The ruler is dropped and then grasped by the other person.
A person sits with their index finger and thumb opened to a gap of about 8 cm.

Worked example

A ruler drop test is conducted five times with the same person. The results show that the five distances fallen are: 0.16 m, 0.17 m, 0.18 m, 0.16 m and 0.15 m.

(a) Calculate the mean distance that the ruler drops. **(3 marks)**

mean = (0.16 + 0.17 + 0.18 + 0.16 + 0.15) ÷ 5 = 0.16 m

(b) Calculate the person's reaction time. **(3 marks)**

$$\text{reaction time} = \sqrt{\frac{2 \times 0.16}{10}} = 0.18\,\text{s}$$

Worked example

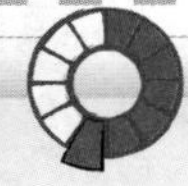

A driver moving at 30 km/h in her car has a reaction time of 0.25 s. Calculate the distance the car travels between seeing a hazard in the road and then applying the brakes. **(4 marks)**

distance (m) = speed (m/s) × time(s)

$$30\,\text{km/h} = \frac{30\,000\,\text{m}}{3600\,\text{s}} = 8.3\,\text{m/s}$$

distance travelled = 8.3 m/s × 0.25 s = 2.1 m

This distance for a driver is called the thinking distance. Read more about it on page 180.

Now try this

1 (a) Define **human reaction time**. **(1 mark)**

(b) Give **three** factors that can cause reaction times to increase. **(3 marks)**

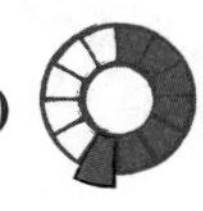

2 Explain why the reaction time doubles when the distance that a ruler falls increases by a factor of 4. **(2 marks)**

Stopping distances

Stopping distance is the **total distance** over which a vehicle comes to rest.

It takes time for a moving car to come to a stop, and the car is still moving during this time. Understanding the factors that affect stopping distance is important for road safety.

Danger appears.

Driver brakes.

Car has stopped.

thinking distance = the distance the car travels while the driver reacts to the danger and applies the brakes

braking distance = the distance the car travels while it is slowing down, once the brakes have been applied

stopping distance = thinking distance + braking distance

Factors affecting thinking distance

Thinking distance increases when the driver's reaction time increases. This can be due to:

- the driver being **tired**
- the driver being **distracted**
- the driver having taken **alcohol** or **drugs**.

Thinking distance also increases if the car's speed increases. The thinking distance at 20mph will be 6 m, and at 40 mph it will double to 12 m.

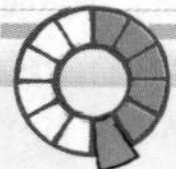

Cars can experience large decelerations when they slow down, and this means that typical forces, of around 10,000 N, can be exerted on the vehicles.

Factors affecting braking distance

Braking distance is the distance the car travels once the brakes have been applied. It increases when:

- the amount of **friction** between the **tyres** and the **road** decreases, such as when the road is **icy** or **wet**
- the **brakes** are worn
- the **tyres** are worn
- the **mass** of the car is bigger.

Braking distance also increases as the speed of the car increases. At 20 mph it will be 6 m, and at 40 mph it will be 24 m.

Worked example

A car is moving at a constant speed. Explain how and why the stopping distance changes when:

(a) the speed of the car increases **(2 marks)**

(b) the car has more passengers **(2 marks)**

(c) the driver has been drinking alcohol. **(2 marks)**

(a) The stopping distance increases because the thinking distance and braking distance increase.

(b) The mass of the car will be greater so the braking distance will increase, which means the stopping distance will increase.

(c) The stopping distance will increase because the thinking distance will increase.

Now try this

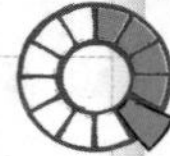

1 State **two** factors that affect:

(a) thinking distance **(2 marks)**

(b) braking distance. **(2 marks)**

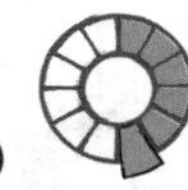

2 Explain why icy and wet roads lead to a greater stopping distance. **(3 marks)**

Extended response – Motion and forces

There will be at least one 6-mark question on your exam paper. For these questions, you will need to think scientifically and structure your answer logically, showing how the points you make are related to each other. You can revise the topics for this question, which is about motion and forces, on pages 167–180.

Worked example

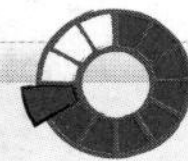

The diagram shows a trolley, ramp and light gates being used to investigate the relationship between the acceleration of a trolley on a slope and the angle that the slope makes with the horizontal.

Describe a simple investigation to determine how the acceleration of the trolley depends on the angle of the slope.

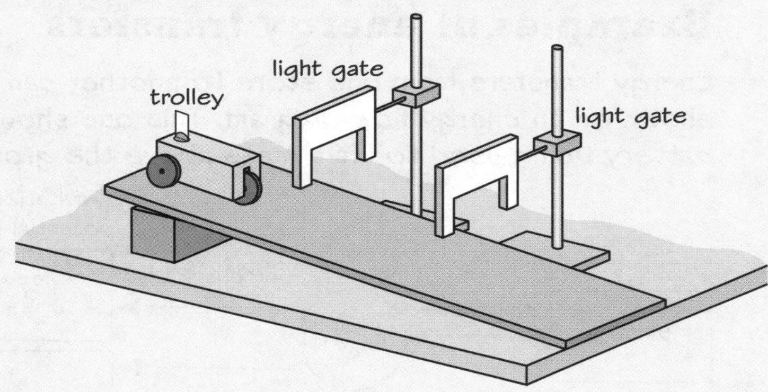

(6 marks)

I will start with the ramp at an angle of 10° and roll the trolley down the slope from the same starting position along the slope each time. The mass of the trolley needs to be kept constant, and a suitable value would be 500 g. I will then increase the ramp angle by 5° each time until I have at least 6 readings. I will repeat readings to check for accuracy and precision.

The speed of the trolley at each light gate is calculated using the equation speed = distance ÷ time, where the distance recorded is the length of the card that passes through the light gates. The acceleration of the trolley is calculated by finding the difference in the speed values at the two light gates and then dividing this by the time it takes the trolley to pass between the gates.

I would expect the acceleration of the trolley to increase as the angle increases, although the relationship may not be linear. This is because as the slope angle increases, you increase the component of the trolley's weight that is acting down the slope, which is the force responsible for the trolley's acceleration.

Command word: Describe

When you are asked to **describe** something, you need to write down facts, events or processes accurately.

Your answer should refer to the equations needed to calculate the velocity and acceleration of the trolley, as well as how you will show the relationship between acceleration and angle of slope.

You need to be careful when stating what you will vary and what you will keep the same. Since the dependent variable is acceleration and the independent variable is angle of slope, you will need to keep the mass of the trolley and its starting point the same each time.

Be clear when referring to how the speed of the trolley will be measured by referring to the card that passes through the light gates, and mention that the length of the card needs to be known if the light gates are recording time values. You need two speed and time values to calculate the acceleration.

You can also refer to the increase in gravitational potential energy as the slope angle increases. This means more is transferred to kinetic energy and so a greater change in velocity per second.

Now try this

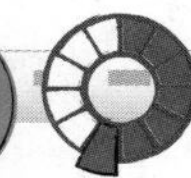

The distance taken for vehicles to stop on roads depends on a number of factors. Discuss what these factors are and how they affect stopping distance. **(6 marks)**

Had a look ☐ Nearly there ☐ Nailed it! ☐

Energy stores and transfers

Energy can be stored in different ways, and can be transferred from one **energy store** to another.

Examples of energy stores

There are eight main energy stores.

Energy store	Example
chemical	fuel, food, battery
kinetic	moving objects
gravitational potential	raised mass
elastic	stretched spring
thermal	hot object
magnetic	two magnets
electrostatic	two charges
nuclear	radioactive decay

Closed systems

When there are **energy transfers** in a closed system, there is no net change to the total energy in the system.

A **closed system** is one where energy can flow in or out of the system, but there is no transfer of mass. An example of a closed system is a pan of water being heated that has a lid on it so that no steam can escape.

Energy transfers

Energy is transferred from one store to another in four different ways:

- **mechanically** – by a force moving through a distance
- **electrically** – by the use of electric current
- **thermally** – because of a difference in temperature
- **radiation** – by waves such as electromagnetic or sound.

Some of the energy will be usefully transferred, and some will be wasted or **dissipated**.

Examples of energy transfers

Energy transfers from one store to another can be shown on an energy flow diagram. This one shows a battery being used to lift a mass above the ground.

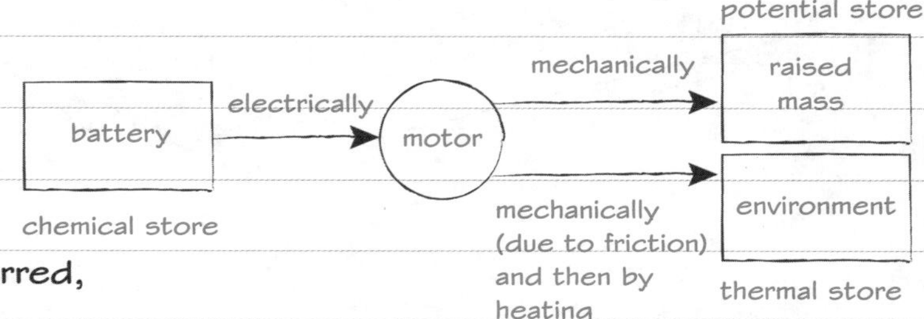

Mechanical processes become wasteful when they lead to a rise in the temperature of the surroundings through heating. In all system changes, energy is stored in less useful ways. For example, the light and sound energy from a TV will eventually be absorbed by the walls and by people, leading to a rise in their temperature.

Worked example

Draw a flow diagram to show the useful and wasted energy transfers for a camping gas stove. **(4 marks)**

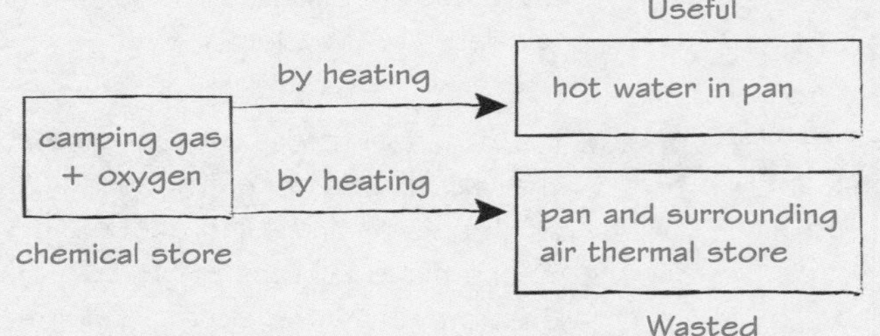

In all flow diagrams, there must be energy stores and energy transfers present to show how the energy is transferred from one store to others.

Based on the law of conservation of energy, the total energy present in the original stores must be equal to the energy present in the stores after the transfers have taken place, since you cannot create or destroy energy in any given system.

Now try this

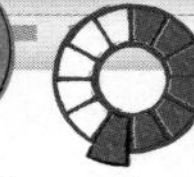

1. Draw an energy store flow diagram for a car driving up a hill. **(4 marks)**
2. When a person lifts a 2 kg mass above the ground, the energy stores change. Draw a possible flow diagram for the transfer and explain how the energy stores change. **(5 marks)**

Efficient heat transfer

The rate at which a material transfers thermal energy depends on a number of factors. The **efficiency** of a device is a measure of how much **useful** energy it transfers.

Thermal energy transfer

The rate at which thermal energy is transferred through a wall of a house depends on:

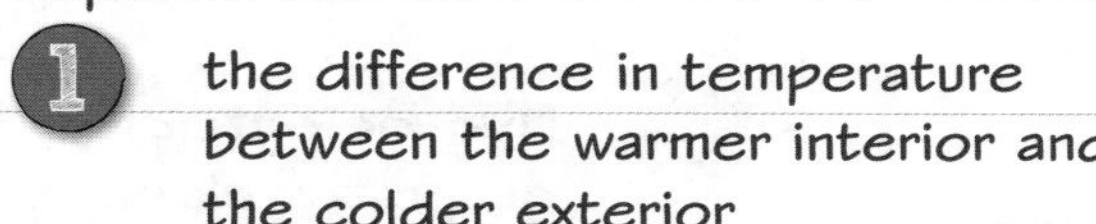

the difference in temperature between the warmer interior and the colder exterior

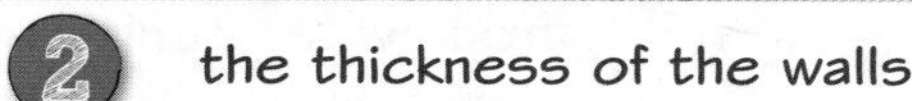

the thickness of the walls

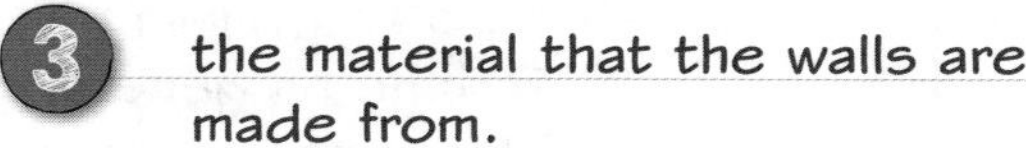

the material that the walls are made from.

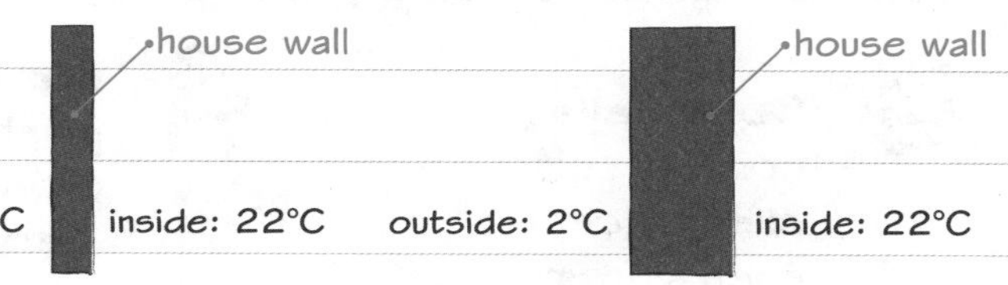

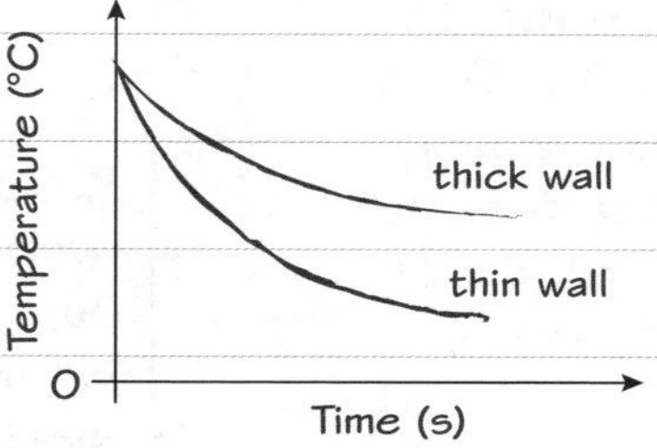

Efficiency

All machines **waste** some of the **energy** they **transfer**. Most machines waste energy as **heat** energy. The **efficiency** of a machine is a way of saying how good it is at transferring energy into **useful** forms.

A very efficient machine has an efficiency that is nearly 100%. The higher the efficiency, the better the machine is at transferring energy to useful forms.

Unwanted energy transfers can be reduced by thermal insulation and lubrication.

Thermal conductivity

A material with a high thermal conductivity is a better conductor of energy than one with a lower thermal conductivity. The rate at which the blue wall transfers energy is greater than that of the red wall. Different materials have different **relative thermal conductivities**.

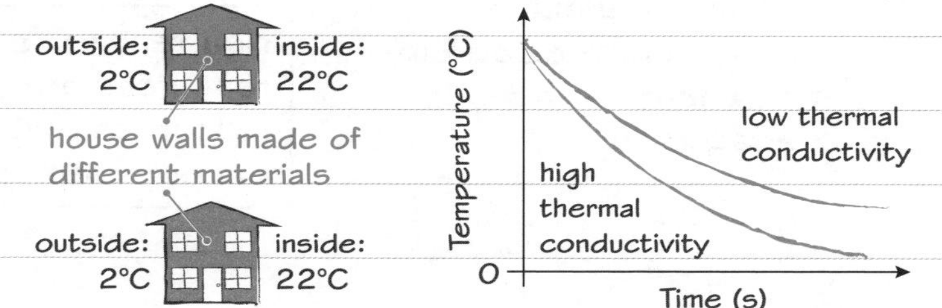

Worked example

A motor transfers 100 J of energy by electricity. 60 J are transferred as kinetic energy, 12 J as sound energy and 28 J as thermal energy. Calculate the efficiency of the motor. **(3 marks)**

$$\text{efficiency} = \frac{\text{useful energy transferred by the machine}}{\text{total energy supplied to the machine}} \times 100\%$$

Efficiency does not have units.

LEARN IT! IT'S NOT ON THE EQUATIONS LIST

total useful energy = 60 J

$$\text{efficiency} = \frac{60\text{ J}}{100\text{ J}} \times 100\%$$

$$= 60\%$$

No machine is ever 100% efficient. If you calculate an efficiency greater than 100% you have done something wrong!

Now try this

1 Calculate the efficiency of a lamp that transfers 14 J of energy into useful light energy for every 20 J of electrical energy input. **(2 marks)**

2 Explain why it is not possible for any device to be 100% efficient. **(3 marks)**

Use an electric appliance such as a kettle as an example.

Had a look ☐ Nearly there ☐ Nailed it! ☐

Energy resources

The main energy resources include both **renewable** and **non-renewable** resources.

Renewable energy resources

Renewable energy resources will not run out. Most do not cause pollution or emit carbon dioxide.

Bio-fuels are animal or plant matter used to produce thermal energy, electrical energy or used to power cars.

Hydroelectricity generates electricity from water behind a dam flowing down a pipe and turning a turbine to generate electricity.

Wind turbines use kinetic energy from the wind to generate electricity.

Tidal power uses the rise and fall of the tide or tidal currents to generate electricity.

Solar cells convert solar energy, or energy from the Sun, directly to electricity. Solar energy can also be used directly to cook food or to heat water.

Non-renewable energy resources

Non-renewable resources are resources that will run out eventually.

Nuclear fuels such as uranium are used:
- to generate electricity
- as energy sources in spacecraft.

Fossil fuels are coal, oil and natural gas. They are used:
- to generate electricity
- to power transport
- to heat homes and for cooking.

They are available all the time, unlike some renewable resources.

Worked example

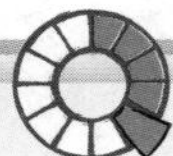

Describe how different renewable and non-renewable resources are used for transport. **(4 marks)**

Renewable and non-renewable energy resources can both be used in power stations, leading to the production of electricity which is used to power trains. Electrical energy can also be supplied from other renewable sources such as solar, hydroelectric and wind.

Biofuel, in the form of methanol, is an example of how renewable fuel can be used to power cars, and solar power is also being developed to do more of this. Petrol and diesel, from oil, are examples of non-renewable fuels that are used to power cars.

Now try this

1 Draw diagrams to show the energy transfers that take place
(a) in hydroelectricity **(3 marks)**
(b) when biofuels are used for transport. **(3 marks)**

2 Suggest the advantages and disadvantages of using biofuels to power cars compared with using solar power to power cars. **(4 marks)**

Patterns of energy use

The patterns and trends in how humans have used energy resources have changed over time.

Patterns and trends in energy use

The use of energy resources has changed over the years. This is because of several factors.

1. **the world's population** In the past 200 years, the population of the world has risen from 1 billion to 7 billion people.
2. **the development of technology** Vehicles such as cars, trains and planes and other devices have increased in number and these all require energy.
3. **electrical energy** Power stations require fuels in order to generate electricity.

Growth in world population

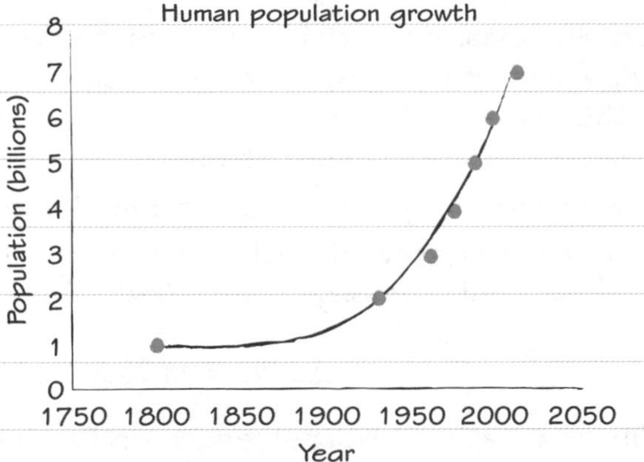

Trends in the world's energy use

As the graph shows, most of the world's energy use has been fossil fuels. Before about 1900, biomass in the form of wood was the major source of fuel and its use has remained constant over time.

In more recent years, there has been an increase in the use of nuclear fuel and hydroelectric power.

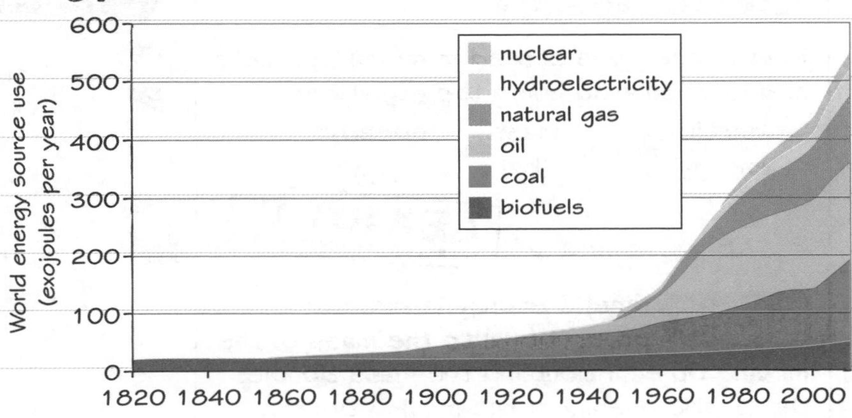

Worked example

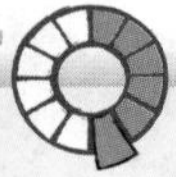

Compare the shape of the world's population graph with that of the energy use graph. **(4 marks)**

They are similar in that as the population increases, the total energy used will increase, assuming that each person uses the same amount of energy. They differ in that the total energy use has increased faster than the population. As societies develop, they use more technology that requires energy to run it. This means that the average amount of energy used per person also increases.

The future?

It is not possible to continue using the Earth's non-renewable energy reserves to the extent that they are being used now. They are a finite reserve, so will run out and not be replaced.

Greater use of fossil fuels will lead to more carbon dioxide in the atmosphere. There will be greater global warming, leading to severe weather, flooding and threats to food supplies.

Now try this

1. Describe the main issues with using the Earth's energy supplies. **(2 marks)**
2. Wood is a biofuel. Suggest why wood has been used constantly for many years, whereas other fuels have only been used much more recently. **(3 marks)**

Had a look ☐ Nearly there ☐ Nailed it! ☐

Potential and kinetic energy

You can calculate the **kinetic** and **gravitational potential energy** of objects using the equations on this page.

Gravitational potential energy

Gravitational potential is the energy possessed by a body due to its height above the Earth. The value of the gravitational potential energy stored depends on:

- the mass of the body
- the gravitational field strength
- the height the body is raised.

The change in gravitational potential energy is given by the equation:

change in gravitational potential energy (J) = mass (kg) × change in gravitational field strength (N/kg) × change in vertical height (m)

$$\Delta GPE = m \times g \times \Delta h$$

LEARN IT! IT'S NOT ON THE EQUATIONS LIST

The gravitational field strength on Earth is 10 N/kg. It can also be given in m/s^2.

Kinetic energy

Kinetic energy is stored in moving objects and is calculated using the equation:

kinetic energy (J) = $\frac{1}{2}$ × mass (kg) × (speed)^2 (m/s)^2

$$KE = \frac{1}{2} \times m \times v^2$$

LEARN IT! IT'S NOT ON THE EQUATIONS LIST

Maths skills Kinetic energy is directly proportional to the mass of the moving object: doubling the mass doubles the kinetic energy of the moving object.

It is directly proportional to the square of the speed, so doubling the speed means the KE increases by a factor of four.

Worked example

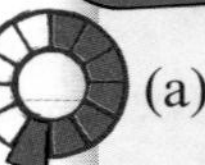

(a) A mass of 800 g is moving at 14 m/s. Calculate its kinetic energy. **(3 marks)**

$KE = \frac{1}{2} \times m \times v^2$

$KE = \frac{1}{2} \times 0.8 \times 14^2 = 78.4$ J

(b) A body of mass 4.8 kg has kinetic energy of 200 J. Calculate the speed it is moving at. **(4 marks)**

$200\text{ J} = \frac{1}{2} \times 4.8 \times v^2$

Rearranging gives $v = \sqrt{400 \div 4.8} = 9.1$ m/s

Remember to square the value of speed when calculating the kinetic energy. The term v^2 means $v \times v$.

Worked example

(a) A body of mass 73 kg is lifted through a vertical height of 26 m. Calculate how much gravitational potential energy it has gained. **(3 marks)**

$\Delta GPE = m \times g \times \Delta h$

$= 73\text{ kg} \times 10\text{ N/kg} \times 26\text{ m} = 18980$ J

(b) The body is now dropped from 26 m above the ground. At what speed will it hit the floor? **(4 marks)**

using the law of conservation of energy:

loss in GPE = gain in KE

$18980\text{ J} = \frac{1}{2} \times 73\text{ kg} \times v^2$

Rearranging gives $v^2 = 520$ (m/s)^2

hence $v = \sqrt{520} = 23$ m/s

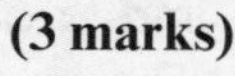

Now try this

1 A mass of 5 kg is raised through a vertical height of 18 m. Calculate the change in gravitational potential energy. **(3 marks)**

2 A motorbike of mass 80 kg is moving at 30 km/h. Calculate its kinetic energy. **(4 marks)**

3 A ball of mass 3 kg falls from 34 m above the ground. Calculate its speed when it lands. **(4 marks)**

Extended response – Conservation of energy

There will be at least one 6-mark question on your exam paper. For these questions, you will need to think scientifically and structure your answer logically, showing how the points you make are related to each other. You can revise the topics for this question, which is about **the principle of the conservation of energy**, on pages 182–185.

Worked example

Figure 1 shows the arrangement of apparatus for a simple pendulum. Figure 2 shows how the kinetic energy of the pendulum changes with time.

Describe the energy transfers taking place during the motion of the simple pendulum. Your answer should refer to the energy stores and energy transfers that are involved.

(6 marks)

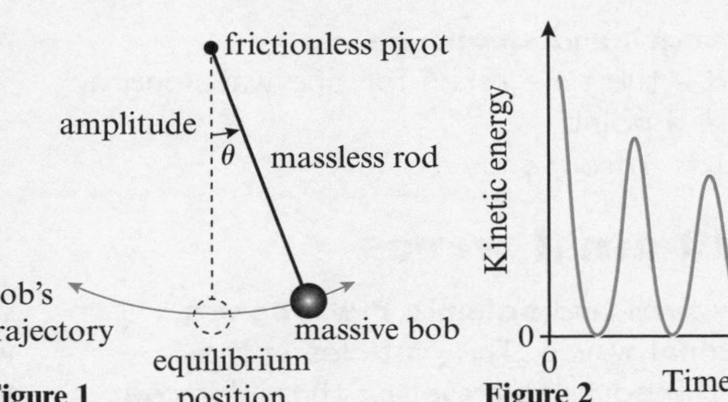

Figure 1 **Figure 2**

Initially, the pendulum has to be raised through a small angle, so it will gain gravitational potential energy. Once released, this gravitational potential energy store will be transferred mainly to a kinetic energy store, with some energy being dissipated as sound and heat by thermal transfer from friction due to air resistance.

The principle of conservation of energy states that the total energy of the system at any time must be the same, since energy cannot be created nor destroyed. Initially, this was all in the gravitational potential energy store, but this is then transferred to the kinetic energy store once the pendulum starts to swing. At the bottom of the swing, the gravitational potential store will be greatly reduced and the energy will be mostly in the kinetic store. Some energy is also transferred to the surroundings as heat and sound by thermal transfer from friction due to air resistance, which must also be taken into consideration as part of the total energy of the system.

Eventually, the pendulum will stop swinging, so the gravitational energy has been transferred to kinetic energy and it has eventually all ended up in the surroundings by friction due to air resistance, leading to the surroundings becoming slightly warmer. The useful energy store has been dissipated to the surroundings.

Command word: Describe

When you are asked to **describe** something, you need to write down facts, events or processes accurately.

It is a good idea to state what the principle of conservation of energy is, so that the examiner knows that you understand it. You can then apply this idea to the changes in the energy stores and transfers in the system. It is also worth noting that the way we talk about energy at GCSE now has changed – you need to refer to energy stores as well as energy transfers. In this example, the main stores are gravitational and kinetic, and the motion of the pendulum is via mechanical transfer, with energy transfer to the surroundings by thermal transfer.

In energy transfers, the energy stores are usually dissipated to the surroundings, which become warmer. In a perfect situation, with no frictional forces or energy wasted, the pendulum would swing forever, but in reality this does not happen.

Refer to **energy stores** and **energy transfers** in your answer. There are eight energy stores and four energy transfers that you need to know. See page 182 to remind yourself of these.

Now try this

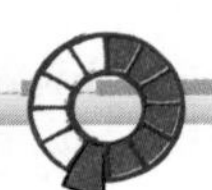

The people on a small island want to use either wind power, on hills and out at sea, or its reserves of coal to generate electrical energy. Compare the advantages and disadvantages of each resource. **(6 marks)**

Had a look ☐ Nearly there ☐ Nailed it! ☐

Waves

Waves transfer energy and information without transferring matter. Evidence of this for a water wave can be seen when a ball dropped into a pond bobs up and down, but the wave energy travels outwards as ripples across the surface of the pond.

Waves can be described by their

- **frequency** – the number of waves passing a point each second, measured in **hertz (Hz)**
- **speed** – measured in **metres per second (m/s)**
- **wavelength** and **amplitude**
- **period** – the time taken for one wavelength to pass a point
- period = 1/frequency

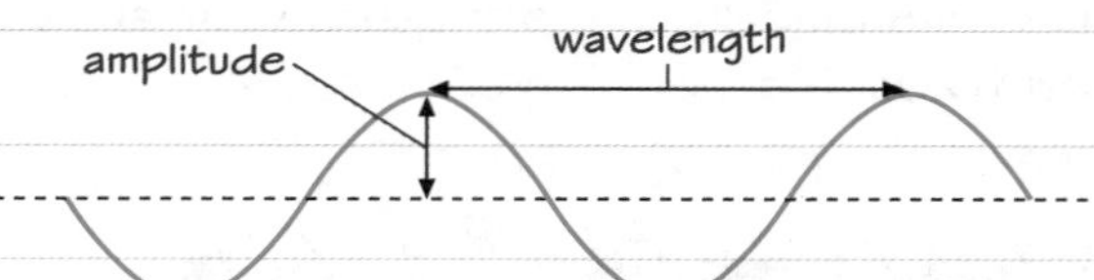

Watch out! Remember that the amplitude is *half* of the distance from the top to the bottom of the wave. Students often get this wrong.

Longitudinal waves

Sound waves and **seismic P waves** are **longitudinal** waves. The particles in the material the sound is travelling through move back and forth along the same direction that the sound is travelling.

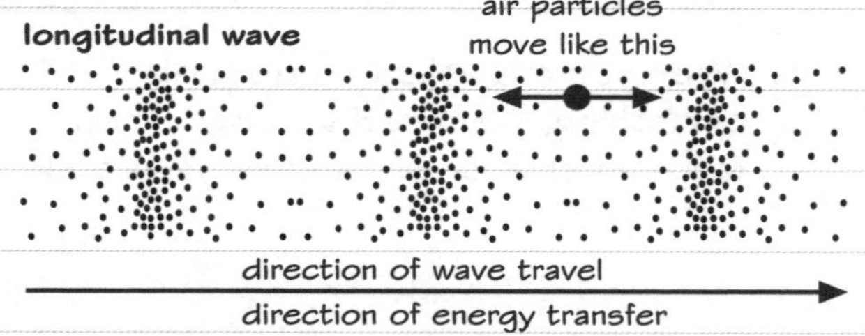

Particles in a **long**itudinal wave move **along** the same direction as the wave.

Transverse waves

Waves on a **water surface, electromagnetic waves** and **seismic S** waves are all **transverse** waves. The particles of water move in a direction at right angles to the direction the wave is travelling.

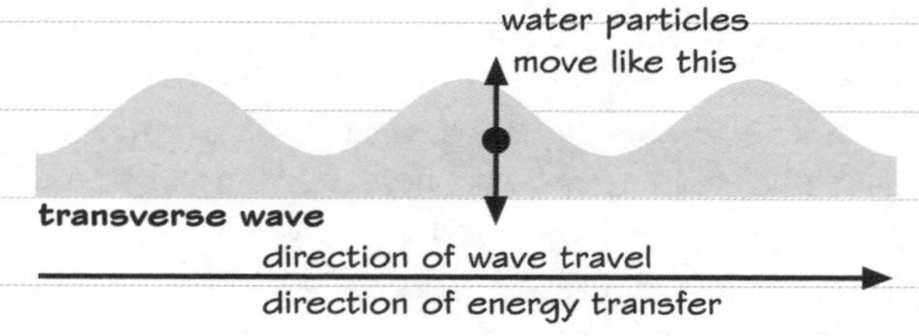

Particles in a tran**s**ver**se** wave move acro**ss** the direction the wave is travelling.

Worked example

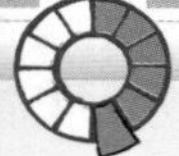

Give two ways in which longitudinal and transverse waves are

(a) similar **(2 marks)**

They both transfer energy without transferring matter and have an amplitude, speed, wavelength and frequency.

(b) different. **(2 marks)**

Particles in longitudinal waves vibrate along the direction of movement, whereas particles in transverse waves move at 90 degrees to the direction of travel. They can also have different speeds, frequencies and wavelengths.

Now try this

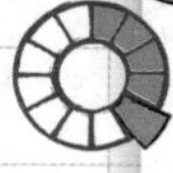

1 (a) Sketch a transverse wave and mark the amplitude and wavelength on it. **(3 marks)**

(b) Draw an arrow to show which way the wave moves. **(1 mark)**

(c) Draw a small particle on the wave, with arrows to show which way it moves. **(1 mark)**

2 The graph shows a wave. Each vertical square represents 1 mm. Work out the amplitude of the wave. **(1 mark)**

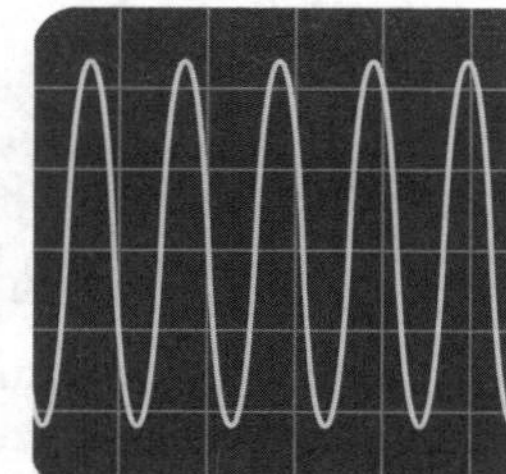

Wave equations

Both of these equations can be used to find the **speed** or **velocity** of a wave.

Speed, frequency and wavelength

LEARN IT! IT'S NOT ON THE EQUATIONS LIST

$v = f \times \lambda$

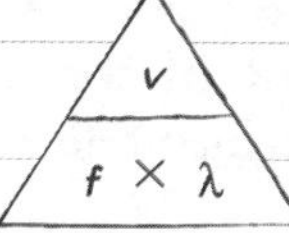

v = wave speed (metres per second, m/s)

f = frequency (hertz, Hz)

λ = wavelength (metres, m)

Speed, distance and time

LEARN IT! IT'S NOT ON THE EQUATIONS LIST

$v = \frac{x}{t}$

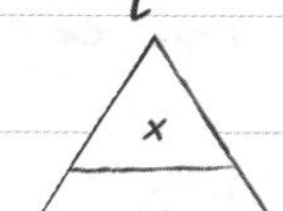

v = wave speed (metres/second, m/s)

x = distance (metres, m)

t = time (seconds, s)

Worked example

A seismic wave has a frequency of 15 Hz and travels at 4050 m/s. Calculate its wavelength. **(4 marks)**

$\lambda = \frac{v}{f}$

$= \frac{4050\,\text{m/s}}{15\,\text{Hz}}$

$= 270\,\text{m}$

λ is the Greek letter 'lambda'.

Start with the equation $v = f \times \lambda$ and rearrange it to make λ the subject.

Divide both sides of the equation by f, to get

$\lambda = \frac{v}{f}$

Substitute the values for v and f to calculate the value for the wavelength, λ.

The units for v are m/s and the units for f are Hz (or per second), so the units for λ are $\frac{\text{m/s}}{1/\text{s}}$ = m.

Always state the correct units as part of your answer.

Worked example

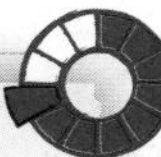

A wave on the sea is travelling at 4 m/s. Calculate how long it takes to travel along a 20 m long pier. **(4 marks)**

$t = \frac{x}{v}$

$= \frac{20\,\text{m}}{4\,\text{m/s}}$

$= 5\,\text{s}$

Rearrange the equation $v = \frac{x}{t}$ to give

$t = \frac{x}{v}$.

You can also find this easily by covering the letter t in the triangle at the top of the page.

When dividing the units, m ÷ (m/s), you get m × s/m = s, which is the correct unit for time.

Again, always include the correct unit in your answer.

Now try this

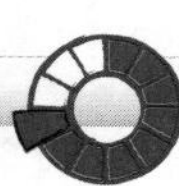

1 A sound wave with a frequency of 100 Hz has a speed of 330 m/s. Calculate its wavelength. **(3 marks)**

2 A wave in the sea travels at 25 m/s. Calculate the distance it travels in one minute. **(4 marks)**

Had a look ☐ Nearly there ☐ Nailed it! ☐

Measuring wave velocity

You need to be able to **calculate** the speed of sound in air or the speed of ripples on the surface of water.

Calculating the speed of sound in air

Method 1: using an echo

1. Measure the distance from the source of the sound to the reflecting surface (the wall).
2. Measure the time interval, with a stopwatch, between the original sound being produced and the echo being heard.
3. Use $\underset{(\text{m/s})}{\text{speed}} = \frac{\text{distance (m)}}{\text{time (s)}}$ to calculate the speed of sound in air.

Repeating the experiment a number of times over a range of distances will allow you to obtain accurate and precise results.

Method 2: using two microphones and an oscilloscope

1. Set up the microphones one in front of the other at different distances in a straight line from a loudspeaker.
2. Set the frequency of the sound from the loudspeaker to a known, audible value.
3. Display the two waveforms on the oscilloscope. Measure the distance between the microphones.
4. Move the microphones apart so that the waveforms move apart by 1 wavelength.
5. Calculate the speed of sound using the equation:
   $\underset{(\text{m/s})}{\text{wave speed}} = \underset{(\text{Hz})}{\text{frequency}} \times \underset{(\text{m})}{\text{wavelength}}$

Calculating the speed of ripples on water surfaces

You can work out the speed of ripples on the surface of water using a ripple tank and a strobe.

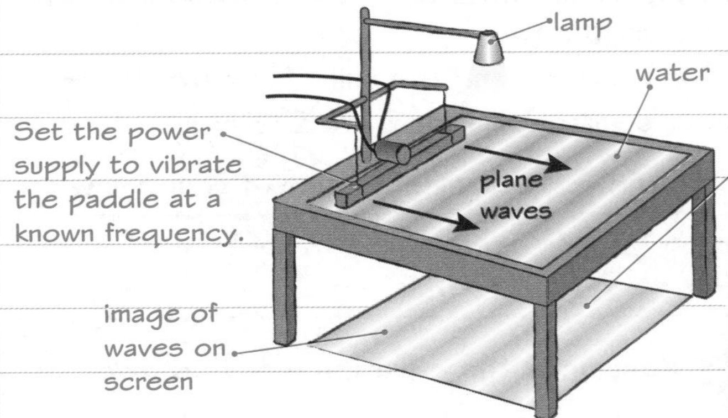
lamp
water
Set the power supply to vibrate the paddle at a known frequency.
plane waves
image of waves on screen

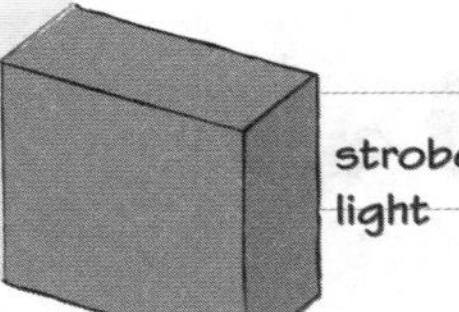
Use a strobe light to 'freeze' the water waves so that you can measure the wavelength.
strobe light
Use the equation *wave speed = frequency × wavelength* to calculate the speed or velocity of the water waves on the surface of the ripple tank.

Worked example

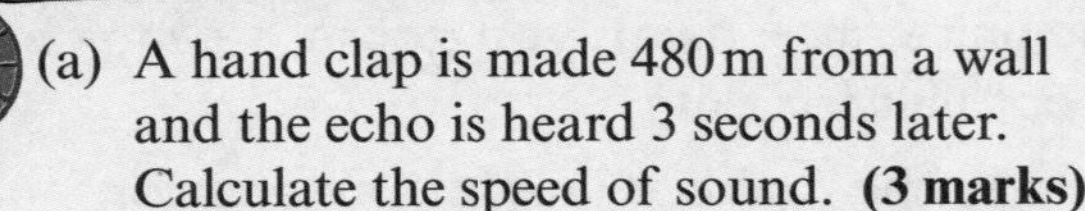

(a) A hand clap is made 480 m from a wall and the echo is heard 3 seconds later. Calculate the speed of sound. **(3 marks)**

$\text{speed} = \frac{\text{distance}}{\text{time}} = \frac{960\,\text{m}}{3\,\text{s}} = 320\,\text{m/s}$

(b) Method 2 is used to determine the speed of sound. The distance between the two microphones is 35 cm and it represents one wavelength. The frequency of the sound is 1 kHz. Calculate the speed of sound in air. **(3 marks)**

wave speed = frequency × wavelength
$= 1000\,\text{Hz} \times 0.35\,\text{m} = 350\,\text{m/s}$

Now try this

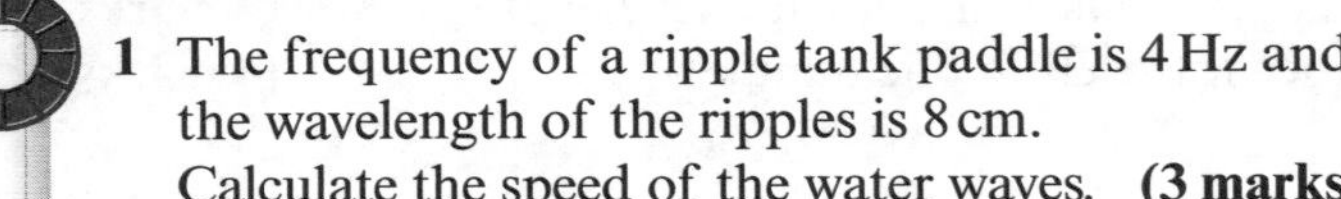

1 The frequency of a ripple tank paddle is 4 Hz and the wavelength of the ripples is 8 cm. Calculate the speed of the water waves. **(3 marks)**

2 Explain how you could obtain a value for the speed of sound in air with a small percentage error. **(4 marks)**

Waves and boundaries

Waves can show different effects when they move from one material to another. These changes can occur at the **boundary** or **interface** between the two materials.

Waves and boundaries

Whenever a sound wave, light wave or water wave reaches the boundary between two materials, the wave can be:

- **reflected**
- **transmitted** or
- **refracted**
- **absorbed.**

Different substances reflect, refract, transmit or absorb waves in ways that vary with the wavelength.

Different wavelengths of radiation are absorbed by molecules in the atmosphere by different amounts.

Wavelength, absorption and transmission

Water molecules **absorb** microwave and infrared wavelength but they **transmit** radio waves which have longer wavelengths.

Infrared radiation is **transmitted** by a black plastic bag, but visible rays are not, as visible rays have a shorter wavelength – they will be **absorbed**.

Shorter-wavelength **X-rays** and **gamma-rays** from outer space are **absorbed** by molecules in the atmosphere, so they cannot cause us harm.

Refraction

Sound waves, water waves and light waves can all be refracted. Refraction can result in a change of both speed and direction. The direction does not change if the wavefronts travel perpendicular to the normal.

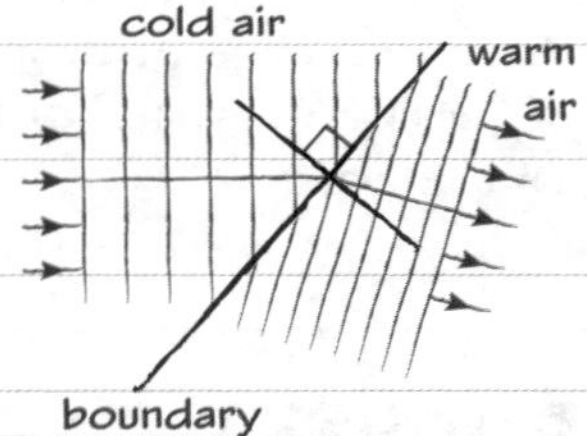

Sound waves travel more slowly in cooler, denser air than in warmer, less dense air.

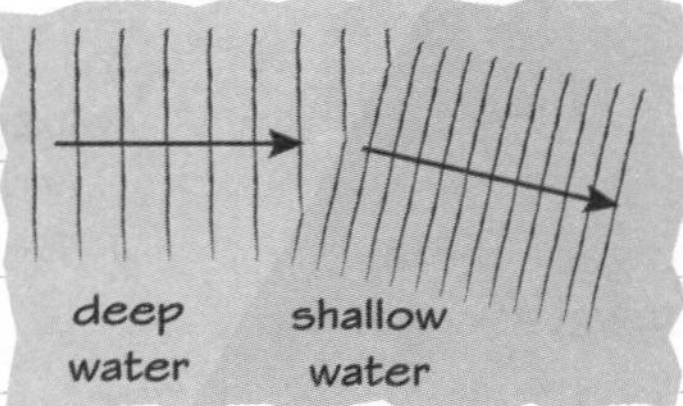

Water waves travel faster in deep water than in shallow water. They can also change direction.

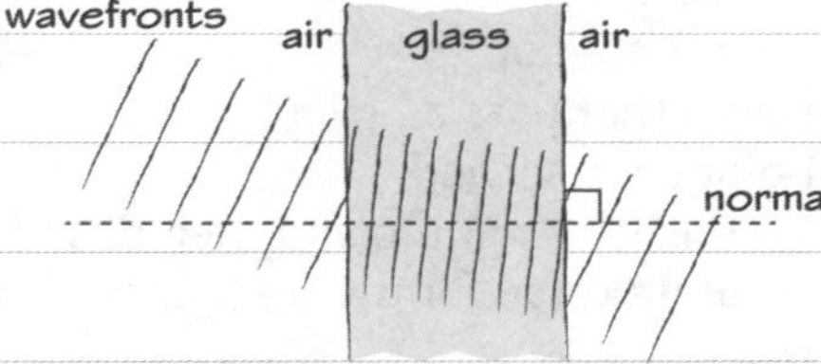

Light waves can slow down and change direction when they pass from air to glass.

Worked example

A person blows a dog whistle. The dog is 200 m away. Explain why the dog will hear the whistle sooner on a warm day than on a cool day. **(3 marks)**

Air temperature affects the speed of sound. Air molecules at a higher temperature have more kinetic energy than air molecules at a lower temperature, so they vibrate faster. This means sound waves travel through warmer air more quickly than through cooler air.

Refraction special case

When light, sound or water waves move from one material into another their **direction does not change** if they are moving along the normal.

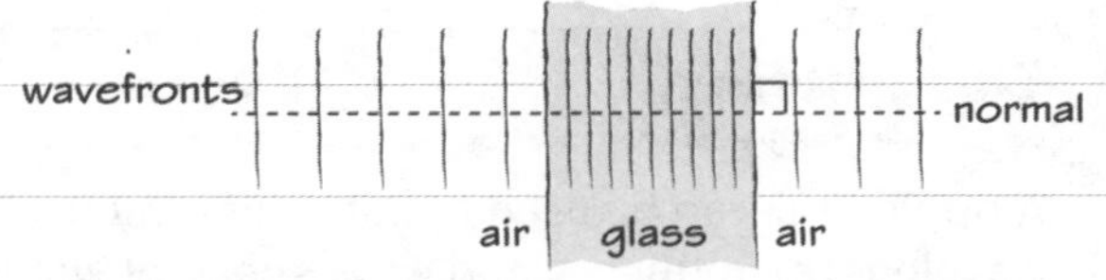

Now try this

1 State **four** things that can happen to a wave at a boundary between two materials. **(2 marks)**

2 Explain the changes that occur to a wave during refraction. **(3 marks)**

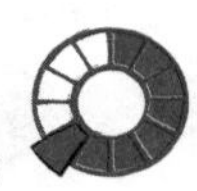

3 Explain why a pencil does not appear to look straight when it is placed vertically into a glass of water. **(3 marks)**

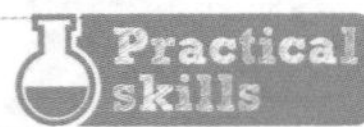

Waves in fluids

You can **determine the wave speed, frequency and wavelength** of waves by using appropriate apparatus.

Core practical

Investigating waves

Aim

To investigate the suitability of apparatus to measure the speed, frequency and wavelength of waves in a fluid.

Apparatus

ripple tank, motor, plane wave generator, stroboscope, ruler, A3 paper and pencil

Method

1 Set up the apparatus as shown.

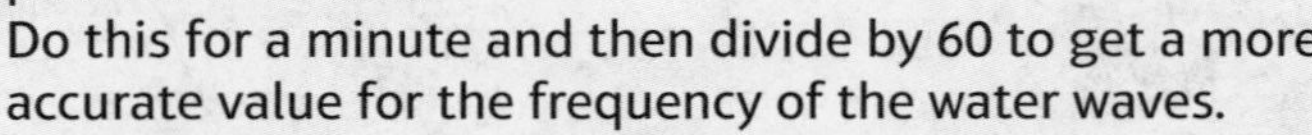

2 Calculate the frequency of the waves by counting the number of waves that pass a point each second. Do this for a minute and then divide by 60 to get a more accurate value for the frequency of the water waves.

3 Use a stroboscope to 'freeze' the waves and find their wavelength by using a ruler. The ruler can be left in the tank or the waves can be projected onto a piece of A3 paper under the tank and the wave positions marked with pencil marks on the paper.

4 Calculate the wave speed.

Results

Wavelength (m)	Frequency (Hz)	Wave speed (m/s)
0.05	10.0	5.0
0.10	5.0	5.0
0.15	3.0	4.5
0.20	2.5	5.0

Conclusion

A ripple tank can be used to determine values for the wavelength, frequency and wave speed of water waves. It is a suitable method, provided that small wavelengths and frequencies are used.

See pages 188–190 for the properties of waves, including the relationship between wave speed, frequency and wavelength.

Water and electricity are being used here, both of which can be dangerous. Be careful to take this into consideration when planning your practical.

To obtain accurate results for the wavelength of the water wave, it is best to find the distance between a large number of waves and then divide this value by the number of waves. This will reduce the percentage error in your value when determining a value for the wavelength of the wave.

Waves in water

Water waves will travel at a constant speed in a ripple tank when generated at different frequencies if the depth of the water is constant at all points. This means that the equation wave speed = frequency × wavelength will give the same wave speed – if the frequency increases, then the wavelength will decrease in proportion.

Now try this

1 Describe the main errors in determining the speed of water waves in this investigation. **(3 marks)**

2 Explain how the investigation described here can be improved. **(4 marks)**

Extended response – Waves

There will be at least one 6-mark question on your exam paper. For these questions, you will need to think scientifically and structure your answer logically, showing how the points you make are related to each other.

You can revise the topics for this question, which is about **waves**, on pages 188–192.

Worked example

Waves transfer energy without there being any overall transfer of matter. These waves can be modelled using a slinky spring, as shown in Figure 1 and Figure 2.

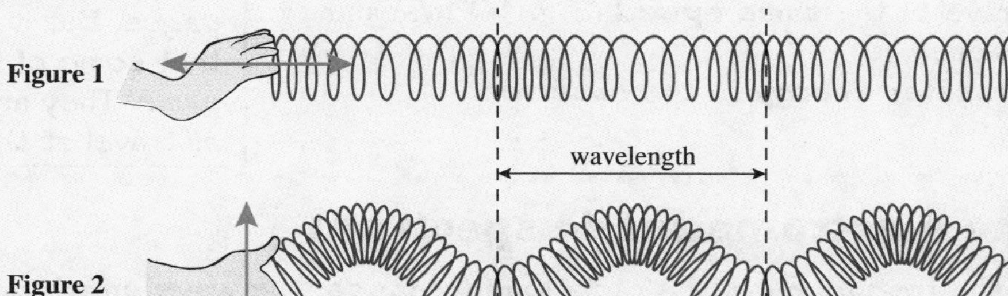

Explain how the wave motion that occurs is related to the disturbance that is causing it and the material that it is moving in.

Your answer should refer to real-life examples of both types of waves. **(6 marks)**

Figure 1 shows a longitudinal wave. A longitudinal wave is produced when the vibration is parallel to the direction of energy transfer.

Figure 2 shows a transverse wave. A transverse wave is produced when the vibration or oscillation is at right angles to the direction of energy transfer.

Sound waves are examples of longitudinal waves, as are seismic P waves. These waves can only transfer energy if there is a medium (a solid, liquid or gas) for them to travel through. Some transverse waves also travel through matter, with examples including water waves and seismic S waves.

Unlike longitudinal waves, some transverse waves can travel through a vacuum and do not need a medium. This is the case for all electromagnetic waves.

The wave in Figure 1 has correctly been identified as a longitudinal wave. A simple description of how it is produced is also given.

The wave in Figure 2 has correctly been identified as a transverse wave. There is also a simple description of how it is produced.

Command word: Explain

When you are asked to **explain** something, it is not enough to just state or describe something. Your answer **must** contain some reasoning or justification of the points you make.

After identifying the waves correctly, examples are given of these waves and the fact that longitudinal waves require a medium. Transverse waves can also do this, but one distinct difference is that transverse electromagnetic waves can travel through a vacuum, for example space.

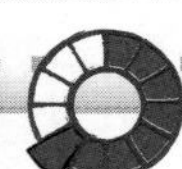

All electromagnetic waves are transverse and all sound waves are longi tudinal in nature. In terms of seismic waves, think of the letter 's' – a **s**ei**s**mic **S**-wave is tran**s**ver**s**e. These **S** waves will only travel through **s**olids. Unlike P waves, they will **not** travel through liquids such as the Earth's liquid outer core.

Now try this

Longitudinal and transverse waves can both be produced in water. Explain how each type of wave is produced and how the speed of each wave can be determined. **(6 marks)**

Had a look ☐ Nearly there ☐ Nailed it! ☐

Electromagnetic spectrum

Infrared radiation, **visible light** and **ultraviolet radiation** are all part of the **electromagnetic spectrum**.

All electromagnetic waves...

- are **transverse waves** (the electromagnetic vibrations are at right angles to the direction the wave is travelling – see page 188)
- travel at the **same speed** (3×10^8 m/s) in a **vacuum**
- **transfer energy** to the observer.

Watch out!

The different parts of the electromagnetic spectrum have different properties, which you will read about on the following pages. But it is important to remember that some of their properties are *the same*. They are *all* transverse waves, and *all* travel at the same speed in a vacuum.

The electromagnetic spectrum

As the **frequency** of the radiation **increases**, the **wavelength decreases**.

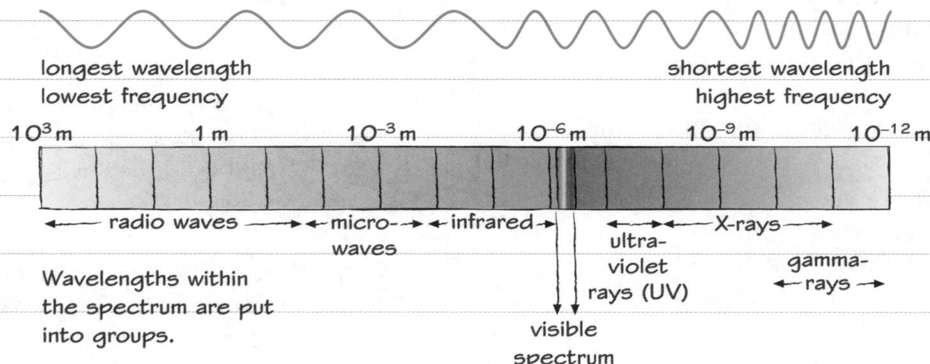

Worked example

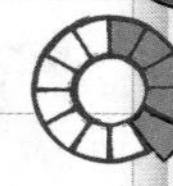

(a) List the main groups of electromagnetic waves from longest to shortest wavelength. **(1 mark)**

radio, microwave, infrared, visible, ultraviolet, X-rays, gamma

(b) Calculate the frequency of an electromagnetic wave travelling in air that has a wavelength of 5×10^{-7} m. **(4 marks)**

$$\text{frequency} = \frac{\text{wave speed}}{\text{wavelength}}$$

$$\text{frequency} = 3 \times \frac{10^8}{5} \times 10^{-7} = 6 \times 10^{14}\,\text{Hz}$$

You can use a mnemonic to help you to remember the order. A mnemonic is a sentence or phrase whose words have the same initial letters as the list you are trying to remember. For example: **R**ed **M**onkeys **I**n **V**ans **U**se **X**-ray **G**lasses.

You can find the wave equation on page 189.

Now try this

1 State the type of electromagnetic radiation that has the (a) longest wavelength, (b) highest frequency. **(2 marks)**

2 The human eye detects wavelengths that range from 4×10^{-7} m to 7×10^{-7} m. Work out the range of frequencies that the eye detects. **(3 marks)**

3 An electromagnetic source produces 4×10^{18} waves in 1 minute.
(a) Calculate the wavelength of the source. **(3 marks)**
(b) State what part of the electromagnetic spectrum this radiation belongs to. **(1 mark)**

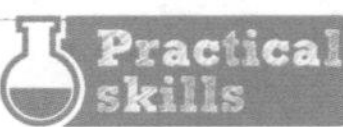

Investigating refraction

You can investigate refraction in rectangular glass blocks in terms of the interaction of light waves with matter.

Core practical

Aim

To investigate the nature of how light waves change direction when they move from air into glass.

Apparatus

ray box and slits, 12 V power supply, glass block, protractor, A3 paper, sharp pencil

Method

1 Place the rectangular block on the A3 paper and draw around it with the sharp pencil.
2 Draw the normal line, which will be at right angles to the side of the block towards which the light ray will be shone.
3 Using the protractor and pencil, mark on the paper angles of incidence of 0° through to 80° in 10° intervals.
4 Starting with the 0° angle (the light ray travelling along the normal line), direct the light ray towards the block and mark its exit point from the block with a sharp pencil dot.
5 Remove the glass block and join the dot to the point of incidence by drawing a straight line. Measure and record this angle, which is the angle of refraction.
6 Repeat for all of the other angles from 10° to 80°.

Results

Angle of incidence	Angle of refraction
0°	0°
10°	7°
20°	13°
30°	19°
40°	25°
50°	31°
60°	35°

Conclusion

When a light ray travels from air into a glass block, its direction changes and the angle of refraction will be less than the angle of incidence unless it is travelling along the normal.

There is more information about the refraction of light on page 191.

When carrying out the investigation, make sure that you direct a thin beam of light towards the point at which the normal makes contact with the glass block. Clearly mark the angles from 0° to 80° with a sharp pencil and ruler.

Refraction

The refraction of a light ray involves a change in:

- the direction of the light ray
- the speed of the light.

Light slows down when it moves from air into glass and speeds up when it moves from glass into air.

The only time when the direction does not change is when the beam is travelling along the normal.

Remember that the angle of incidence is measured with respect to the normal line.

Light will slow down more when it travels from air into glass than it will when it travels from air into water. This is because the 'optical density' of glass is greater than that of air.

Now try this

1 State the **two** things that can change when a light ray is refracted. **(2 marks)**

2 Draw a table of results for a light ray travelling from air into a block of water instead of a block of glass. **(2 marks)**

3 Explain how the (a) frequency and (b) wavelength of a light ray are affected when it enters glass from air. **(4 marks)**

Wave behaviour

The behaviour of **electromagnetic waves (EM)** in different materials depends on their **wavelength** and **velocity**.

Electromagnetic waves

All electromagnetic waves travel at the **same speed in a vacuum**. This is the **speed of light**, 3×10^8 m/s. Our eyes detect the part of the electromagnetic spectrum that is visible light, but there are other wavelengths that behave differently.

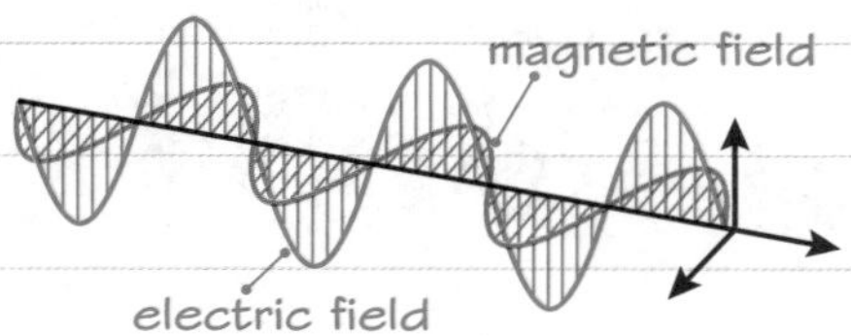

Electromagnetic waves are transverse. They are composed of an electric field and a magnetic field at right angles to one another.

Behaviour of EM waves

Electromagnetic waves may be:

1. reflected off a surface
2. refracted when they move from one material to another
3. transmitted when they pass through a material
4. absorbed by different materials. For example, UV is absorbed by the skin but not by the Earth's atmosphere.

The extent to which these four things happen depends on the material and the wavelength of the EM waves.

Radio waves and microwaves

Radio waves and microwaves are both used for communicating. They have different wavelengths and behave differently in different materials.

X-rays and gamma-rays cannot reach the surface of the Earth as they are absorbed by the upper atmosphere.

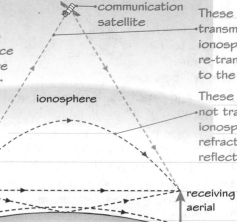

communication satellite

These microwaves are transmitted by the ionosphere and then re-transmitted back to the receiver.

ionosphere

These radio waves are not transmitted by the ionosphere; they are refracted and then reflected to the receiver.

transmitting aerial

receiving aerial

Earth

Worked example

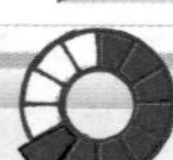

Explain why it is not possible to get a suntan if you are inside a house, but it is possible to listen to the radio. **(4 marks)**

The ultraviolet rays that cause suntans are not transmitted by the walls of houses, they are absorbed by them. Radio waves are transmitted by the walls of houses so can be detected by a radio receiver inside the house.

Wave velocities

Electromagnetic waves have different velocities in different materials. This is linked, to an extent, to the density of the material. The speed of light in a diamond is around 40% of the speed of light in a vacuum or air, due to the high optical density of diamond.

Now try this

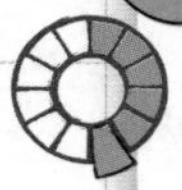

1 State **four** things that can happen to electromagnetic waves. **(2 marks)**

2 Explain why we do not need a communication satellite to relay some radio waves. **(2 marks)**

Dangers and uses

The amount of **energy** that is transferred by an electromagnetic wave is dependent on its **wavelength** or **frequency**. The **highest frequencies** (shortest wavelengths) are the most energetic and the **most dangerous** waves.

Dangers	Type (high frequency → low frequency)	Uses
X-rays and gamma-rays can cause **mutations** (changes) to the **DNA** in cells in the body. This may kill the cells or cause **cancer**.	gamma-rays	• to sterilise food and medical equipment • in scanners to detect cancer • to treat cancer
(see above)	X-rays	• to look inside objects, including medical X-rays to look inside bodies • in airport security scanners, to see what people have in their luggage
UV in sunlight can damage skin cells, causing **sunburn**. Over time, exposure to UV can cause **skin cancer**. UV can also damage the **eyes** leading to eye conditions.	ultraviolet	• to detect security marks made using special pens • inside fluorescent lamps • to detect forged banknotes (real banknotes have markings that glow in UV light) • to disinfect water
	visible light	• allows us to see, lights up rooms and streets, buildings and roads (illumination) • photography
Infrared radiation transfers thermal energy. Too much infrared radiation can cause **skin** burns.	infrared	• in cooking (by grills and toasters) • to make thermal images (images using heat), used by police and rescue services • in short-range communications, such as between laptops or other small computers • in remote controls for TVs and other appliances, where the signal only has to travel short distances • to send information along optical fibres • in security systems such as burglar alarms, to detect people moving around
Microwaves heat water – so they can heat the water inside our bodies. Heating cells can damage or kill them.	microwaves	• in mobile phones, and to communicate by satellite transmissions • for cooking (in microwave ovens)
	radio waves	• broadcasting radio and TV programmes • communicating with ships, aeroplanes and satellites

Worked example

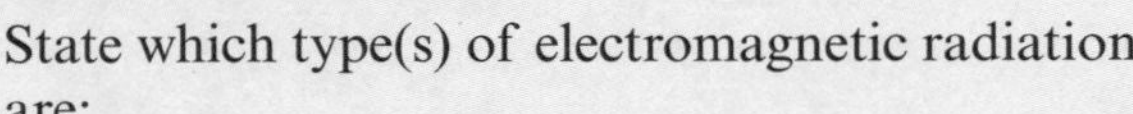

State which type(s) of electromagnetic radiation are:

(a) used for communicating **(4 marks)**

radio waves, microwaves, infrared and visible light

(b) used for cooking **(2 marks)**

microwave and infrared

(c) used for cleaning things **(2 marks)**

ultraviolet and gamma

(d) most dangerous. **(3 marks)**

ultraviolet, X-rays and gamma

Radio waves can be produced by, or can themselves induce, oscillations in electrical circuits.

The different types of electromagnetic radiation share uses and dangers. For example, more than one type can be used for communicating and cooking, and there are three main types of electromagnetic radiation that can cause humans serious harm.

Now try this

1 Explain how X-rays are useful to doctors but can also be harmful to patients. **(3 marks)**

2 Explain how gamma-rays can be used to detect cancer even though they can also cause it. **(2 marks)**

3 Compare the use of microwaves and infrared radiation to cook food. **(5 marks)**

Changes and radiation

Radiation is **absorbed** or **emitted** when **electrons** jump between **energy levels**.

Energy levels

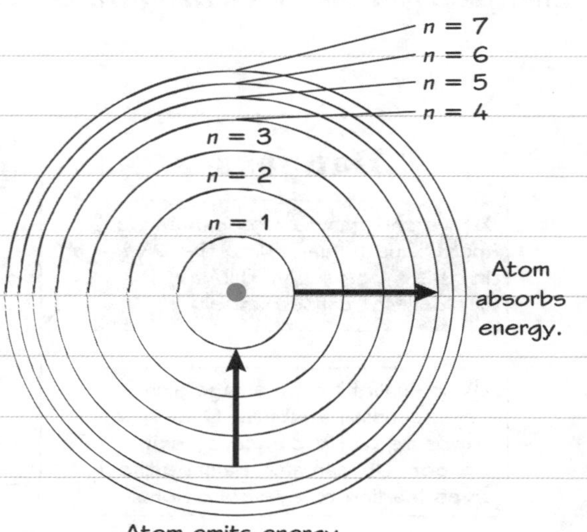

Electrons can only exist in atoms at certain, well-defined energy levels. These **energy levels** depend on the atom, and the electrons inside the atom can move between the shells or leave the atom completely.

Electromagnetic radiation is **emitted** or **absorbed** by atoms based on whether energy is given out or taken in.

Electrons move up energy levels when they absorb energy and they fall down to lower energy levels when they emit energy.

Electrons move from a lower energy level to a higher energy level when the correct amount of energy is absorbed.

Electromagnetic radiation is emitted when electrons fall down from a higher to a lower energy level.

The electromagnetic radiation can have a **wide frequency range**, from **infrared** through to **ultraviolet** and beyond.

Nuclear radiation

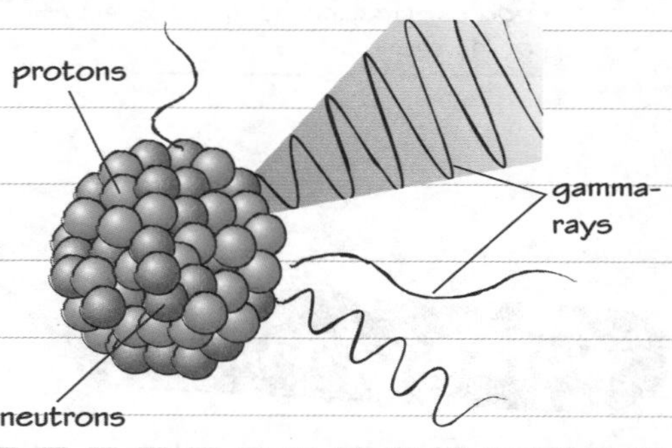

Energy is also emitted from the **nuclei** of **unstable** atoms. **Protons** and **neutrons** also occupy **energy levels** in the nucleus, in the same way that electrons do in the atom.

When energy changes occur in the nucleus, high-energy **gamma-rays** are emitted.

Gamma-rays can be emitted over a large range of frequencies, depending on the energy levels within the nucleus.

Worked example

Explain the energy changes that occur within the electron levels of the atom. **(3 marks)**

Electrons can move between energy levels if they absorb or emit electromagnetic radiation. When they absorb electromagnetic radiation they move up energy levels and when they emit electromagnetic radiation they move down energy levels.

The energy of electromagnetic radiation that is emitted or absorbed has to **match** the difference between the **energy levels** that the **electron** is moving between.

Energy changes in the **nucleus** are far greater than those seen between electron energy levels.

Now try this

1 Describe what happens when electrons:
(a) absorb electromagnetic radiation **(2 marks)**
(b) emit electromagnetic radiation. **(2 marks)**

2 Explain why visible light is emitted when electrons move between shells but gamma-rays are emitted when there are changes in the nucleus. **(4 marks)**

Extended response – Light and the electromagnetic spectrum

There will be at least one 6-mark question on your exam paper. For these questions, you will need to think scientifically and structure your answer logically, showing how the points you make are related to each other. You can revise the topics for this question, which is about **the electromagnetic spectrum**, on pages 194–197.

Worked example

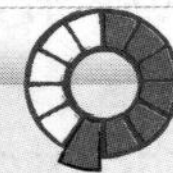

Exposure to certain types of electromagnetic radiation can have harmful effects on the human body.

Describe how exposure to electromagnetic radiation can be harmful.

Your answer should refer to the types of electromagnetic radiation and the damage that they may cause. **(6 marks)**

long wavelength, low frequency

short wavelength, high frequency

radio waves, microwaves, infrared, visible light, ultraviolet, X-rays, gamma-rays

Apart from radio waves and visible light, which are not deemed to cause damage to humans, the other five members of the electromagnetic spectrum do cause harmful effects.

Microwaves are absorbed by water molecules and can lead to internal heating of body cells, which can be harmful. Infrared radiation can cause burns to the skin.

Ultraviolet radiation can damage the eyes, leading to eye conditions. It can also cause skin cancer as surface cells are affected. X-rays and gamma-rays can both lead to cellular damage and mutations, which can lead to diseases such as cancer.

The initial sentence states which types of electromagnetic wave are dangerous and which are not. This will help to structure the rest of the answer.

Two of the low-energy members of the electromagnetic spectrum are commented on next.

The three higher-energy electromagnetic waves are then commented on. The energy of electromagnetic radiation is related directly to its frequency, so as frequency increases, so does the energy and so does the extent of the damage.

UV, X-rays and gamma-rays are examples of **ionising radiation**. This can lead to tissue damage and possible mutations in cells. Although not mentioned here, you could be asked a question that relates specifically to the dangers of ionising radiation. Of the three types mentioned here, only gamma rays come from the nucleus of the atom. You can read more about this on pages 200–211.

Command word: Describe

When you are asked to **describe** something, you need to write down facts, events or processes accurately.

Now try this

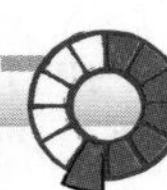

Describe how the electromagnetic spectrum is useful to humans. Refer to specific members of the EM spectrum in your answer. **(6 marks)**

Had a look ☐ Nearly there ☐ Nailed it! ☐

Structure of the atom

Th[illegible] very big difference in the sizes of **atoms, nuclei** and **molecules.**

Structure of the atom

Atoms have a **nucleus** containing **protons** and **neutrons. Electrons** move around the nucleus of an atom. An atom has the same number of protons and electrons, so the **+** and **– charges** balance and the atom has no overall charge.

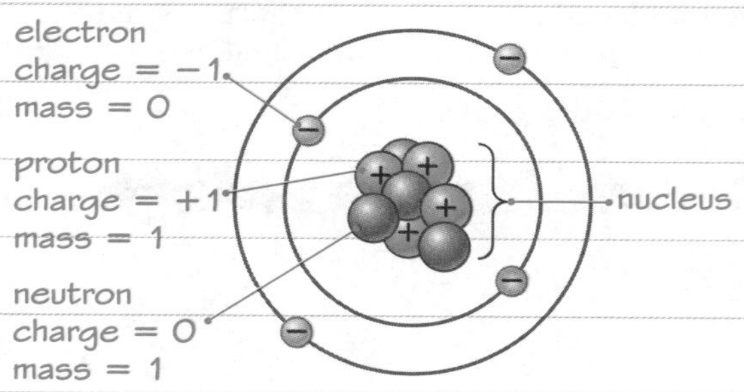

The atom and the nucleus

1. All atoms have a **nucleus.** The **nucleus** is always **positively charged** as it contains **protons** which have a positive charge and neutrons which do not have a charge.
2. The **nucleus** contains more than **99%** of the **mass** of the atom.
3. The total number of **protons** in an atom's **nucleus** must be the same as the total number of **electrons** in the shells.
4. Electrons in atoms always orbit the **nucleus** and have a **negative charge.**
5. Atoms are always **neutral** because the positive charge from the **protons** cancels out the negative charge from the **electrons.**
6. The **nucleus** of an atom of an element may contain **different numbers of neutrons.**

Atom, nucleus and molecule

A **molecule** is two or more atoms bonded together. A **molecule** is about 10 times the diameter of an atom and about 10^6 times the diameter of a nucleus.

10^{-9} m

molecule

10^{-10} m

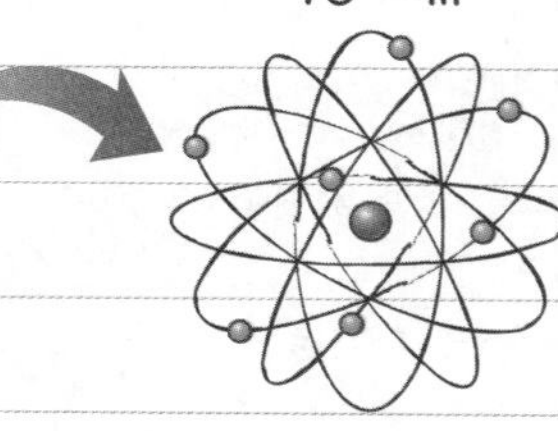

atom

10^{-15}–10^{-14} m

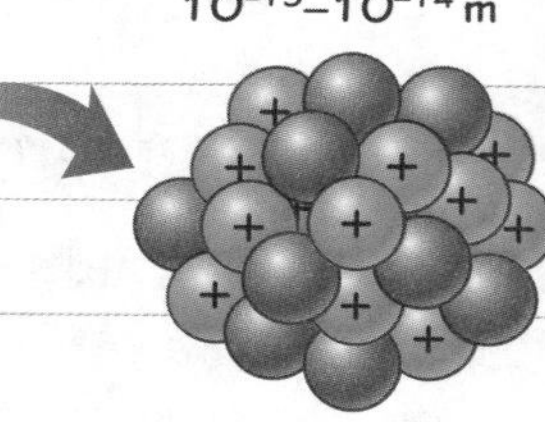

nucleus

Gases such as oxygen and carbon dioxide are **molecules.** Water is also a **molecule.**

The diameter of an atom is about 10^{-10} m. The diameter of the nucleus is about 10^{-15} m.

Diagrams of atoms are always incorrect in terms of the scale. In reality, if the nucleus were the size shown here, the nearest electron would be at least 1000 m away.

Worked example

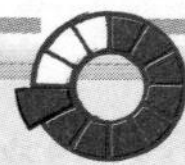

A molecule has a length of 4.8×10^{-9} m. A nucleus has a diameter of 1.2×10^{-15} m. How many nuclei would need to be placed side by side in a line to have the same length as the molecule? **(3 marks)**

$$\frac{4.8 \times 10^{-9}\,\text{m}}{1.2 \times 10^{-15}\,\text{m}} = 4 \times 10^6 \text{ or 4 million nuclei.}$$

Maths skills When dividing values in standard form, remember that you need to:

- divide the numbers at the front
- subtract the powers.

i.e. $\frac{A \times 10^x}{B \times 10^y} = \left(\frac{A}{B}\right) \times 10^{(x-y)}$.

Remember, −9 − (−15) is equal to 6 because two minuses make a plus, so −9 + 15 = 6.

Now try this

A poster shows a diagram of an atom that is not drawn to scale. The overall diameter of the atom on the poster is 60 cm. Calculate the diameter of the nucleus on the poster if it was properly drawn to scale. **(4 marks)**

Atoms and isotopes

Atoms of the same **element** can be **different** based on the composition of their **nuclei**.

Atoms

Atoms are made up of **protons**, **neutrons** and **electrons**. The **protons** and **neutrons** are in the nucleus of the atom, and the **electrons** move around the outside.

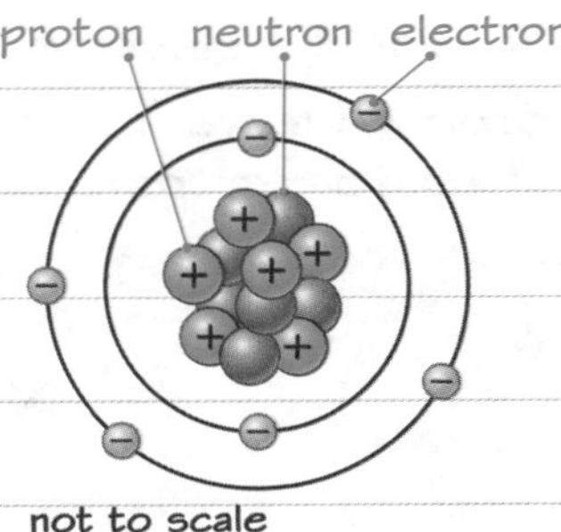

Describing atoms

All the atoms of a particular element have the same number of **protons**. The number of protons in each atom of an element is called the **atomic number**, or **proton number**.

The total number of protons and neutrons in an atom is the **mass number**, or **nucleon number**.

The atomic number and mass number of an element can be shown like this:

mass number → 16

atomic number → 8

$^{16}_{8}O$

Isotopes

Atoms of a particular element always have the **same number of protons**, but they can have **different numbers of neutrons**. Atoms with the **same number of protons** but **different numbers of neutrons** are **isotopes** of the same element.

As the number of protons determines the element, you can also say that the nuclei of different elements have their own characteristic charge. In this example, a carbon nucleus always has a charge of +6.

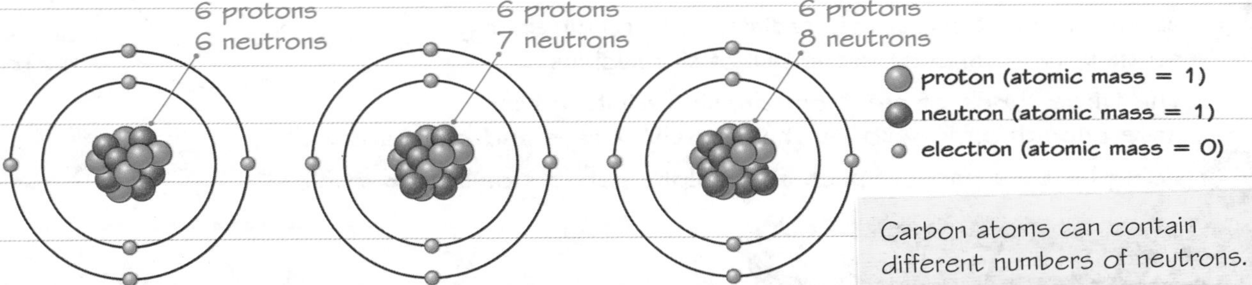

Carbon atoms can contain different numbers of neutrons.

Worked example

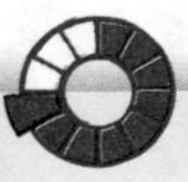

The symbol represents an atom of sodium.

$^{23}_{11}Na$

Explain what these numbers tell you about the structure of the sodium nucleus. **(2 marks)**

The atomic number shows that sodium contains 11 protons. The mass number shows that it contains (23 − 11) = 12 neutrons.

Worked example

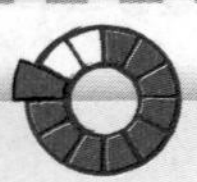

Explain how these atoms are similar and different:

$^{5}_{3}Li$, $^{7}_{3}Li$, $^{6}_{3}Li$ **(3 marks)**

They all contain the same number of protons and the same number of electrons (three of each). They have different numbers of neutrons, so they are different isotopes of lithium.

Now try this

1 An atom of boron (B) has 5 protons and 6 neutrons. Show the atomic number and mass number as a symbol for the isotope. **(1 mark)**

2 Explain how these atoms are similar and different: $^{14}_{7}N$ $^{15}_{7}N$. **(3 marks)**

Had a look ☐ Nearly there ☐ Nailed it! ☐

Atoms, electrons and ions

Electrons can **move** between energy levels, or they can **leave** the atom completely.

Electrons in orbit

In all atoms, electrons orbit the nucleus in different orbits or energy levels, at different, fixed distances from the nucleus.

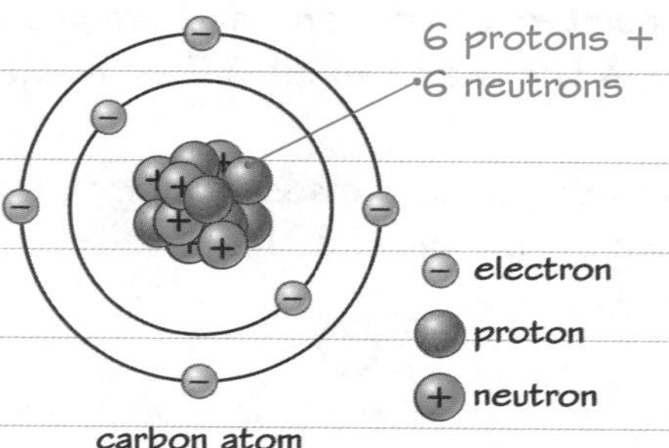

Moving between energy levels

An **electron** will move from a **lower to a higher orbit** if it **absorbs** electromagnetic radiation.

An **electron** will move from a **higher to a lower orbit** if it **emits** electromagnetic radiation.

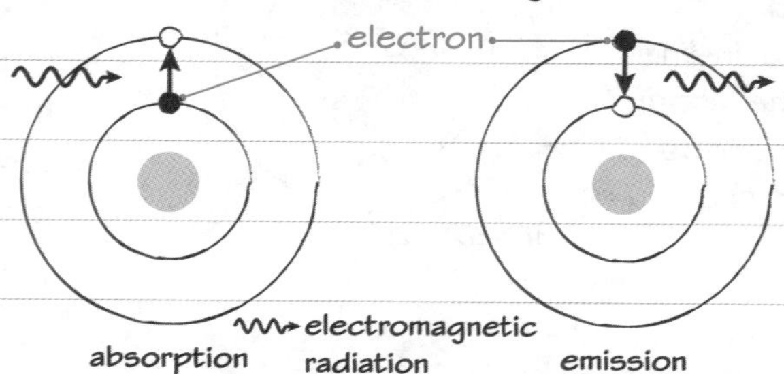

Forming positive ions

Atoms become positively charged particles called **positive ions** when they **lose electrons**. There are now more protons present in the nucleus than there are electrons in the shells.

Electrons can leave an atom by:

1. absorbing electromagnetic radiation of enough energy so that they can escape the pull of the nucleus.

   The three types of electromagnetic radiation that have enough energy to do that are UV, X-rays and gamma-rays.

2. being hit by a particle such as an alpha particle or beta particle.

Worked example

State the overall charge of

(a) an atom **(1 mark)**

neutral or 0

(b) a sodium atom that has lost an electron **(1 mark)**

+1

(c) an atom where an electron has moved from a higher to lower orbit. **(1 mark)**

neutral or 0

All atoms have a neutral overall charge. Positive ions are formed when an atom loses an electron.

Atoms can also gain electrons. They become negative ions with an overall negative charge.

No electron has been lost from the atom so there is no net change in the amount of charge present within the atom.

Now try this

1 State where
(a) protons **(1 mark)**
(b) neutrons **(1 mark)**
and
(c) electrons **(1 mark)**
are found in atoms.

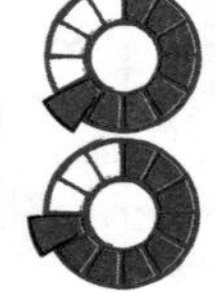

2 Explain how ions are different from atoms. **(4 marks)**

3 Suggest how you can tell, by experiment, if an atom has become a positive or a negative ion. **(2 marks)**

Ionising radiation

Alpha, beta, gamma and **neutron** radiation are **emitted** by **unstable nuclei.** The process is **random**, which means it is not possible to determine exactly **when** any nucleus will decay next.

Alpha, beta, gamma and neutron radiation

Some elements are **radioactive** because their nuclei are **unstable.** This means that they will undergo radioactive decay and change into other elements. Unstable nuclei will decay when alpha, beta, gamma or neutron radiation is emitted.

An **alpha particle** is a helium nucleus. It is composed of two protons and two neutrons. It has a charge of +2 and is the heaviest of the particles emitted by unstable atoms. A **beta particle** is an electron. It has a charge of −1. A **positron** is an anti-electron and has a charge of +1.

A **gamma-ray** is a form of high-energy electromagnetic radiation. It has no mass or charge.

A **neutron** has zero charge.

Properties of radiation

Alpha and beta particles and gamma-rays can collide with atoms, ionising them by causing them to lose electrons.

Neutrons:

- are not directly ionising
- have a very high penetrating power due to them having no charge and not interacting strongly with matter
- can travel through humans and buildings for long distances before being stopped.

1 (α) Alpha particles:
- will travel around 5 cm in air
- very ionising
- can be stopped by a sheet of paper.

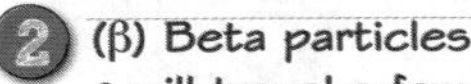

2 (β) Beta particles:
- will travel a few metres in air
- moderately ionising
- can be stopped by aluminium 3 mm thick.

3 (γ) Gamma-rays:
- will travel a few kilometres in air
- weakly ionising
- need thick lead to stop them.

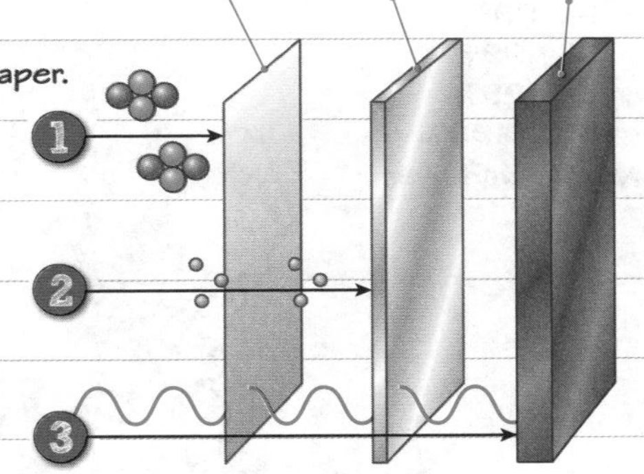

Worked example

Complete the table to show the properties of alpha, beta, gamma and neutron radiation. **(5 marks)**

Type of radiation emitted	Relative charge	Relative mass	Ionising power	Penetrating power	Affected by magnetic fields?
alpha, α (helium nucleus, two protons and two neutrons)	+2	4	heavily ionising	very low, only ~5 cm in air	yes
beta, $\beta-$ (an electron from the nucleus)	−1	$\frac{1}{1840}$	weakly ionising	low, stopped by thin aluminium	yes, but in the opposite direction to α and $\beta+$
positron, $\beta+$ (a particle with same size as electron but an opposite charge)	+1	$\frac{1}{1840}$	weakly ionising	low, stopped by thin aluminium	yes
neutron, n	0	1	not directly ionising	high	no, since neither are charged
gamma, γ (waves)	0	0	not directly ionising	very high, stopped only by thick lead	

Now try this

1 Explain why alpha particles are the most ionising radiation. **(3 marks)**

2 Explain why radioactive decay is a random process. **(2 marks)**

Background radiation

Low levels of radiation are present around you all the time. This radiation is both **natural** and **man-made** and is called **background radiation.**

Background radiation

We are always exposed to ionising radiation. This is called **background radiation.** This radiation comes from different sources, as shown in the pie chart.

Radon is a radioactive gas that is produced when **uranium** in rocks decays. Radon decays by emitting an alpha particle. The radon can build up in houses and other buildings. The amount of radon gas varies from place to place, because it depends on the type of rock in the area.

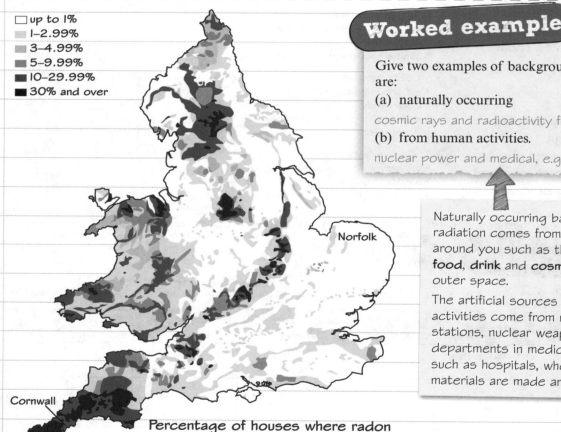

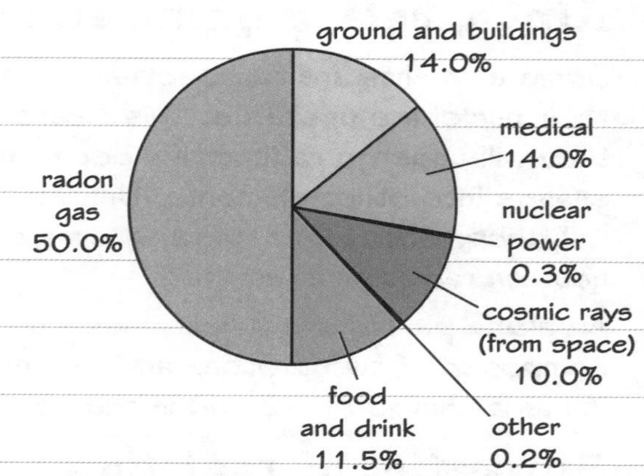

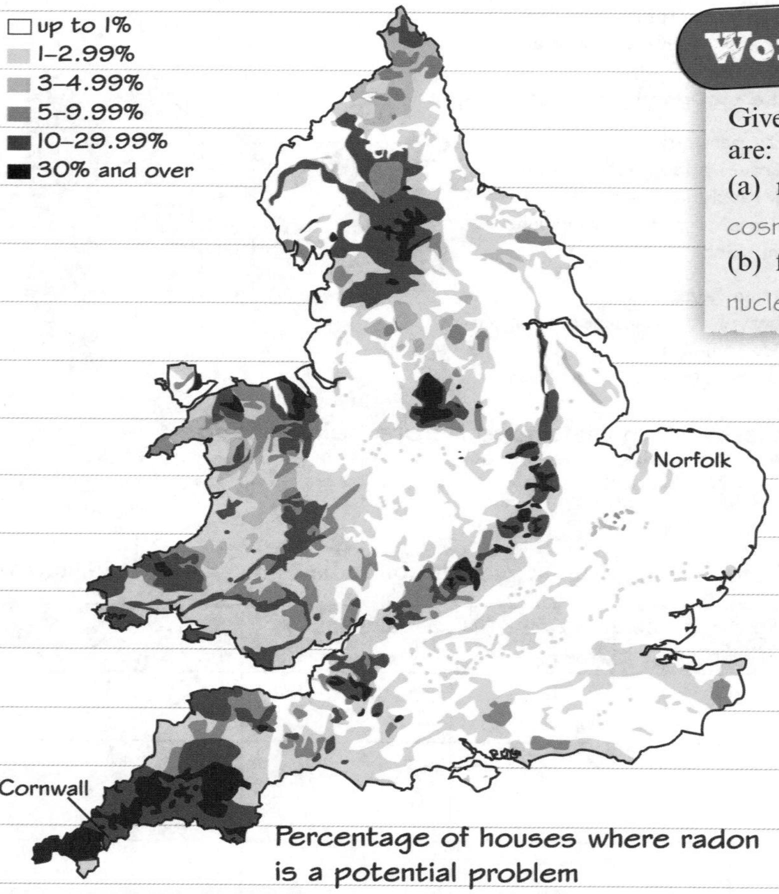

☐ up to 1%
1–2.99%
3–4.99%
5–9.99%
10–29.99%
30% and over

Norfolk

Cornwall

Percentage of houses where radon is a potential problem

Worked example

Give two examples of background radiation that are:

(a) naturally occurring **(2 marks)**

cosmic rays and radioactivity from the ground

(b) from human activities. **(2 marks)**

nuclear power and medical, e.g. hospitals

Naturally occurring background radiation comes from the environment around you such as the **soil**, **rocks**, **food**, **drink** and **cosmic rays** from outer space.

The artificial sources from human activities come from nuclear power stations, nuclear weapons and from departments in medical settings, such as hospitals, where radioactive materials are made and used.

Now try this

1 Which one of these is a source of artificial background radiation?

☐ **A** soil
☐ **B** rock
☐ **C** cosmic rays
☐ **D** medical X-rays **(1 mark)**

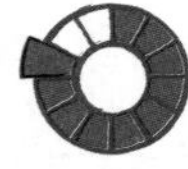

2 Look at the map above. How would the pie chart above change if you lived in
(a) Norfolk
(b) Cornwall? **(2 marks)**

3 Explain why radon is more dangerous inside the body than outside the body. **(3 marks)**

Measuring radioactivity

Photographic film and a **Geiger–Müller tube** can both be used to **measure** and **detect** radiation.

The Geiger–Müller tube

The Geiger–Müller (GM) tube is used to detect nuclear radiation. It is connected to a counter or ratemeter which shows the amount of radiation that has been detected.

The text in red shows what is happening inside the GM tube, but you do not need to remember this in detail.

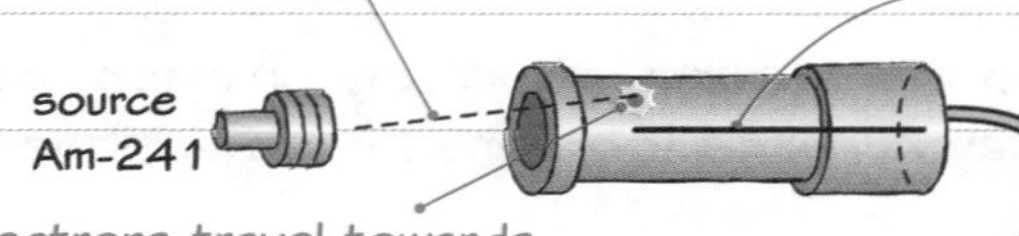

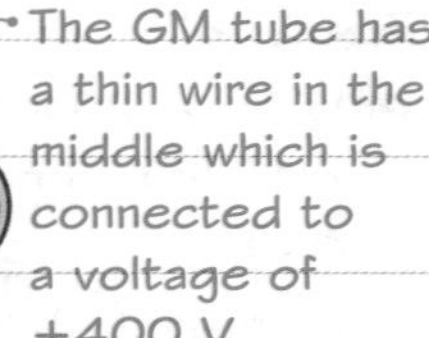

The GM tube contains argon gas. When radiation enters the tube, the atoms of argon are ionised and electrons travel towards the thin wire.
source
Am-241
As the electrons travel towards the thin wire, more electrons are knocked from atoms and an avalanche effect occurs.

The GM tube has a thin wire in the middle which is connected to a voltage of +400 V.

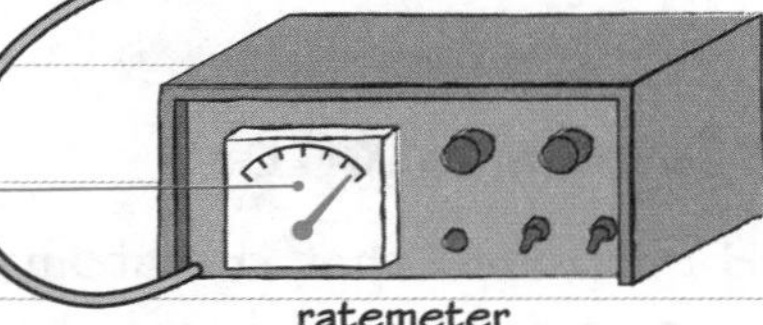

The amount of radiation detected is shown by the ratemeter.

Detecting radiation

In 1896, Henri Becquerel discovered that uranium salts would lead to the darkening of a photographic film, even if it was wrapped up so that no light could reach it.

Radiation was being emitted from the uranium nuclei and this was responsible for the darkening of the film. This is now made use of in the nuclear industry as workers will wear a film badge to determine if they are being exposed to different forms of nuclear radiation.

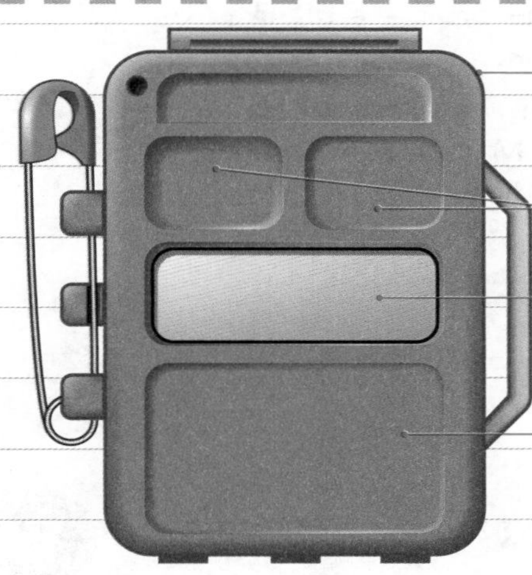

photographic film inside
thin and thick plastic windows or aluminium – stop some beta particles

lead between the plastic case and the film – stops beta and most gamma radiation

This dosimeter is a film badge used to monitor the radiation received by its wearer.

Worked example

(a) Explain why there is an open window in the film badge. **(3 marks)**

Light rays will darken the film. Since the film badge is designed to detect nuclear radiation, this acts as a control so that you know that the film is working.

(b) Explain why the film badge has aluminium and lead sheets inserted in it. **(3 marks)**

Aluminium absorbs beta particles and lead absorbs gamma-rays. If there is darkening behind these sheets, then high-energy beta- or gamma-rays may have got through and so the nature of the radiation can be determined and the risk identified.

You should use your knowledge of the penetrating powers of the different types of radioactivity to answer the question.

Now try this

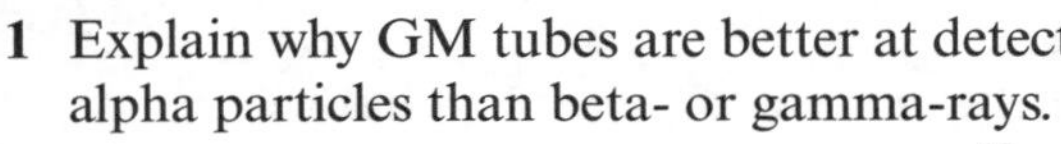

1 Explain why GM tubes are better at detecting alpha particles than beta- or gamma-rays. **(3 marks)**

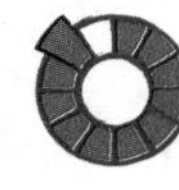

2 Explain how a film from a film badge would tell you if there was a potential risk to humans. **(3 marks)**

Models of the atom

The **model of the atom** has changed over time, based on **evidence** of its structure that became available from **experiments**.

Plum pudding model

J. J. Thomson's model of the atom was that the atom was like a 'plum pudding' with negatively charged 'electron plums' embedded in a uniform, positively charged 'dough' – a bit like the way currants look when in a Christmas pudding.

This model showed that both positive and negative charges existed in atoms and accounted for the atom being neutral.

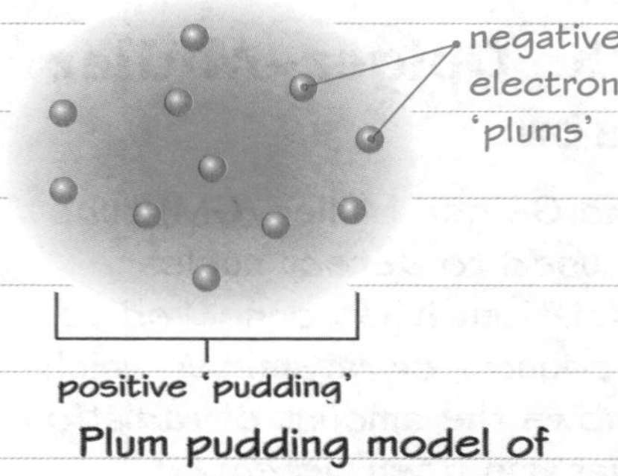

Plum pudding model of the atom

Rutherford's model

Rutherford proposed that the atom must contain a very **small, positively charged nucleus** which electrons orbit – a bit like planets orbiting the Sun.

Rutherford's hypothesis was proved to be correct by Geiger and Marsden, who fired alpha particles at gold film.

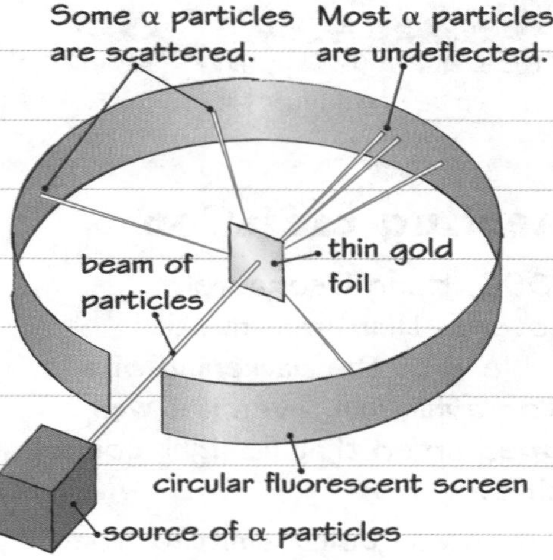

Rutherford scattering

The Bohr model

Niels Bohr showed that **electrons** had to orbit a **positive nucleus** in well-defined **energy levels** or orbits, but could move between energy levels if they gained or lost energy.

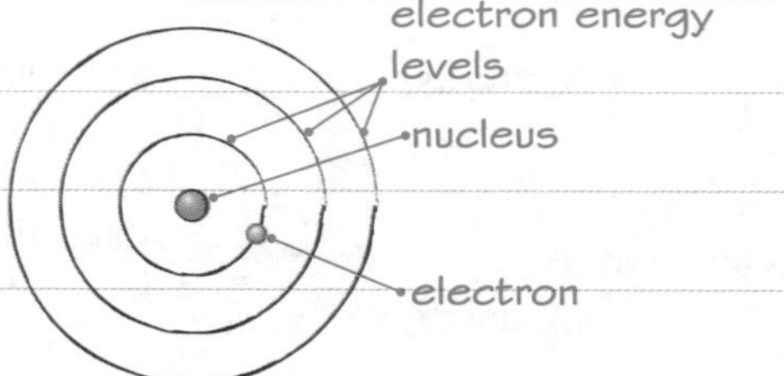

Worked example

Which model of the atom is being described?

(a) The atom is mostly empty space and has a positive nucleus which contains most of the mass of the atom. **(1 mark)**

Rutherford's model

(b) Electrons can only orbit the atom in certain energy levels or orbits. **(1 mark)**

the Bohr model of the atom

(c) The atom is neutral with negative 'plums' embedded in a positive 'dough'. **(1 mark)**

the plum pudding model

Now try this

Refer to page 202 for more information on electron levels.

1 Describe the three models of the atom. **(3 marks)**

2 Explain how Rutherford's work led to the conclusion that the atom had a small, positive nucleus. **(4 marks)**

3 Suggest how the Bohr model allows astronomers to determine which gases are present in the outer surfaces of stars. **(4 marks)**

Beta decay

One way in which unstable nuclei can undergo radioactive decay is by **beta decay.** There are two types – one where an **electron is emitted**, the other where a **positron is emitted**.

Beta-minus (β^-) decay

In β^- decay, a **neutron** in the nucleus of an unstable atom decays to become a **proton** and an **electron**. The proton stays within the nucleus, but the **electron**, which is the β^- **particle**, is emitted from the nucleus at high speed as a fast-moving electron.

$$n \longrightarrow p + e^-$$

The decay of **carbon-14** into **nitrogen-14** by the emission of a β^- particle is an example of β^- decay. The mass number does not change, but the proton number **increases** by 1.

$$^{14}_{6}C \longrightarrow {}^{14}_{7}N + {}^{0}_{-1}e$$

Beta-plus (β^+) decay

In β^+ decay, a **proton** in the nucleus decays to become a **neutron** and a **positron.** The neutron stays in the nucleus. The **positron**, which is the β^+ **particle**, is emitted from the nucleus at a very high speed, carrying away a positive charge and a very small amount of the nuclear mass.

$$p \longrightarrow n + e^+$$

The decay of **sodium-22** into **neon-22** by the emission of a positron is an example of β^+ decay. The mass number does not change, but the proton number **decreases** by 1.

$$^{22}_{11}Na \longrightarrow {}^{22}_{10}Ne + {}^{0}_{+1}\beta$$

Worked example

Complete each equation and state whether it is β^- or β^+ decay:

(a) $^{131}_{53}I \longrightarrow {}^{131}_{54}Xe + ?$ **(2 marks)**

Add $^{0}_{-1}e$. β^- decay

(b) $^{23}_{12}Mg \longrightarrow {}^{23}_{11}Na + ?$ **(2 marks)**

Add $^{0}_{1}e$. β^+ decay

Uses of beta decay

Carbon-14 is an unstable isotope of carbon and emits β^- particles – it has a half-life of over 5700 years. This means that it can be used for radiocarbon dating, which involves finding the ages of materials that are thousands of years old. Positrons can be used in hospitals to form images of patients by the use of PET scans.

Now try this

1 Describe the changes that take place in the nucleus when it undergoes:
(a) β^- decay **(2 marks)**
(b) β^+ decay. **(2 marks)**

2 How is a β^- particle different from an electron that is found in a stable atom? **(2 marks)**

3 Write a balanced equation for nickel-66 which undergoes β^- decay. **(3 marks)**

Had a look ☐ Nearly there ☐ Nailed it! ☐

Radioactive decay

When unstable nuclei decay, the changes that occur depend on the **radiation** that is emitted from the **nucleus**.

Changes to the nucleus

Type of radiation	Effect on the mass of the nucleus	Effect on the charge of the nucleus
alpha α	nuclear mass reduced by 4 [−4]	positive charge reduced by 2 [−2]
beta $\beta-$	no change [0]	positive charge increased by 1 [+1]
beta $\beta+$	no change [0]	positive charge reduced by 1 [−1]
gamma	no effect on either the mass or the charge of a nucleus	
neutron	mass reduced by 1	no change of nuclear charge

Balancing nuclear decay equations

In any nuclear decay, the total mass and charge of the nucleus are conserved – they are the same before and after the decay. So the masses and charges on each side of the equation must balance.

When uranium-238 undergoes alpha decay, mass and charge are conserved. Nuclei that have undergone radioactive decay also undergo a rearrangement of their protons and neutrons. This involves the loss of energy from the nucleus in the form of gamma radiation. The mass and charge of the gamma-ray must be zero for the equation to balance:

$$^{238}_{92}\text{U} \longrightarrow \, ^{234}_{90}\text{Th} + \, ^{4}_{2}\alpha + \gamma$$

uranium thorium alpha particle gamma-ray

When neutron decay occurs, mass and charge are also conserved:

$$^{13}_{4}\text{Be} \longrightarrow \, ^{12}_{4}\text{Be} + \, ^{1}_{0}\text{n}$$

Neutron decay results in another isotope of the same element being formed.

When beryllium decays by neutron emission, the mass and charge are both conserved. The nuclear charge will be +4 on both sides of the decay equation and the total mass will be 13.

Worked example

Balance the nuclear equations for these decays:

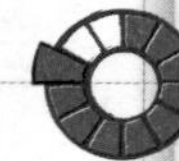

(a) Iodine-121 undergoes β^+ decay to form tellurium. **(2 marks)**

$^{121}_{53}\text{I} \rightarrow \, ^{121}_{52}\text{Te} + \, ^{0}_{+1}\text{e}$

$^{0}_{+1}\text{e}$ is the emitted β^+ particle.

(b) An isotope of carbon undergoes β^- decay to form nitrogen-14. **(2 marks)**

$^{14}_{6}\text{C} \rightarrow \, ^{14}_{7}\text{N} + \, ^{0}_{-1}\text{e}$

$^{0}_{-1}\text{e}$ is the emitted β^- particle.

(c) Radon-220 undergoes α decay to form an isotope of polonium. **(2 marks)**

$^{220}_{86}\text{Rn} \rightarrow \, ^{216}_{84}\text{Po} + \, ^{4}_{2}\text{He}$

$^{4}_{2}\text{He}$ is the emitted α particle.

Now try this

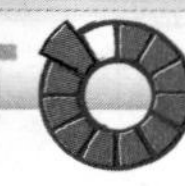

1 A thorium-232 nucleus undergoes α decay to become radium. Write a balanced equation to show this. **(2 marks)**

2 A nucleus decays by α decay, followed by two β^- decays and then neutron decay. If the original nucleus had a mass number of A and proton number of Z, what would the mass number and the proton number be of the resulting nucleus? **(4 marks)**

Half-life

The **activity** of a radioactive source is the number of atoms that **decay** every second. The unit for activity is the **becquerel (Bq)**. When an atom decays it emits radiation but changes into a more stable isotope.

Unstable atoms

The activity of a source depends on how many **unstable** atoms there are in a sample, and on the particular isotope. As more and more atoms in a sample decay, there are fewer unstable ones left, so the activity decreases. The **half-life** of a radioactive isotope is the time it takes for **half** of the **unstable** atoms to **decay**. This is also the time for the activity to go down by half.

We cannot predict when a particular nucleus will decay. But when there is a very large number of nuclei, the half-life gives a good prediction of the proportion of nuclei that will decay in a given time.

Radioactive decay and half-life

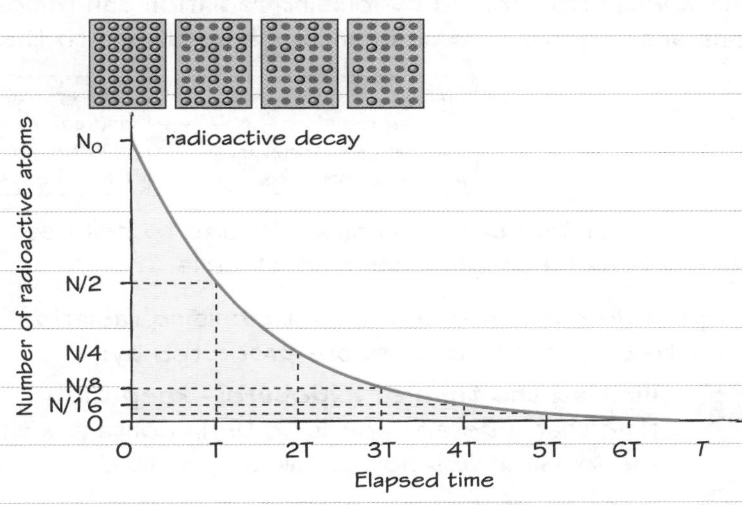

The number of radioactive nuclei present in a sample will halve after each successive half-life. After 1 half-life there will be 50% of the radioactive atoms left; after 2 half-lives there will be 25% of the radioactive atoms left, and so on.

Half-life example

The activity of a radioactive source is 240 Bq and its half-life is 6 hours.

After 1 half-life, the activity will halve to 120 Bq.

After 2 half-lives it will halve again to 60 Bq.

After 3 half-lives it will halve again to 30 Bq.

After 4 half-lives (one day) it will be 15 Bq.

Maths skills

Be careful when multiplying fractions:

After **1 half-life**, $\frac{1}{2}$ of the initial radioactive atoms are left.

After **2 half-lives**, $\frac{1}{2} \times \frac{1}{2}$ or $\frac{1}{4}$ are left.

After **3 half-lives**, $\frac{1}{2} \times \frac{1}{2} \times \frac{1}{2}$ or $\frac{1}{8}$ are left.

Worked example

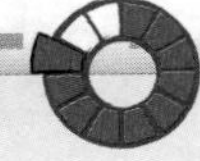

The graph shows how the activity of a sample changes over 24 hours. What is the half-life of the sample? **(3 marks)**

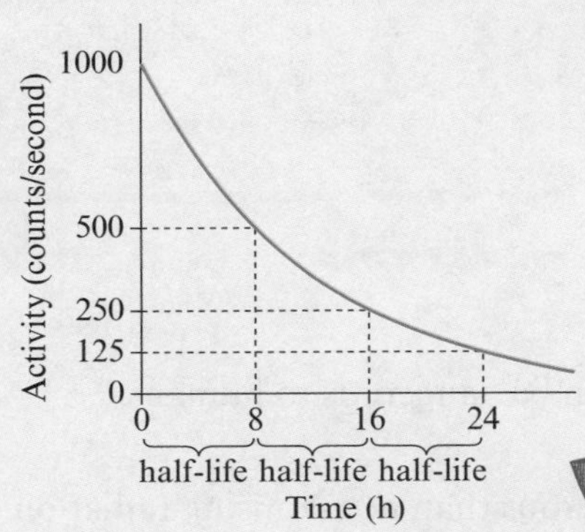

activity at time 0 = 1000 Bq

Half of this is 500 Bq.

The activity is 500 Bq at 8 hours.

The half-life is 8 hours.

Now try this

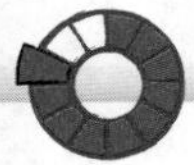

1. The activity of a source is 120 Bq. Four hours later it is found to be 30 Bq. Calculate the half-life of the source. **(2 marks)**
2. The half-life of a source is 15 minutes. Calculate the fraction that will remain after one hour. **(3 marks)**

As the activity of a radioactive source decreases, the gradient of the graph will get less and less steep.

Dangers of radiation

Ionising radiation can knock electrons out of atoms, turning the atoms into **ions**. This can be very **harmful** to humans.

Ionisation and cellular mutation

The energy transferred by ionising radiation can remove electrons from atoms to form **ions**. Ions are very reactive and can cause mutations to the **DNA** in **cells**. This can lead to **cancer**.

Energy transferred by ionising radiation removes electrons from atoms to form ions. → Ions are reactive and can cause mutations to the DNA in cells. → Damaged DNA can lead to cancer.

Ionising radiation can also cause damage to cell tissue in the form of radiation burns. When its energy is high enough it can also kill cells.

People who come into contact with ionising radiation need to be protected. They are protected by:

1. **limiting the time of exposure** – keep the time that a person needs to be in contact with the ionising radiation as low as possible.
2. **wearing protective clothing** – wearing a lead apron will absorb much of the ionising radiation.
3. **increasing the distance between the person and the radioactive source** – the further the person is from the ionising radiation, the less damage it will do.

To determine how much radiation a person has been exposed to, a film badge may be worn. See page 205 for more details.

Precautions

People using radioactive material take precautions to make sure they stay safe.

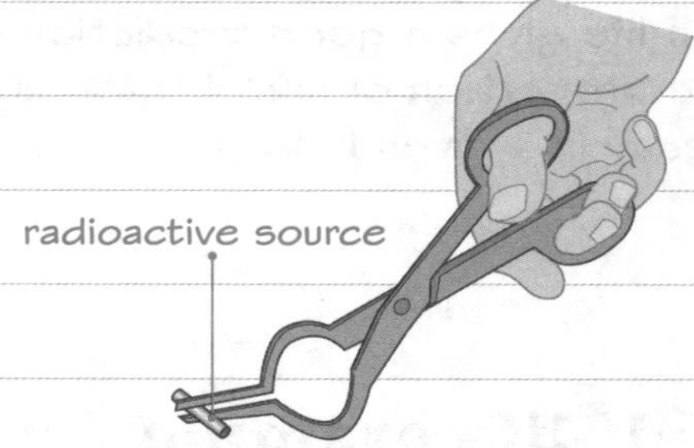

The radioactive source is being moved using tongs to keep it as far away from the person's hand as possible. The source is always kept pointing away from people.

Worked example

Describe two precautions taken by dentists and dental nurses to reduce their exposure to ionising radiation while taking an X-ray. **(2 marks)**

1. They go out of the room in which the X-ray is taking place.
2. They keep the X-ray pulse as short as possible. (This also minimises the patient's exposure.)

The amount of energy that the human body is exposed to from a radioactive source is referred to as the **dose**. The dose needs to be big enough to obtain the X-ray image, but low enough to be safe for the patient and the dentist.

Now try this

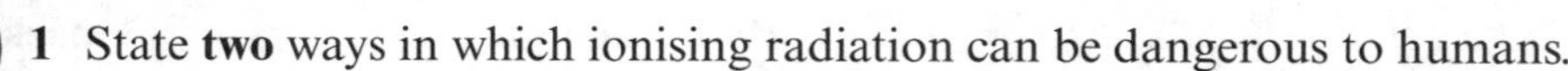

1 State **two** ways in which ionising radiation can be dangerous to humans. **(2 marks)**

2 Explain why ionising radiation is more dangerous than non-ionising radiation. **(3 marks)**

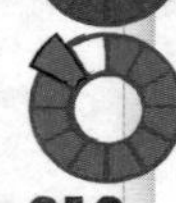

3 Film badges have been superseded in some places by the use of semiconductor devices that measure levels of ionising radiation. Suggest why these devices are preferred to film badges. **(3 marks)**

Contamination and irradiation

You may be exposed to the effects of radioactive materials by being **irradiated** or **contaminated.**

Irradiation

Irradiation is ionising radiation from an external radioactive source travelling to the body – it is not breathed in, eaten or drunk.

Irradiation does not refer to non-harmful rays from televisions, light bulbs or other non-ionising sources.

Irradiation is the exposure of a person to ionising radiation from outside the body. This could be in the form of harmful gamma-rays, beta particles or X-rays.

Alpha particles are unlikely to be harmful outside the body as they have a very short range in air (5 cm) and are unlikely to reach a person.

When ionising radiation reaches the body, cells may be damaged or killed, but you will not become radioactive.

Contamination

External contamination occurs when radioactive materials **come into contact with a person's hair, skin or clothing.**

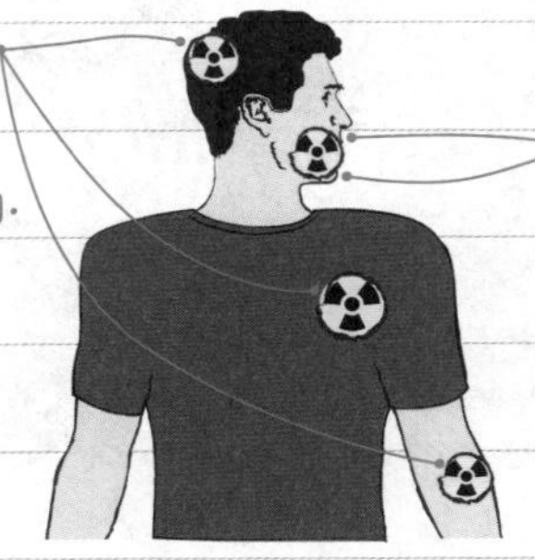

Internal contamination occurs when a **radioactive source is eaten or drunk**. Some nuts, plants, fruits and alcoholic drinks have low levels of radioactivity. This is due to the radioactive minerals that they are exposed to during their growth or manufacture.

Worked example

State what type of contamination the following examples are describing:

(a) eating radioactive strontium-90 that is found in some foods **(1 mark)**

internal contamination

(b) being exposed to cosmic rays from the Sun **(1 mark)**

irradiation

(c) having an X-ray to find if a bone is broken **(1 mark)**

irradiation

(d) radioactive dust comes into contact with the skin **(1 mark)**

external contamination.

Remember that with contamination, the radioactive source comes into contact with the skin or is taken into the body. This can be through the mouth or nose, or through a cut in the skin. With irradiation, the source does not come into contact with the skin.

Now try this

1 Define
   (a) irradiation **(2 marks)**
   (b) contamination. **(2 marks)**

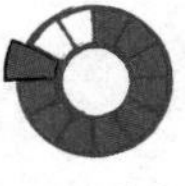

2 Explain why alpha particles are more likely to cause damage inside the body rather than outside the body. **(3 marks)**

3 Explain why background radiation is a mixture of contamination and irradiation. **(2 marks)**

Extended response – Radioactivity

There will be at least one 6-mark question on your exam paper. For these questions, you will need to think scientifically and structure your answer logically, showing how the points you make are related to each other.

You can revise the topics for this question, which is about **radioactivity**, on pages 200–211.

Worked example

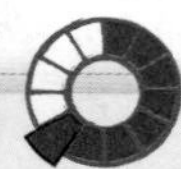

Unstable nuclei emit **ionising** radiation which can cause damage to human tissue.

Explain the precautions that need to be taken in order to reduce the potential danger of ionising radiation to humans.

Your answer should refer to the nature of ionising radiation and how it may be dangerous both inside and outside the body. **(6 marks)**

Ionising radiation is radiation that has enough energy to knock electrons from the outer shells of atoms, turning them from neutral atoms to positively charged ions. There are three types of ionising radiation from the nucleus – alpha, beta and gamma radiation. Alpha radiation is the most ionising and gamma-rays are the least ionising.

Since ionising radiation can damage human tissue and cause mutations in cells, leading to cancer, it is necessary to take precautions. These precautions include wearing lead aprons to absorb the radiation, being exposed to as low a radiation energy as possible, being as far from the radioactive source as possible and being exposed for as short a period of time as possible. All of these ensure the dose is not too high.

Inside the body, alpha particles are easily absorbed and can cause cellular mutations which can lead to cancer. Beta and gamma radiation tend to pass out of the body. Outside the body, alpha particles are not dangerous as they will not reach the body due to their short range, but beta and gamma radiation can reach the body and can cause damage to cells.

The answer starts with an explanation of what ionising radiation is as well as a description of the three main types of ionising radiation and their relative ionising abilities.

Having explained the nature of ionising radiation, the second paragraph addresses the dangers and the precautions that need to be taken to ensure that exposure does not lead to adverse health issues.

The final paragraph looks at which type of radiation is most dangerous inside the body as well as which types of radiation are most dangerous when the radioactive source is outside the body.

Command words: Explain

When you are asked to **explain** something, it is not enough to just state or describe something. Your answer **must** contain some reasoning or justification of the points you make based on the physics being addressed. This needs to be communicated clearly with good spelling, punctuation, grammar and sentence structure.

Contamination and irradiation

Although not mentioned here specifically in the answer, you could state that material emitting ionising radiation can be breathed into the body or eaten, in which case the person is **contaminated**. The radiation could also reach them as beta or gamma radiation from an external source, in which case the person is being **irradiated**.

Now try this

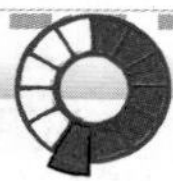

Describe how and why the atomic model has changed over time. Your answer should refer to three different models of the atom. **(6 marks)**

Work, energy and power

The **work** done by a force is the same as the **energy** transferred by the force. **Power** is the **rate of work** being done or the **rate at which energy is transferred.**

Work done and energy

Work done is the amount of energy transferred, and is measured in **joules (J).**

The work done by a force is calculated using this formula:

work done (J) = force (N) × distance moved in the direction of the force (m)

$$E = F \times d$$

LEARN IT! IT'S NOT ON THE EQUATIONS LIST

(Triangle: E over F × d)

You can measure the work done by recording the size of the force and the distance moved in the direction of the force. Having recorded these values, multiply them together to find the value for the work done.

Power

Power is the rate of doing work (how fast energy is transferred). Power is measured in **watts (W).** 1 watt is 1 joule of energy being transferred every second.

$$\text{power (W)} = \frac{\text{work done (J)}}{\text{time taken (s)}}$$

$$P = \frac{E}{t}$$

LEARN IT! IT'S NOT ON THE EQUATIONS LIST

(Triangle: E over P × t)

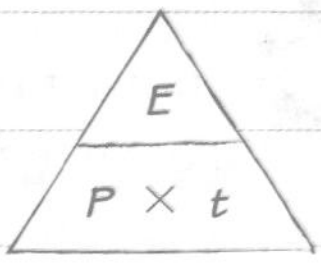

A hairdryer with a power of 1800 W transfers 1800 J each second.

Changing the energy of a system

The energy of a system can be changed by:

1. **work done through forces:** a body can be lifted through a vertical height by a force. This will increase the store of gravitational potential energy.
2. **electrical equipment:** a cell in a circuit provides a potential difference so that components in the circuit can transfer energy into other forms, such as thermal energy.
3. **heating a material:** supplying thermal energy to a system will increase the kinetic energy of the particles in the material. Its temperature will increase if it is not changing state.

Worked example

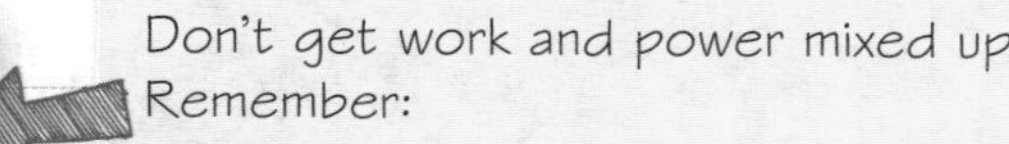

Don't get work and power mixed up. Remember:

work = energy transferred, measured in joules

power = *rate* of energy transfer, measured in watts

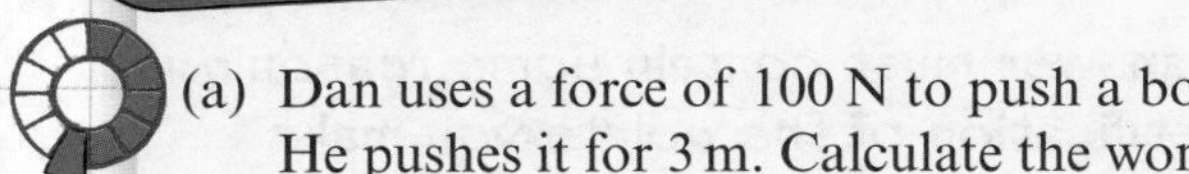

(a) Dan uses a force of 100 N to push a box across the floor. He pushes it for 3 m. Calculate the work done. **(3 marks)**

work done = 100 N × 3 m = 300 J

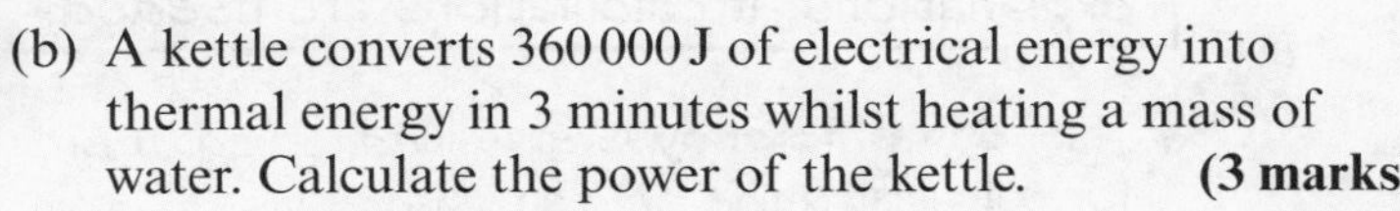

(b) A kettle converts 360 000 J of electrical energy into thermal energy in 3 minutes whilst heating a mass of water. Calculate the power of the kettle. **(3 marks)**

$$\text{power} = \frac{\text{energy}}{\text{time}} = \frac{360\,000\text{ J}}{3 \times 60\text{ s}} = 2000\text{ W}$$

Always make sure that energy is in J and time is in s before substituting to find an answer in W.

Now try this

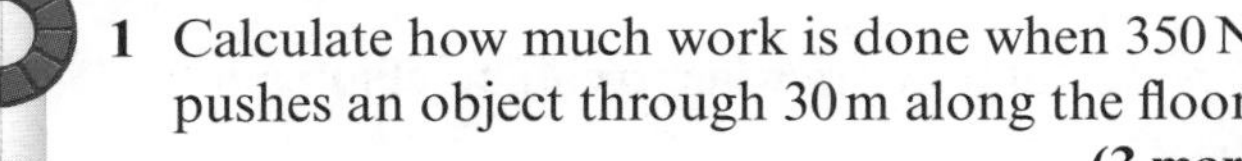

1 Calculate how much work is done when 350 N pushes an object through 30 m along the floor. **(3 marks)**

2 Calculate how much energy an 1800 W hairdryer transfers when used for 8 minutes. **(3 marks)**

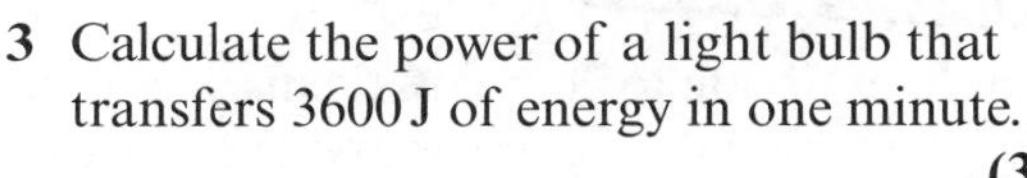

3 Calculate the power of a light bulb that transfers 3600 J of energy in one minute. **(3 marks)**

4 Calculate the temperature rise of a 36 kg mass of water when it is heated by a 4 kW heater for 46 minutes and 40 seconds. **(5 marks)**

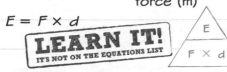

Had a look ☐ Nearly there ☐ Nailed it! ☐

Extended response – Energy and forces

There will be at least one 6-mark question on your exam paper. For these questions, you will need to think scientifically and structure your answer logically, showing how the points you make are related to each other. You can revise the topics for this question, which is about **energy and forces**, on pages 182, 183, 186 and 213.

Worked example

You can find more information on efficiency on page 183.

Energy stores and energy transfers can be shown on energy transfer diagrams.

An energy transfer diagram is shown for a battery-operated fan.

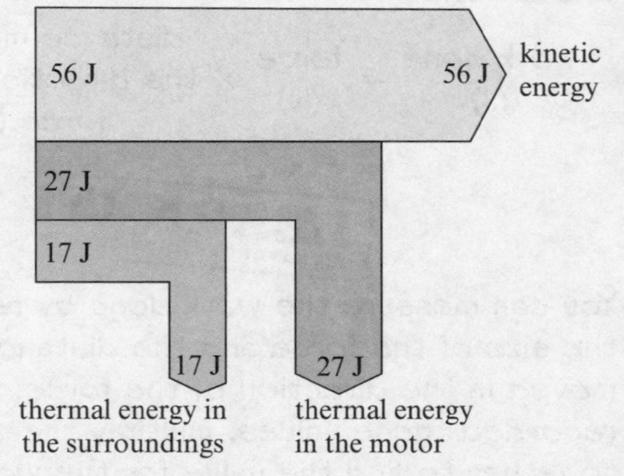

Explain how this energy transfer diagram can be used to determine the efficiency of the fan.

Your answer should refer to the term **efficiency**, the equation for efficiency and a calculation. **(6 marks)**

For the energy transfer diagram shown, the input is in the form of a chemical energy store, which is transferred to kinetic energy, sound energy and thermal energy.

For the energy transfer diagram shown, for every 100 J of energy transferred from the chemical energy store of the battery, there will be 56 J transferred to the kinetic energy, 17 J to the thermal energy store in the surroundings and 27 J to the thermal energy store in the motor.

Since a battery-operated fan is designed to transfer the chemical energy store to kinetic energy as a useful form, 56 J of the 100 J of chemical energy can be deemed to be useful, and 44 J can be assumed to be wasted.

The efficiency of a device is the proportion of the input energy that is usefully transferred and is calculated using the equation:

$$\text{efficiency} = \frac{\text{(useful output energy)}}{\text{(total input energy)}} \times 100\%$$

For the battery-operated fan, the efficiency is $\frac{56\text{ J}}{100\text{ J}} \times 100\%$ which is an efficiency of 56%.

The answer starts by simply stating what the energy transfer diagram is showing in terms of energy stores and transfers.

This part of the answer explains which energy store is useful as an output energy store and which of the stores are wasted energy stores. It also states the amounts that are useful and wasted.

Command word: Explain

When you are asked to explain something, it is not enough just to state or describe it.

Your answer must contain some reasoning or justification of the points you make.

Your explanation can include mathematical explanations, if calculations are needed.

The term **efficiency** is explained, an equation for calculating the efficiency is provided and a value is calculated for the fan's efficiency.

The question expects you to talk about **energy stores** and **energy transfers**. There are eight types of energy store and 4 types of energy transfer (which you can read more about on page 182).

Now try this

Two athletes decide to compare themselves in terms of how powerful they are when trying to climb a hill. One of the athletes is a climber and the other is a runner.

Describe an experiment to see which of the two athletes is the most powerful.

Your answer should refer to the equations for calculating changes in gravitational potential energy and power. **(6 marks)**

Interacting forces

Pairs of forces can interact **at a distance** or by **direct contact**.

Non-contact forces

Forces can be exerted between objects without them being in contact with one another. There are three **non-contact forces** that you need to know about.

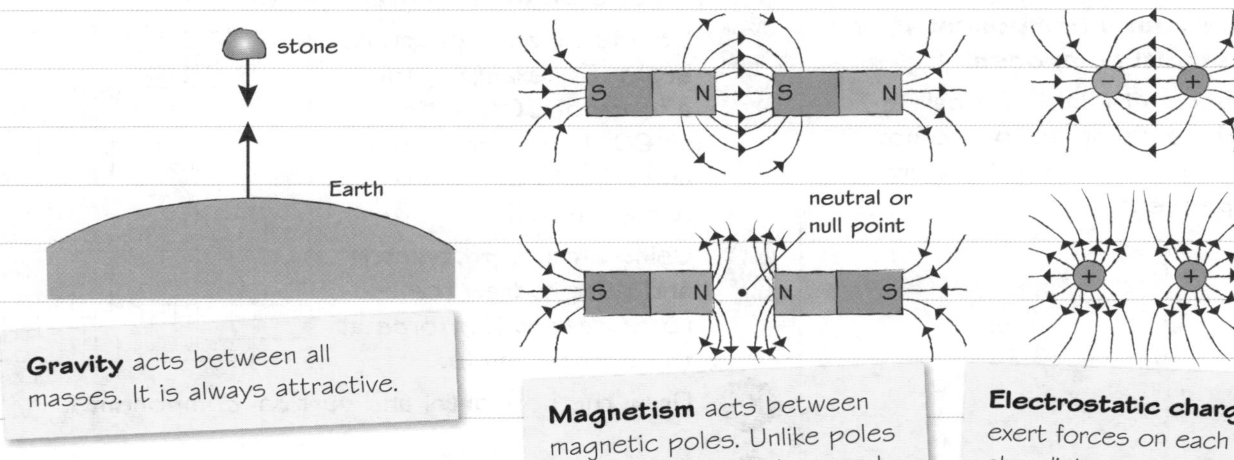

Gravity acts between all masses. It is always attractive.

Magnetism acts between magnetic poles. Unlike poles attract and like poles repel.

Electrostatic charges exert forces on each other at a distance. Unlike charges attract and like charges repel.

Contact forces

Forces can be exerted between objects due to them being in contact. In each case there is an interaction pair of forces that act in opposite directions. The forces can be represented by vectors.

The **normal contact force** acts upwards in opposition to the weight of the object. The interaction pair is the force of the ground on the box and the force of the box on the ground.

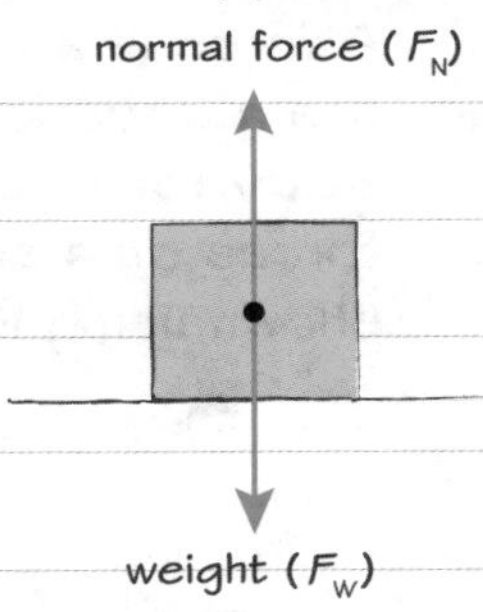

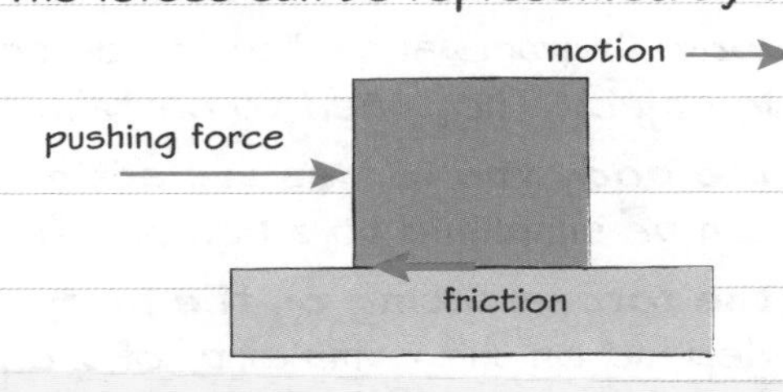

The **force of friction** acts in opposition to the pushing force that is trying to change its motion. Friction always acts to slow a moving object down. The interaction pair is the force of the object on the surface and the force of the surface on the object.

Worked example

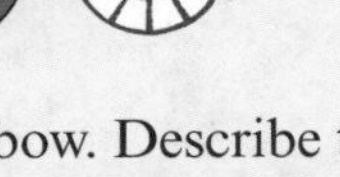

An archer fires an arrow from a bow. Describe the contact and non-contact forces involved in firing the arrow. **(2 marks)**

Contact forces include the normal force acting on the arrow, friction as it moves through the air and the tension in the string. Non-contact forces include gravity once it has been released.

Other examples of contact forces include tension and drag.

Now try this

1 State how gravity is different from magnetism and the electrostatic force. **(1 mark)**

2 State how the forces of friction and drag are similar. **(1 mark)**

3 Describe the contact and non-contact forces acting on a moving ship. **(3 marks)**

Free-body force diagrams

A **resultant force** can be resolved into its **horizontal** and **vertical** component forces. **Free-body force diagrams** can be drawn to show all the forces acting on a body.

Components of a force

Any force can be resolved into its **horizontal component** F_x and its **vertical component** F_y.

The horizontal and vertical components of motion do not affect one another – they are independent.

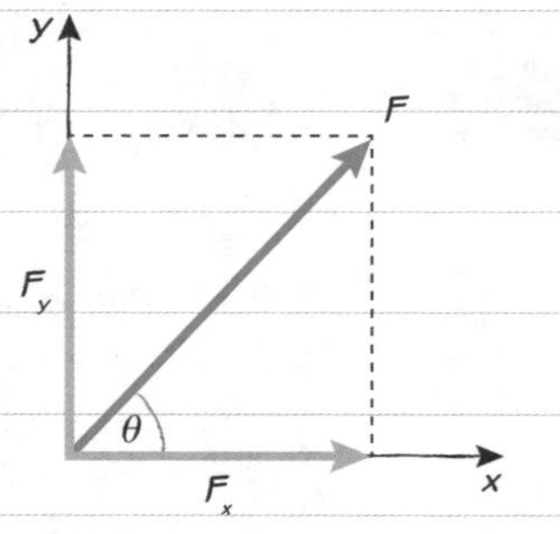

Resolving the forces by scale drawing

You can resolve a force into its horizontal components by doing a scale drawing on graph paper.

1. Decide on an appropriate scale. For example, for a force of 50 N acting at 60° to the horizontal use a scale of 1 cm to represent 10 N.
2. Using a ruler, protractor and pencil, draw the line to represent the force at the correct angle.
3. Draw the horizontal and vertical components.
4. Measure the lengths and convert to a force using the same scale.

F = 50 N
F_y = 43 N
F_x = 25 N

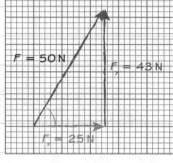

Free-body diagrams

Free-body diagrams show the forces acting on a single object. They need to contain:

1. **the body the forces are acting on** This can be simplified to a box or a dot.
2. **the forces acting on the body** This will depend on the complexity of the problem and the assumptions being made.

Free-body diagrams use arrows to show the **size** and **direction** of the forces involved. If these arrows **balance** then there is **no resultant force.** If the arrows do **not** balance then the body will be **accelerating.**

Free-body force diagrams

A force is a vector quantity, because it has a direction as well as a size. A **free-body force diagram** represents all the forces on a single body. Larger forces are shown using longer arrows.

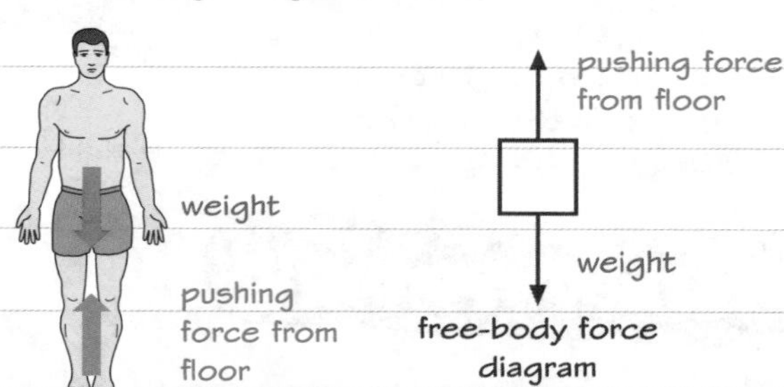

Worked example

Each square on this 1 cm grid represents a force of 10 N. Use this to find:

c
b
a

(a) the resultant force **(2 marks)**

The length of the resultant force is 5.4 cm, so the resultant force is 54 N.

(b) the horizontal component of the force **(2 marks)**

The length of *a* is 5 cm, so the force is 50 N.

(c) the vertical component of the force. **(2 marks)**

The length of *b* is 2 cm, so the force is 20 N.

Now try this

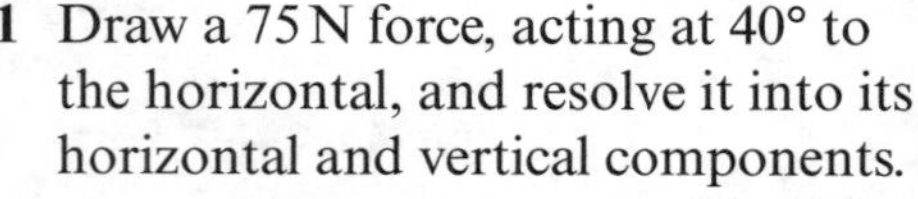

1 Draw a 75 N force, acting at 40° to the horizontal, and resolve it into its horizontal and vertical components. **(4 marks)**

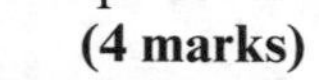

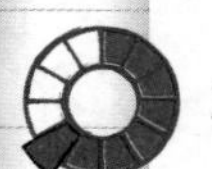

2 Draw a free-body diagram for a sky diver who has just jumped out of a plane from 3000 m. **(3 marks)**

Resultant forces

Resultant forces determine whether a body will be **stationary**, moving at a **constant speed** or **accelerating**.

Resultant forces

Forces are **vectors**, so they have a **size** and a **direction**. Both of these need to be taken into consideration when finding the resultant force.

The resultant force is the single force that would have the same effect as all of the other forces acting on the object. There can be many of these, but they always simplify to just one resultant force.

A resultant force of zero means a body is either stationary or moving at a constant speed.

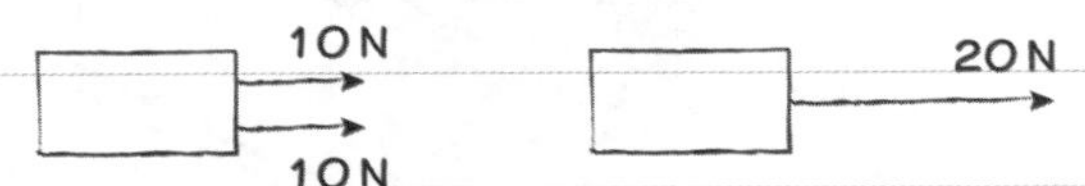

Forces acting in the same direction are added.

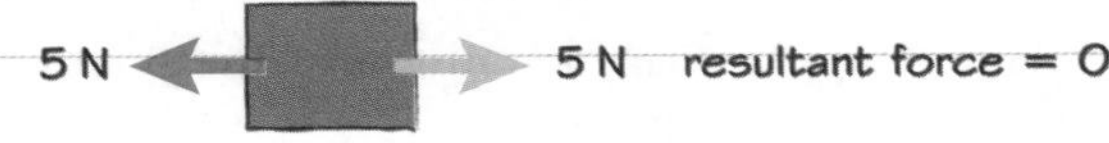

3 N 5 N resultant force = 2 N

Forces acting in opposite directions are subtracted.

Resultant forces in 2D

Sometimes, forces act at right angles to each other. This means that they cannot simply be added or subtracted like parallel forces can be.

The resultant force can then be found from a scale drawing on graph paper.

When the forces are at right angles, you can check the answer using Pythagoras' theorem.

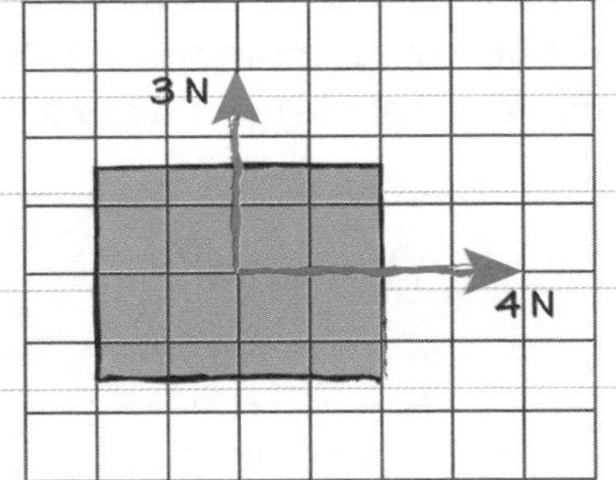

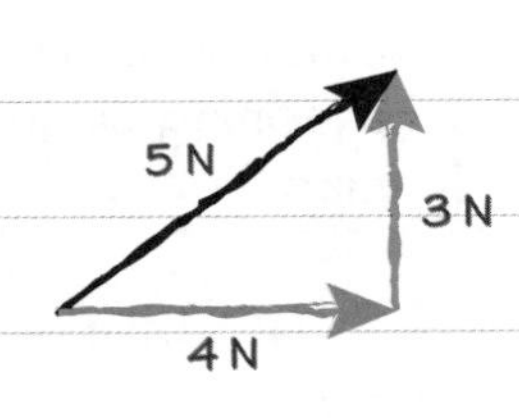

3 N acting up and 4 N acting to the right produce a 5 N resultant force in the direction shown.

Check using Pythagoras' theorem:

$$a^2 + b^2 = c^2$$

$$3^2 + 4^2 = 25$$

$$\text{resultant} = \sqrt{25} = 5\text{ N}$$

angle of resultant $= \tan^{-1}\left(\frac{3}{4}\right) = 37°$ to the horizontal

Worked example

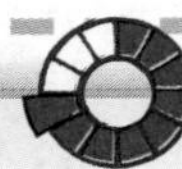

Work out the size and direction of the resultant force. **(5 marks)**

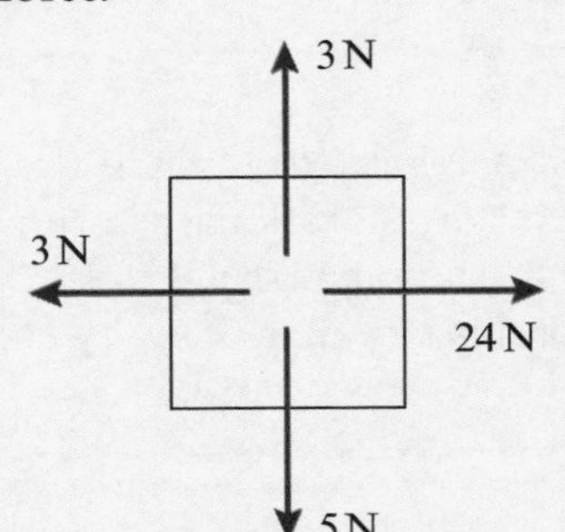

resultant down = 2 N

resultant right = 21 N

overall resultant force = $\sqrt{21^2 + 2^2} = 21.1$ N

Angle of resultant is below the horizontal, so is $-\tan^{-1}\left(\frac{2}{21}\right) = -5.4°$

Now try this

1 Draw a diagram to show how two forces of 20 N can have a resultant force of
   (a) 40 N upwards **(1 mark)**
   (b) 0 N. **(1 mark)**

2 Describe the motion of a body that has a resultant force of 8 N acting on it towards the ground. **(2 marks)**

3 A force of 5 N acts to the left and a force of 12 N acts upwards on a ball. Draw a scale diagram to work out the size and direction of the resultant force. **(4 marks)**

You can check your answer using Pythagoras' theorem.

Extended response – Forces and their effects

There will be at least one 6-mark question on your exam paper. For these questions, you will need to think scientifically and structure your answer logically, showing how the points you make are related to each other. You can revise the topics for this question, which is about **forces and their effects**, on pages 215–217.

Worked example

The free-body diagram shows the forces acting on a skier. The skier is moving down the slope.

Explain what the possible motion of the skier could be with reference to the diagram.

Your answer should refer to the size, nature and direction of the forces acting on the skier. **(6 marks)**

There are three forces shown on the diagram – the weight of the skier acting vertically downwards, the normal contact force of the skier, acting at right angles to the slope, and the force of friction acting up the slope.

The three forces are identified and stated. Their directions are also provided.

The weight is caused by the gravitational force, a non-contact force, acting on the skier's mass. The reaction force is a contact force acting perpendicular to the slope due to the slope pushing upwards on the skier. The force of friction is a contact force caused when the skis are in contact with the snow or ice. This force will act parallel to the slope and up the slope in the opposite direction to the skier's motion.

Having stated the names of the three forces, a description is then given as to how they have been caused. Reference to contact and non-contact forces is also provided in the answer.

The skier will accelerate down the slope if the component of his weight acting down the slope is greater than the force of friction acting up the slope. If the component of the weight acting down the slope is less than the friction, the skier will slow down. If these two forces are balanced then the skier will either be stationary or will move at a constant speed.

Having stated the nature and direction of the three forces, the motion of the skier can be explained by referring to the size of any balanced or unbalanced forces.

Command word: Explain

When you are asked to **explain** something, it is not enough to just state or describe something. Your answer must contain some reasoning or justification of the points you make.

A body will be stationary or move at a constant speed if the forces acting on it are balanced. Since the question states that the skier is moving down the slope, do not state that the skier is stationary in the answer or make it clear that you have discounted the possibility because of what the question states. The acceleration of any body, including the skier, is explained by Newton's second law of motion. You can read about this on page 173.

Now try this

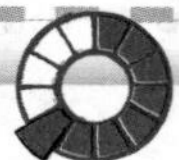

Describe what can happen when two objects interact with one another. Your answer should refer to contact and non-contact forces. **(6 marks)**

Circuit symbols

Electric circuit diagrams are drawn using agreed symbols and conventions that can be understood by everybody across the world.

Circuit symbols

You need to know the symbols for these components.

Component	Symbol	Purpose
cell	positive terminal negative terminal	provides a potential difference
battery		provides a potential difference
switches		allows the current flow to be switched on or off
voltmeter	V	measures potential difference across a component
ammeter	A	measures the current flowing through a component
fixed resistor		provides a fixed resistance to the flow of current
variable resistor		provides a variable (changeable) resistance
filament lamp		converts electrical energy to light energy as a useful form
motor	M	converts electrical energy to kinetic energy as a useful form
diode		allows current to flow in one direction only
thermistor		resistance decreases when the temperature increases
LDR		resistance decreases when the light intensity increases
LED		a diode that gives out light when current flows through it

Worked example

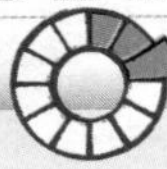

Name four components that are commonly used to change or control the amount of resistance in a circuit. **(2 marks)**

thermistor, LDR, fixed resistor, variable resistor

2 marks for all correct, 1 mark for 3 correct and 0 marks for less than 3 correct

Worked example

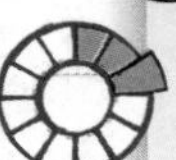

Describe how an LED is different from a filament lamp. **(1 mark)**

An LED will only allow electrical current to pass through it in one direction.

Now try this

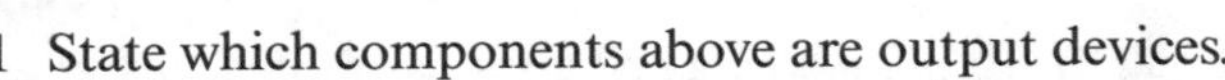

1 State which components above are output devices. **(3 marks)**

2 State how ammeters and voltmeters should be arranged with components in circuits. **(2 marks)**

3 Design a circuit to turn on a lamp when it goes dark. **(3 marks)**

Series and parallel circuits

Components in circuits can be arranged in **series** or **parallel**. The rules for current and potential difference in these two types of circuits are different.

Series circuits

A **series circuit** contains just one loop, around which an electric current can flow.

Electrons do not flow through any of the voltmeters.

Electrons flow clockwise in this circuit, through the cell, wires, three ammeters and two resistors.

Ammeters are always connected in series with components.

Ammeters have a very low resistance so the current can flow through them and be measured accurately.

The size of the current in a series circuit is the same at every point in the circuit. All three ammeters in this circuit will show the same value for the current.

The potential difference across components that are arranged in series must add up to give the cell voltage, so $V_1 = V_2 + V_3$.

Voltmeters have a very high resistance so that no current will flow through them.

Parallel circuits

A **parallel circuit** contains more than one loop and the current will split up or recombine at the junctions.

The potential difference across the components in each branch of a parallel circuit must add to give the cell voltage. The voltmeter readings across the two resistors will both have the same value as the potential difference across the cell.

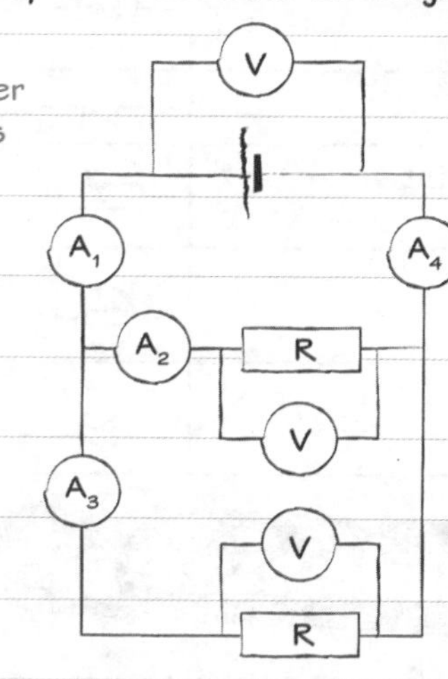

The sum of the currents in each of the branches must equal the current leaving the cell. For this example, $A_1 = A_4 = A_2 + A_3$.

Worked example

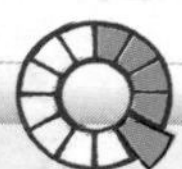

What are the missing values in this parallel circuit if the bulbs are identical? **(2 marks)**

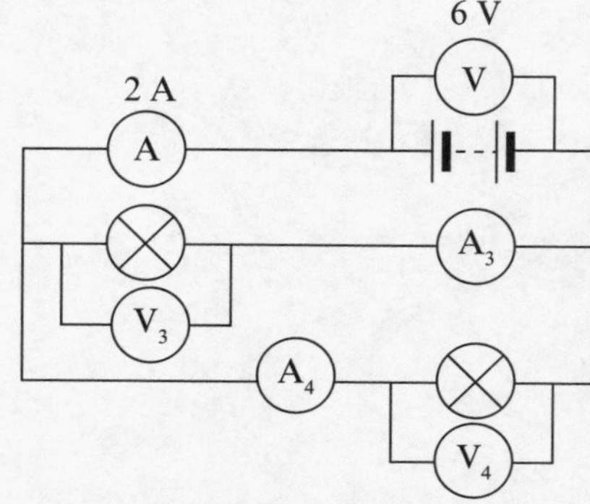

Both the ammeter readings A_3 and A_4 = 1 A. The voltmeter readings V_3 and V_4 will both be 6 V.

Now try this

1 Describe the differences for current and potential difference in series and parallel circuits. **(4 marks)**

2 Explain why car lights are connected in parallel and not in series. **(3 marks)**

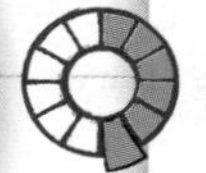

3 Explain why a cell goes flat more quickly when lamps are arranged in parallel. **(4 marks)**

Current and charge

An **electric current** is the **rate of flow of charge.** In a metal, electric current is the flow of electrons.

Charge and current

The size of a current is a measure of how much charge flows past a point each second. It is the rate of flow of charge. The unit of charge is the **coulomb** (C). One ampere, or amp (A), is one coulomb of charge per second. You can calculate charge using the equation:

charge (C) = current (A) × time (s)

$$Q = I \times t$$

LEARN IT! IT'S NOT ON THE EQUATIONS LIST

Watch out! Don't get the units and quantities confused. The units have sensible abbreviations (C for coulombs, A for amps). The symbols for the quantities are not as easy to remember (Q stands for charge, I for current).

You also need to know how to rearrange the equation using the triangle shown so that you can calculate Q, I and t using values provided in questions.

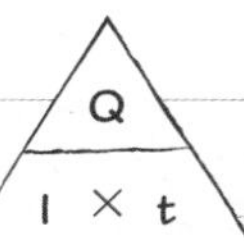
Q
I × t

Measuring current

Electric current will flow in a closed circuit when there is a source of potential difference. To measure the size of the current flowing through a component, an ammeter is connected in series with the component.

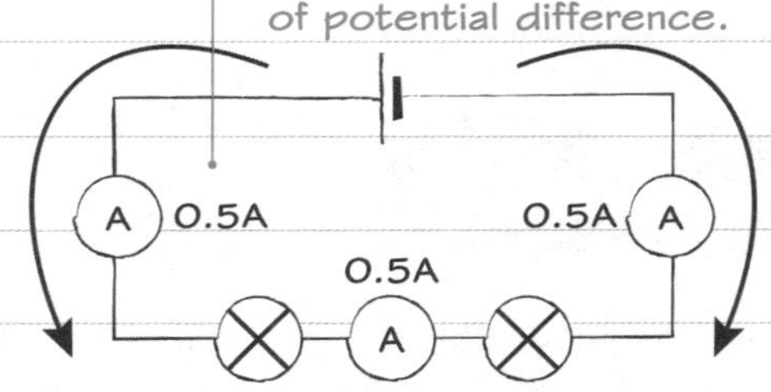
The current flowing is the same at all points in a series circuit. In this example, the current is 0.5 A.
The cell is the source of potential difference.
Conventional current flows this way, from + to – in a circuit.
Electrons flow this way, from – to +.
A
0.5A
0.5A
A
0.5A
A
The three ammeters are connected in series with the two filament lamps.

Worked example

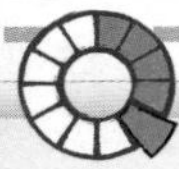

(a) Describe how ammeters should be connected with components in circuits. **(1 mark)**

Ammeters should always be arranged in series with the component to be measured.

(b) Describe what happens to the size of the current in a series circuit. **(1 mark)**

The current is the same value at any point in a series circuit.

Worked example

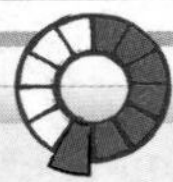

(a) A current of 1.5 A flows for 2 minutes. Calculate how much charge flows in this time. **(3 marks)**

$Q = I \times t$

$= 1.5\,\text{A} \times 120\,\text{s} = 180\,\text{C}$

(b) A charge of 1200 C flows through a filament lamp for 4 minutes. Calculate the average current in the filament lamp. **(3 marks)**

$I = \frac{Q}{t} = \frac{1200\,\text{C}}{240\,\text{s}} = 5\,\text{A}$

Now try this

1 State the units of current and charge. **(2 marks)**

2 Calculate the charge that flows in one hour when there is a current of 0.25 A in a circuit. **(3 marks)**

3 A charge of 3×10^4 C is transferred by a current of 250 mA. Calculate the length of time the current flows. **(4 marks)**

Had a look ☐ Nearly there ☐ Nailed it! ☐

Energy and charge

Energy, **charge** and **potential difference** are closely related when dealing with electrical circuits.

Energy, charge and potential difference

Energy, charge and potential difference are related by the equation:

energy transferred (J) = charge moved (C) × potential difference (V)

$$E = Q \times V$$

LEARN IT! IT'S NOT ON THE EQUATIONS LIST

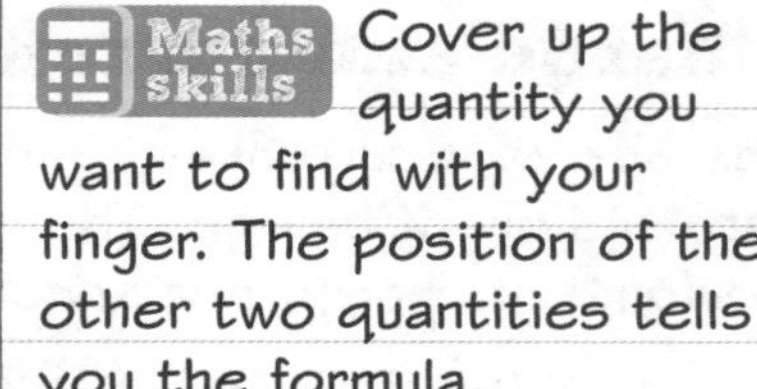

Maths skills: Cover up the quantity you want to find with your finger. The position of the other two quantities tells you the formula.

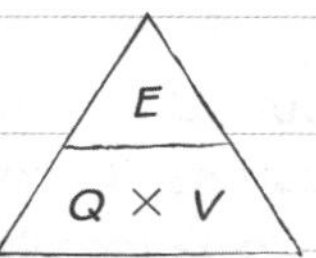

Calculate the charge, Q, from the current reading on the ammeter using the equation $Q = I \times t$ (see page 221).

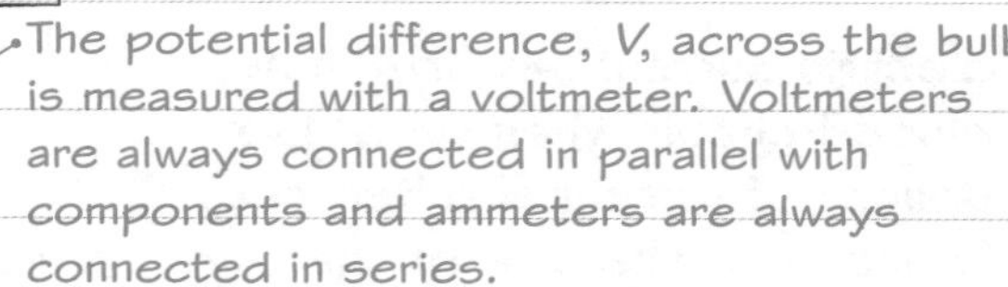

Remember that for this circuit, the value obtained for the energy transferred will be for this bulb, since the potential difference being measured is across this bulb only.

The potential difference, V, across the bulb is measured with a voltmeter. Voltmeters are always connected in parallel with components and ammeters are always connected in series.

Worked example

(a) A charge of 25 C passes through a motor. The potential difference across the motor is 6 V. Calculate how much energy is transferred. **(3 marks)**

$E = Q \times V$, so $E = 25\,C \times 6\,V = 150\,J$

For some calculations, you just need to substitute the numbers straight into the equation $E = Q \times V$.

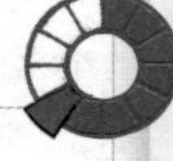

(b) A current of 1.2 A flows through a bulb for 2 minutes. The potential difference across the bulb is 12 V. Work out how much energy is transferred. **(3 marks)**

$E = Q \times V$ and $Q = I \times t$, so $E = I \times t \times V$

$E = 1.2\,A \times 120\,s \times 12\,V = 1728\,J$

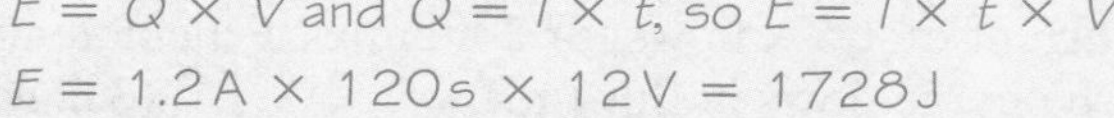

For other calculations, there may be two steps. Here you need to calculate Q first by using $Q = I \times t$ and then substitute this value into $E = Q \times V$.

Worked example

A charge of 18 C transfers 364 J of electrical energy through a resistor. Calculate the potential difference across the resistor. **(3 marks)**

Rearrange $E = Q \times V$ to give $V = \frac{E}{Q}$

$V = \frac{364\,J}{18\,C} = 20.2\,V$

Now try this

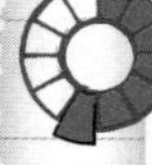

1 A charge of 24 C flows through an LED and the potential difference across it is 6 V. Calculate how much energy is transferred. **(3 marks)**

2 Explain why the volt, V, can also be described as a 'joule per coulomb'. **(3 marks)**

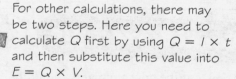

3 The potential difference across a resistor is 18 V and it transfers 500 J in 4 minutes. Calculate the current passing through the resistor. **(4 marks)**

Had a look ☐ Nearly there ☐ Nailed it! ☐

Ohm's law

Ohm's law states how the **current** through a component relates to its **resistance** and the **potential difference** across it.

Resistance

The **resistance** of a component is a way of measuring how hard it is for electricity to flow through it. The units for resistance are **ohms (Ω)**.

The resistance of a whole circuit depends on the resistances of the different components in the circuit. The higher the total resistance, the smaller the current.

resistance UP, current DOWN

The resistance of a circuit can be changed by putting different **resistors** into the circuit, or by using a **variable resistor**. The resistance of a variable resistor can be changed using a slider or knob.

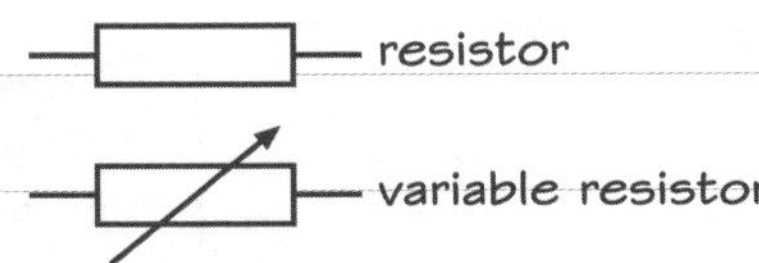

Ohm's law

Ohm's law states that the size of the current, I, flowing through a component of resistance, R, is directly proportional to the potential difference, V, across the component at constant temperature.

The equation for Ohm's law is:

potential difference (V) = current (A) × resistance (Ω)

$$V = I \times R$$

LEARN IT! IT'S NOT ON THE EQUATIONS LIST

Components that obey Ohm's law are said to be ohmic conductors whereas those that do not obey Ohm's law are non-ohmic. Examples of both of these can be found on page 225.

Worked example

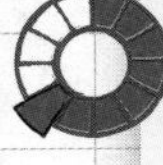

(a) Resistor A has a current of 3 A flowing through it when the potential difference across it is 15 V. What is the size of its resistance? **(3 marks)**

From $V = I \times R$ you obtain $R = \frac{V}{I}$

Substituting gives $R = \frac{15\,V}{3\,A} = 5\,\Omega$

(b) Sketch a circuit that could be used to plot current against potential difference for this resistor. **(4 marks)**

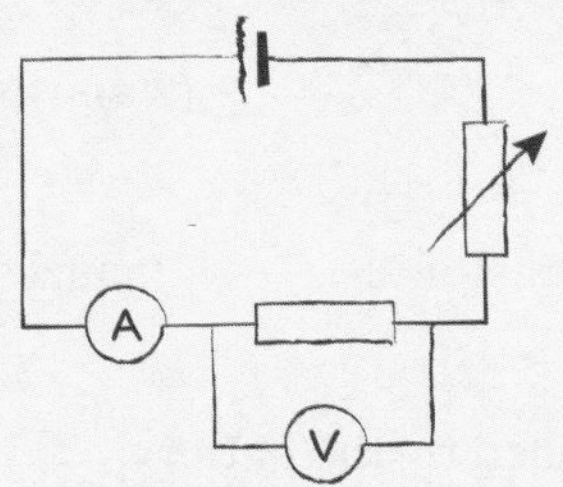

The variable resistor can be used to collect numerous values for I and V so that an I–V graph can be plotted.

(c) Resistor A is then replaced with resistor B. A graph of current against potential difference is plotted.

Explain what the graph tells you about the two resistors. **(2 marks)**

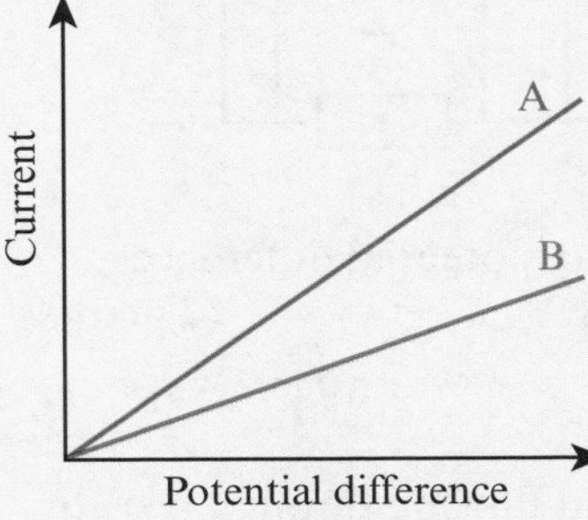

The steeper line shows that resistor A has a lower resistance than resistor B.

Both resistors are ohmic because the graphs are straight lines.

Now try this

1 A current of 3.2 A flows through a lamp of resistance 18 Ω. Calculate the size of the potential difference across the lamp. **(3 marks)**

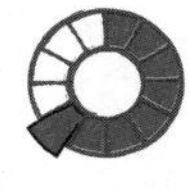

2 The potential difference across a resistor is 28 V when the current flowing through it is 0.4 A. Calculate the resistance of the resistor. **(3 marks)**

Resistors

…t of resistors in **series** or **parallel** determines whether the **resistance** in a …ses or **decreases**.

…rs in series

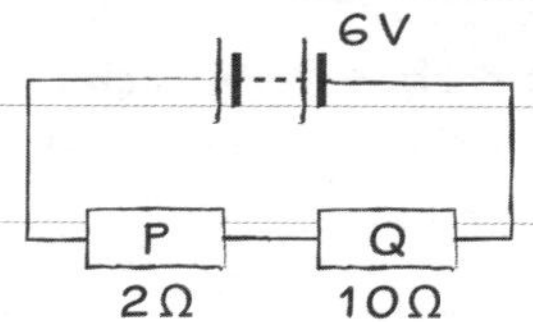

Calculating series resistance

For resistors arranged in series:

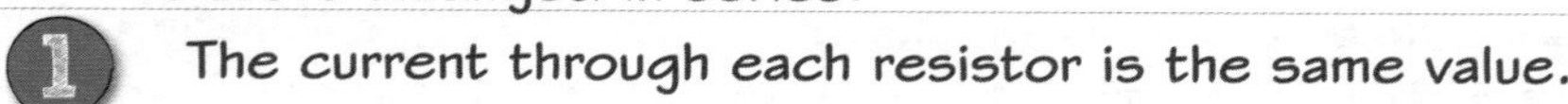
1. The current through each resistor is the same value.

2. The sum of the voltages across the resistors must add up to equal the cell voltage.

The total resistance in this series circuit is 12Ω. The total potential difference from the battery is 6V. Using $I = V \div R$ the current, I, flowing in resistors P and Q, is 6V ÷ 12Ω = 0.5A.

Resistors in parallel

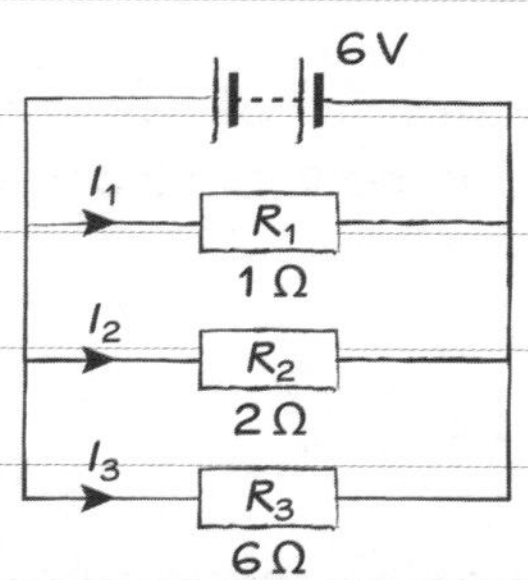

Calculating parallel resistance

For resistors arranged in parallel:

1. The total current leaving the battery is equal to the sum of the current flowing in the separate branches.

2. The potential difference across the resistors in each branch is equal to the potential difference of the battery.

Worked example

In the series circuit in the diagram, resistor X is known to have a resistance of 16Ω.

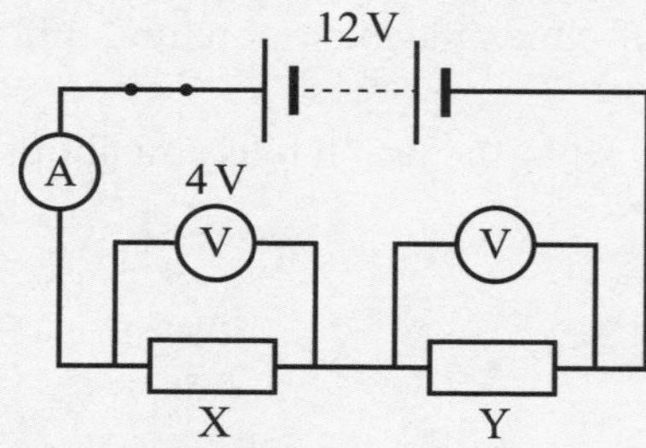

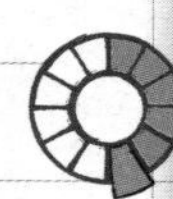

(a) Calculate the missing potential difference value across resistor Y. **(2 marks)**

8V, since the values across X and Y must add up to 12V.

(b) Calculate the current flowing through resistor X. **(3 marks)**

$I = \frac{V}{R} = \frac{4V}{16\Omega} = 0.25A$

Worked example

Two resistors are arranged in parallel as shown in the circuit below. Calculate:

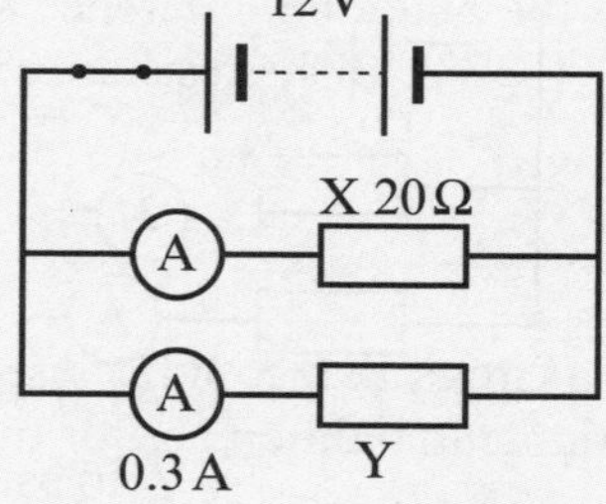

(a) the resistance of Y **(2 marks)**

$R = \frac{V}{I} = \frac{12V}{0.3A} = 40\Omega$

(b) the current through resistor X **(2 marks)**

$I = \frac{V}{R} = \frac{12V}{20\Omega} = 0.6A$

(c) the total current supplied by the battery. **(2 marks)**

total current = 0.3A + 0.6A = 0.9A

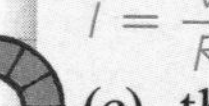

Now try this

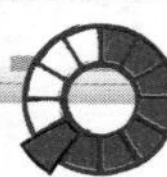

Two resistors, each of resistance 60 ohms, are arranged in series and connected to a 12 V cell. Calculate the size of the current flowing in the circuit. **(3 marks)**

I–V graphs

An **I–V graph** shows how the **current** flowing through a component varies as the **potential difference** across it varies. You need to know the characteristics of these I–V graphs.

1 Fixed resistor

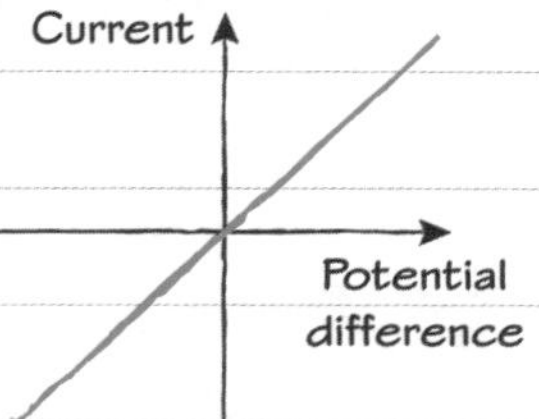

The temperature remains constant so the resistance remains constant. The slope or gradient is constant and the line remains straight throughout.

 Filament lamp

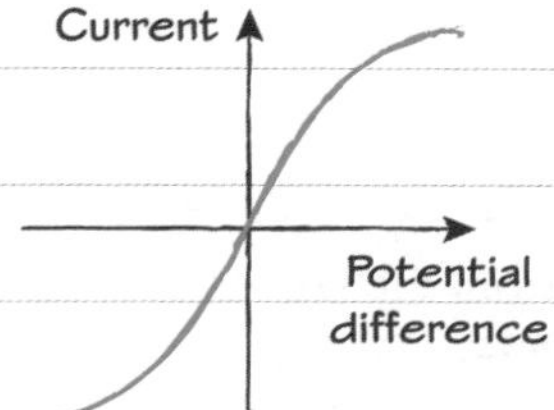

As the potential difference increases, the filament gets hotter, and atomic vibration increases. This leads to greater resistance. The slope or gradient decreases as the potential difference increases since resistance is increasing.

 Diode

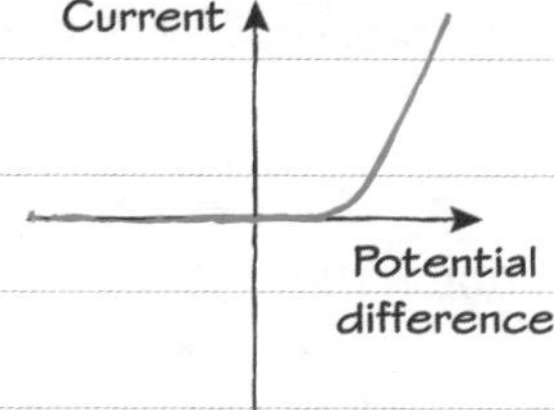

The current only flows in one direction. There is a **threshold** in the forward direction, which is why the graph is flat initially. A diode behaves like a fixed resistor – the resistance does not change. A diode has a very high resistance in the reverse direction.

Drawing an I–V graph

You can use a circuit like this to collect data about current and potential difference.

- ✓ When the switch is closed you can read current from the ammeter and potential difference from the voltmeter.
- ✓ Varying the value of the variable resistor allows you to record the current for different potential differences.
- ✓ You can reverse the cell to obtain negative values for potential difference.

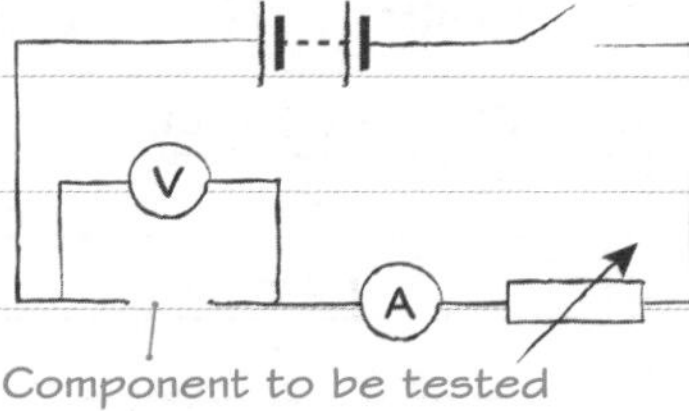

V
A
Component to be tested is placed here.

Worked example

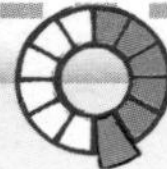

A student is testing the resistance of a component. She draws an I–V graph.

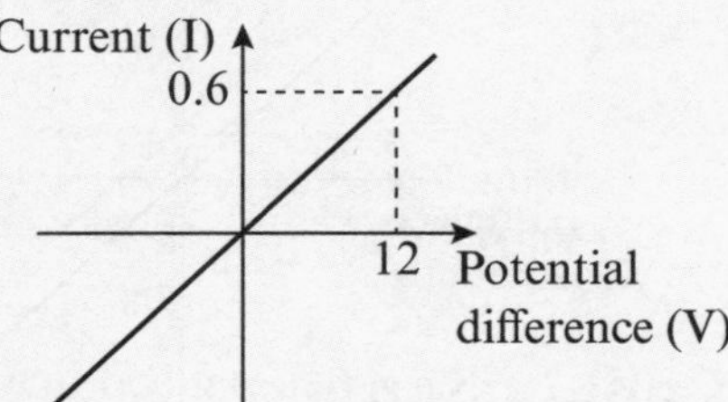

The component being tested is a:

- ☐ **A** diode
- ☐ **B** filament lamp
- ☒ **C** fixed resistor
- ☐ **D** cell

(1 mark)

An I–V graph always has the current (I) plotted on the vertical *y*-axis and the potential difference (V) plotted on the horizontal *x*-axis. To find the resistance of a component from an I–V graph, you will need to divide the *x*-axis value by the corresponding *y*-axis value. That is, the gradient is not the resistance, the gradient = 1/resistance.

Now try this

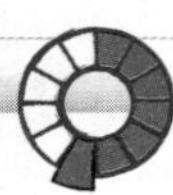

Amy is testing the resistance of a filament lamp. She varies the potential difference across the lamp and records the current that flows.

(a) Sketch a graph of potential difference (V) against current (I) for Amy's experiment. **(2 marks)**

(b) Explain the shape of your graph. **(3 marks)**

Electrical circuits

You can investigate the relationship between **potential difference, current** and **resistance** for a filament lamp and a resistor by arranging them in series and parallel circuits.

Core practical

Aim

To determine the relationship between potential difference, current and resistance for a filament lamp and a resistor when arranged in series and parallel.

Apparatus

cell or battery, ammeters, voltmeters, filament lamps, resistors, connecting wires

Method 1: Investigating V, I and R

Set up the circuit so that the current through the filament lamp and the potential difference across it can be measured. Adjust the variable resistor setting to obtain a number of readings for current and potential difference. Plot a graph of I against V. Repeat for the resistor.

Results

Your results can be recorded in a table.

Potential difference (V)	Current (I)
−2.0	−0.20
−1.5	−0.18
−1.0	−0.15
−0.5	−0.10
0	0
0.5	0.10
1.0	0.16
1.5	0.18
2.0	0.20

Method 2: Testing series and parallel circuits

Set up the circuits in both parallel and series arrangements and use these to verify the rules for the behaviour of current and potential difference in both types of circuits.

series circuit
3 V
0.14 A
R_1 0.14 A R_2
0.14 A
1.5 V 1.5 V

parallel circuit
3 V
0.6 A
3 V 0.3 A
3 V 0.3 A

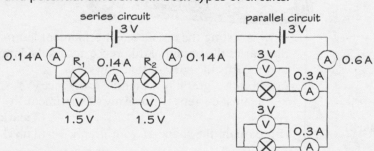

There is more information relating to current, potential difference, resistance and series and parallel circuits on pages 220–225.

Be careful when using electrical circuits as electric current can produce heat and cause burns.

Data collected in an investigation are accurate if close to the true value. The data are precise if the repeat readings of a certain variable are close to one another. The best data are both accurate and precise.

In order to improve the accuracy of your results, place a switch in your circuit so that the current does not get too large. Too large a current leads to a greater value of the resistance being recorded, due to thermal energy causing extra vibrations of the ions in the metal lattice.

Now try this

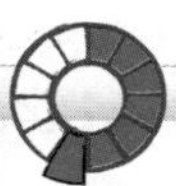

1 Explain why a series or parallel arrangement will still lead to the same shape of I–V graph for both components. **(4 marks)**

2 Explain why the I–V graph for a filament lamp has the shape that it does. **(5 marks)**

Conclusion

Current is the same at any point in a series circuit, but will split up at a junction in a parallel circuit.

The sum of the potential difference across components in any loop in a series or parallel circuit equals the potential difference of the cell.

Had a look ☐ Nearly there ☐ Nailed it! ☐

The LDR and the thermistor

The resistance of **light-dependent resistors (LDRs)** and **thermistors** changes according to light conditions (LDR) or temperature (thermistor).

Light-dependent resistors

The resistance of a **light-dependent resistor (LDR)** is large in the dark. The resistance gets less if light shines on it. The brighter the light, the lower the resistance.

brightness UP, resistance DOWN

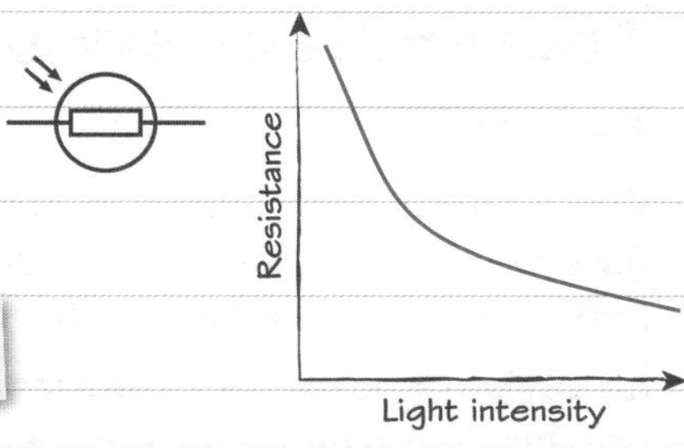

Thermistors

The resistance of a **thermistor** depends on its temperature. The higher the temperature, the lower the resistance.

temperature UP, resistance DOWN

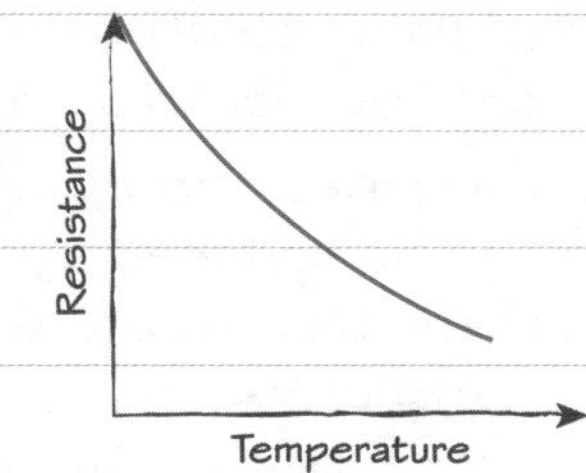

Worked example

A 6 V battery is connected in series with an LDR, a 300 Ω fixed resistor and an ammeter, as shown.

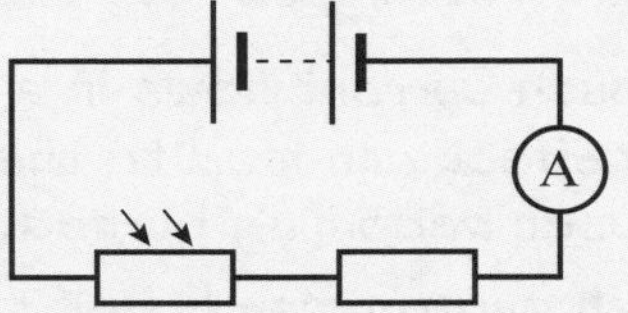

(a) Describe how the circuit shown can be used to explore the variation in resistance as the light levels change. **(2 marks)**

The amount of light being shone on the LDR can be changed by varying the intensity of the light it is exposed to from a torch or a lamp. The size of the current flowing in the circuit can then be recorded for each of these light levels.

(b) Describe the results you would expect. **(2 marks)**

The output current will be highest when the light level is greatest. As the brightness increases, the LDR's resistance decreases.

(c) How could a voltmeter be used in the circuit to show how the resistance changes with light levels? **(3 marks)**

Connect a voltmeter across the LDR. As the light level increases, the potential difference across the LDR will decrease since the resistance has decreased and a greater share of the 6 V will be across the fixed resistor.

Worked example

A 3 V battery is connected in series with a thermistor and an LDR, as shown.

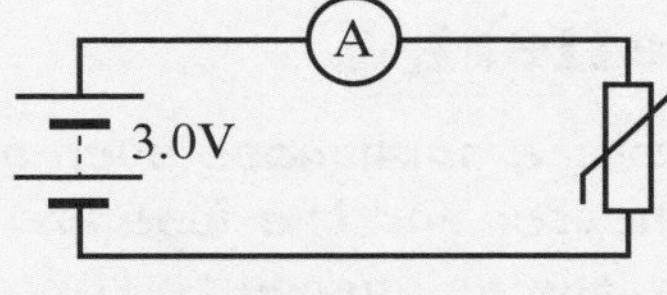

(a) Explain how the circuit can be used to explore how the resistance of the thermistor changes. **(2 marks)**

The thermistor could be immersed in a beaker of cold water, or a water bath, which is then heated from 10 °C to 80 °C. The current flowing in the circuit is recorded at each temperature over this range.

(b) Describe what your results would show. **(2 marks)**

As the temperature increases, the current increases. This is because the thermistor's resistance decreases as it gets warmer.

If a buzzer was placed in this circuit then it could be used as a simple fire alarm, with the buzzer sounding when the current reached a certain value. This could be controlled or calibrated by using a variable resistor.

Now try this

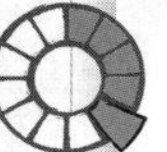

1 State what affects the resistance of (a) a thermistor and (b) an LDR. **(2 marks)**

2 Explain how an LDR and a thermistor could be used together in a circuit. **(4 marks)**

Had a look ☐ Nearly there ☐ Nailed it! ☐

Current heating effect

[...]ent flows in a circuit it has a **heating** effect. This can have **advantages** and disa[...]ges.

Energy transfers in a resistor

When there is an electric current flowing through a resistor, energy is transferred which heats the resistor.

Electrical energy is **dissipated** as thermal energy in the surroundings when an electrical current does work against electrical resistance.

Unwanted thermal energy transfers can be reduced by using low-resistance wires. Wires that are better conductors, shorter or thicker will waste less energy as heat compared with longer, thinner wires of a poorer electrical conductor.

Energy and collisions

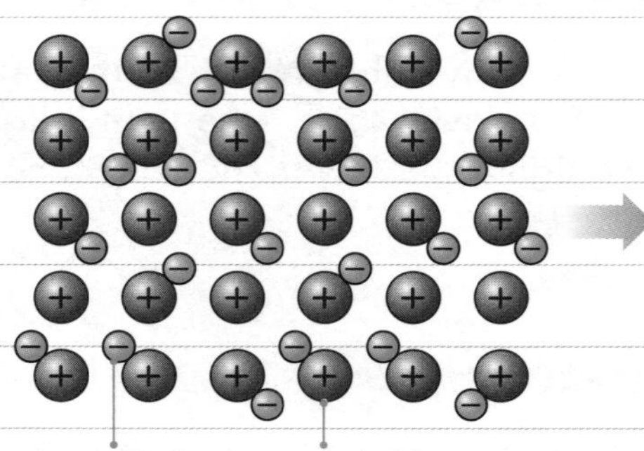

Electrons flow through a metal lattice when a potential difference is applied across the ends of the metal. Collisions between the electrons and the ions in the lattice lead to the kinetic energy of the electrons being dissipated as thermal energy.

Advantages of the heating effect

Some appliances such as the iron, the heater and the fuse are designed to transfer energy to thermal energy by the heating effect of a current.

The heating effect of a current passing through a resistor is useful for:

- heating water in a kettle
- radiant heaters
- toasters, grills and ovens
- underfloor heating.

Disadvantages of the heating effect

If too much current flows in a circuit, then the heating effect can lead to the appliance catching fire or the user becoming burned.

Too much current can flow if too many appliances are being used at the same time. Safety features such as correct fuses, circuit breakers and earthing need to be in place. See page 231.

In many devices, the heating effect of a current means that some energy is wasted. For example, a light bulb and a TV screen transfer some energy to thermal energy by the heating effect of a current.

Worked example

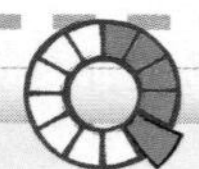

Give three examples of appliances in the home where the heating effect of the current causes energy to be wasted as thermal energy. **(1 mark)**

a computer, computer monitor and a mobile phone screen

Worked example

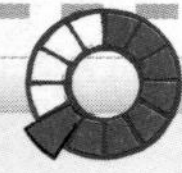

Explain how energy is transferred to thermal energy in a resistor. **(3 marks)**

Collisions between the moving electrons and the fixed metal ions in a lattice result in the kinetic energy of the electrons being transferred to thermal energy. The thermal energy is dissipated into the surroundings.

Now try this

1 Name **three** household devices that usefully transfer electrical energy to thermal energy. **(3 marks)**

2 Explain why it might be dangerous to have many electrical appliances being used at the same time. **(3 marks)**

3 Explain why it is better to use metals with a lower resistance in radiant heaters instead of metals with a higher resistance. **(3 marks)**

Energy and power

Electrical energy is **transferred** in circuits to do **work**. The amount of energy transferred depends on the **current**, **time** and **potential difference**.

Calculating energy

The total energy transferred by a device depends on the current in it, the potential difference across it and how long it is switched on.

Energy transferred (J) = current (A) × potential difference (V) × time (s)

$$E = I \times V \times t$$

Maths skills You can rearrange the energy transferred equation on the left to give the following:

$V = E \div (I \times t)$

$I = E \div (V \times t)$

$t = E \div (V \times I)$

Calculating power

Power is the energy transferred per second or the rate at which energy is transferred. It is measured in **watts**. For electrical devices, power depends on the current in the device and the potential difference across it, or the resistance of the device. It also depends on the total energy transferred and the time taken to transfer the energy.

There are three equations you need to know that can be used to calculate power:

1 electrical power (W) = current (A) × potential difference (V)

$$P = I \times V$$

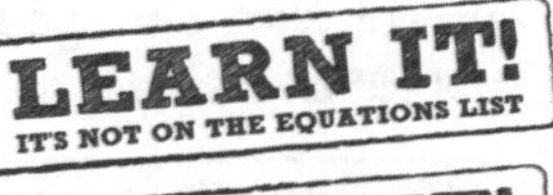
LEARN IT!
IT'S NOT ON THE EQUATIONS LIST

2 power (W) = $\frac{\text{energy transferred (J)}}{\text{time taken (s)}}$

$$P = \frac{E}{t}$$

LEARN IT!
IT'S NOT ON THE EQUATIONS LIST

3 electrical power (W) = current squared (A^2) × resistance (Ω)

$$P = I^2 \times R$$

LEARN IT!
IT'S NOT ON THE EQUATIONS LIST

Maths skills Equation triangles can be used to rearrange equations. For example this is the equation triangle for $P = I \times V$.

P / I × V

Worked example

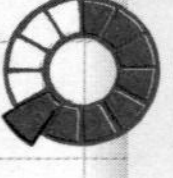
(a) A current of 8 A flows for 4 minutes in a kettle that is connected to a potential difference of 230 V. Calculate how much energy is transferred. **(4 marks)**

$E = V \times I \times t$

$E = 230\,\text{V} \times 8\,\text{A} \times (4 \times 60)\,\text{s} = 441\,600\,\text{J}$

(b) A 2.2 kW hairdryer is used for 5 minutes. Calculate how much energy is transferred. **(4 marks)**

$E = P \times t$

$= 2200\,\text{W} \times (5 \times 60)\,\text{s} = 660\,000\,\text{J}$

Worked example

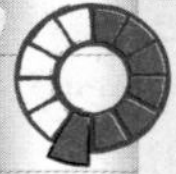
(a) Calculate the power of a microwave that transfers 30 000 J of energy in 35 s. **(4 marks)**

$P = \frac{E}{t} = \frac{30\,000\,\text{J}}{35\,\text{s}} = 857\,\text{W}$

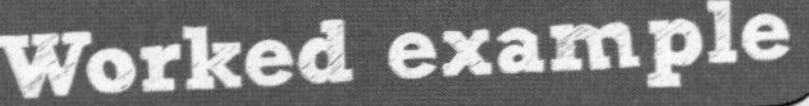

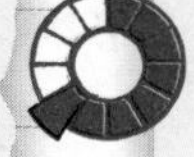
(b) Calculate the current that is taken by a device with a power rating of 3400 W when its resistance is 120 Ω. **(4 marks)**

$I = \sqrt{\frac{P}{R}} = \sqrt{\frac{3400\,\text{W}}{120\,\Omega}} = 5.3\,\text{A}$

Power can be measured in watts (W) or in joules per second (J/s). A power of 1 W means that 1 J of energy is being transferred each second.

Now try this

1 Calculate the current that flows in a 3000 W oven when connected to the 230 V mains supply. **(3 marks)**

2 A device transfers 100 000 J over 3 minutes. The potential difference across the device is 230 V. Calculate the resistance of the device. **(4 marks)**

Had a look ☐ Nearly there ☐ Nailed it! ☐

a.c. and d.c. circuits

Circuits can be operated using **alternating current (a.c.)** or **direct current (d.c.)**.

Direct current

An **electric current** in a wire is a flow of electrons. The current supplied by **cells** and **batteries** is **direct current (d.c.)**. In a direct current the electrons all flow in the same direction.

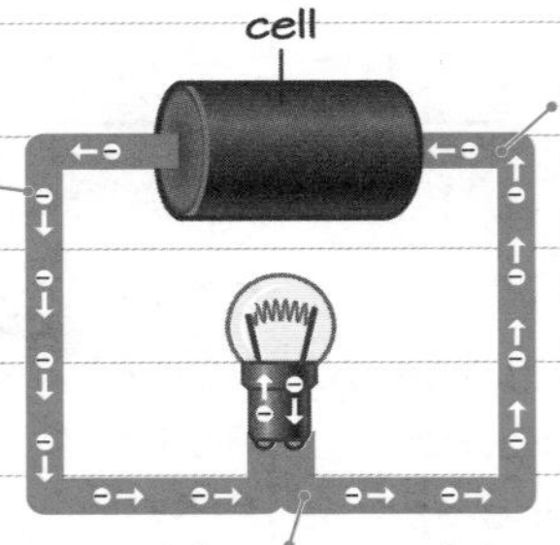

cell
Electrons are pushed out of one end of the cell.
Electrons flow round to the other end of the cell.
There must be a complete circuit for the electrons to flow.

The oscilloscope trace for d.c. from a battery is a horizontal line. The potential difference can be read off the vertical scale.

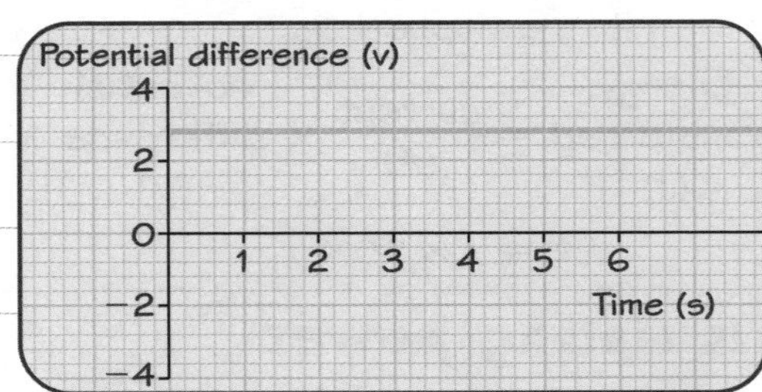

The potential difference above has a constant value of 2.8 V.

Alternating current

Mains electricity supplied to homes and businesses is an **alternating current (a.c.)**. Alternating current is an electric current that changes direction regularly, and its potential difference is constantly changing.

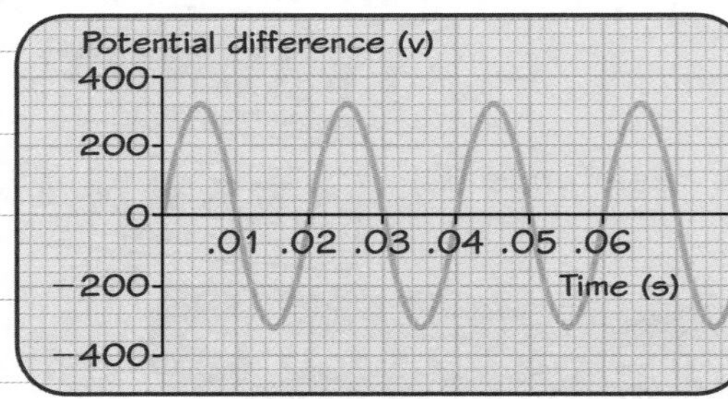

For an a.c. supply, the movement of charge is constantly changing direction. The mains supply has an average working value of 230 V and a frequency of 50 Hz. This means that the electric current and voltage change direction 100 times every second.

d.c. and a.c. supply in the home

Both d.c. and a.c. can be used in the home. D.c. is supplied in the form of cells and batteries. A.c. is the mains supply. Both can be used by devices to transfer energy to motors and heating devices.

Different electrical appliances have different power ratings. The power rating tells you how much energy is transferred by the appliance each second.

Appliance	Power rating
kettle	2200 W
hairdryer	1500 W
microwave	850 W
electric oven	3000 W
electric shaver	15 W

Worked example

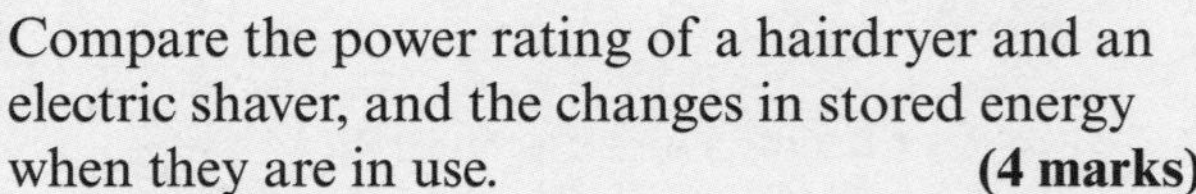

Compare the power rating of a hairdryer and an electric shaver, and the changes in stored energy when they are in use. **(4 marks)**

The hairdryer has a power rating of 1500 W, which means that it transfers 1500 J per second to stores of kinetic energy and thermal energy. The electric shaver has a power rating of one-hundredth of the hair dryer. It transfers only 15 J per second, mainly to a kinetic energy store.

Now try this

1 Give three examples of devices that use:
(a) d.c. **(3 marks)**
(b) a.c. **(3 marks)**

2 Explain how a.c. and d.c. are:
(a) similar **(1 mark)**
(b) different. **(2 marks)**

3 Explain how and why alternating current is generated in power stations instead of direct current. **(3 marks)**

Mains electricity and the plug

Electrical energy enters UK homes as **mains electricity** at **230 V a.c.** A mains plug has three wires, which have different roles in the operation of a.c. in the home.

The 13 A plug

The earth (yellow and green) wire does not form part of the circuit but acts as a safety feature along with the fuse.

The neutral (blue) wire completes the circuit with the appliance. It is at a potential difference of 0 V with respect to the earth wire.

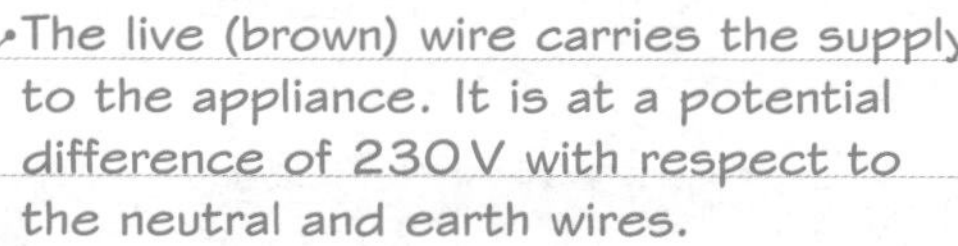

The live (brown) wire carries the supply to the appliance. It is at a potential difference of 230 V with respect to the neutral and earth wires.

The fuse is connected to the live wire and it contains a wire that will melt if the current gets too high. If the fuse blows, the device will be at 0 V and not at 230 V. A switch should also be connected in the live wire for the same reason.

The live, neutral and earth wires are made from copper, but covered in colour-coded insulating plastic so that the consumer is not exposed to a dangerous voltage.

Earthing and fuses

Earthing, circuit breakers and fuses are safety features that are used to ensure that the user does not get electrocuted if a fault occurs.

1 The live wire inside the appliance may come loose and touch a metal part of the device's casing.

2 The large current heats and melts the wire in the fuse, making a break in the circuit.

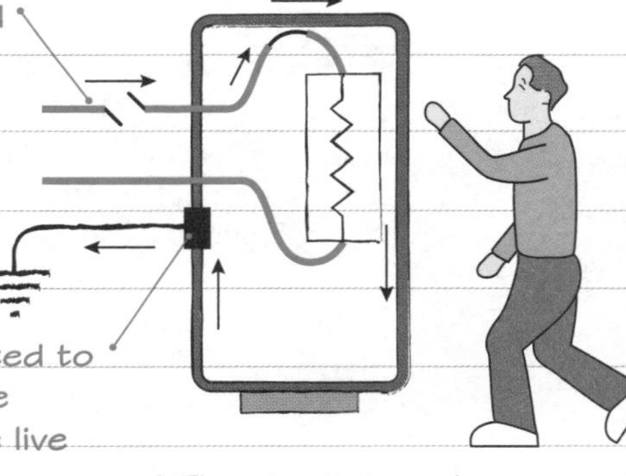

3 The earth wire is connected to the metal casing and a large current flows in through the live wire and out through the earth wire.

4 The circuit is no longer complete, so there is no chance of electric shock or fire.

5 By having an earth wire connected to the metal casing, the user is not at risk if the live wire comes loose and touches anything metallic.

Worked example

(a) State the potential differences of the live, neutral and earth wires in a plug. **(3 marks)**

live: 230 V; neutral and earth: both at 0 V

(b) Name three safety features used by appliances. **(3 marks)**

earth wire, fuse and circuit breaker

Circuit breakers are also used as safety features in addition to fuses and earth wires. When circuit breakers detect that a current is too high, they use a magnetic field to open a switch and isolate the appliance, making it safe. They are easily reset by pressing a switch if they trip.

Now try this

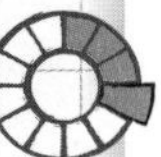

1 State the colours of the:
(a) live wire **(1 mark)**
(b) neutral wire **(1 mark)**
(c) earth wire. **(1 mark)**

2 Give **one** advantage of a fuse and **one** advantage of a circuit breaker. **(2 marks)**

3 Explain how an earth wire and fuse protect the user when the live wire comes loose. **(4 marks)**

Had a look ☐ Nearly there ☐ Nailed it! ☐

Extended response – Electricity and circuits

There will be at least one 6-mark question on your exam paper. For these questions, you will need to think scientifically and structure your answer logically, showing how the points you make are related to each other. You can revise the topics for this question, which is about **electricity and circuits**, on pages 219–231.

Worked example

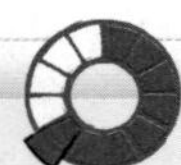

Two identical bulbs can be connected to a 6 V cell in series or parallel as shown in the diagram.
Compare the brightness of the bulbs in the two circuits when the switches are closed.
Your answer should refer to the current, potential difference, energy stores and rates of energy transfer in both circuits.

(6 marks)

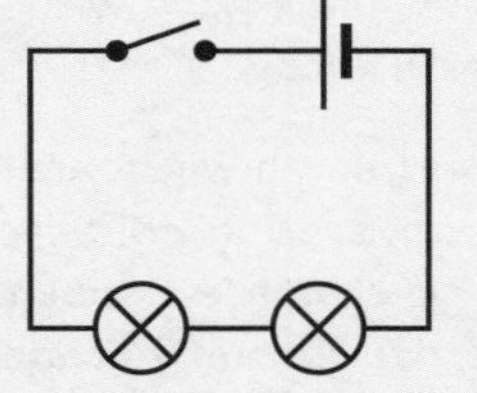

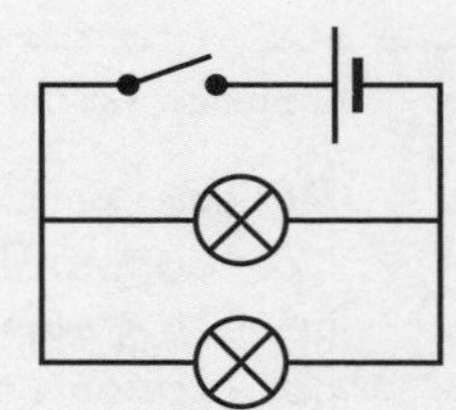

The brightness of the bulbs in the series circuit will be much less than the brightness of the bulbs in the parallel circuit.

The electrical energy that is transferred to light is directly proportional to the potential difference across each bulb and the size of the current flowing through each bulb. The brightness of the bulbs will be greatest when the energy transferred per second is greatest.

In the series circuit, the potential difference across each bulb will be 3 V, whereas in the parallel circuit the potential difference across each bulb will be 6 V. Also, the current flowing through each of the bulbs in the parallel circuit will be greater than the current flowing through the bulbs in the series circuit. Since the current and the potential difference in the parallel bulbs are greater, more energy will be transferred per second and the bulb will be brighter.

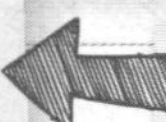

The opening sentence simply states how the brightness of the lamps compares in each of the two circuits.

The answer then states the relationship between electrical energy, current and potential difference before relating brightness to the energy transferred per second, or power, of the bulbs.

The comparison of the bulbs' brightness is related to the relative sizes of any current, potential difference and energy transferred per second.

The rate at which energy is transferred from one form to another is called power. When dealing with electrical circuits, the power is calculated using the equation $P = IV$. You can read more about this on page 229.

Now try this

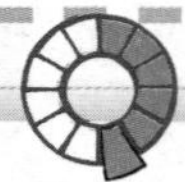

The UK mains electricity supply operates at 230 V a.c. Explain how safety features help to keep users safe. Your answer should refer to the structure and function of a plug. **(6 marks)**

Magnets and magnetic fields

Certain materials can be magnetised to become **permanent** or **temporary magnets**.

Magnets and magnetic fields

Like magnetic poles **repel**. **Unlike** magnetic poles **attract**.

The magnetic field is strongest at the poles. The field lines are shown to be closer together and more concentrated.

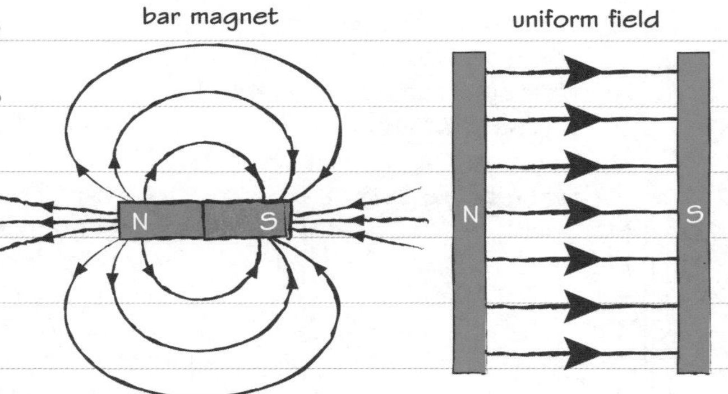

Magnetic field lines are always from N to S.

The field is constant. This is shown by parallel, equally spaced field lines.

Permanent and induced magnets

Magnetic materials include **cobalt**, **steel**, **iron**, **nickel** and magnadur.

A **permanent magnet** has poles which are N and S all of the time. They are made from steel or magnadur.

A **temporary magnet** can be magnetised by bringing a permanent magnet near to it. When the permanent magnet is removed, the temporary magnet loses its magnetism.

Uses of magnets

Permanent	Temporary
fridge magnets	electromagnets
compasses	circuit breakers
motors and generators	electric bells
loudspeakers	magnetic relays
door closers on fridges	

Plotting compasses

A plotting compass is used to plot the shape and direction of magnetic lines of force. You can do this by laying a bar magnet on a piece of paper. Put the plotting compass by one pole of the magnet. Draw a dot by the needle away from the magnet. Move the plotting compass so that the other point is by the dot you have just drawn. Continue doing this until you have mapped all the way around the magnet.

The Earth's magnetic field has the same pattern as that of a bar magnet and can be plotted using a plotting compass. The behaviour of compasses is evidence that the Earth has a magnetic field.

Worked example

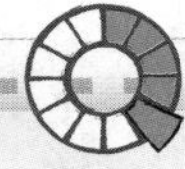

What will the plotting compass needle look like when it is placed at point 1? **(1 mark)**

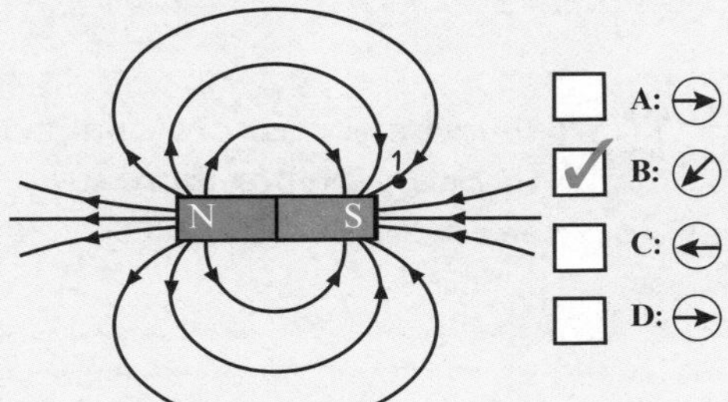

- ☐ A: (→)
- ☑ B: (↙)
- ☐ C: (←)
- ☐ D: (→)

Explain how you can tell if a magnetic material is a permanent or a temporary magnet. **(2 marks)**

A temporary magnetic material will always be attracted by a magnet, but a permanent magnet can be both attracted and repelled.

A magnadur magnet can be used to do this. Magnadur magnets are ceramic and have their poles on the larger flat faces.

Now try this

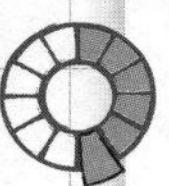

1 Draw the field lines for:
(a) a weak bar magnet **(2 marks)**
(b) a strong uniform field. **(2 marks)**

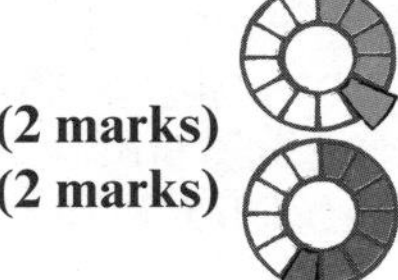

2 Explain why unmagnetised iron is always attracted by a permanent magnet. **(3 marks)**

3 Suggest why motors use permanent magnets. **(2 marks)**

Current and magnetism

An **electric current** will create a **magnetic field**. The **shape, direction** and **strength** of the field depend on a number of factors.

The magnetic field around a long straight conductor

Be careful! Do not confuse the direction of the current with electron flow. You need to use the direction of conventional current which is taken to be from + to −.

The shape of the magnetic field around a long straight conductor can be thought of as a series of concentric circles.

I

I

B

If you point the thumb of your right hand in the direction of the flow of conventional current, the direction that your fingers curl will be the direction of the magnetic field.

The strength of the magnetic field depends on:

- the size of the current in the wire: it is directly proportional to the current.
- the distance from the wire: it is inversely proportional to the distance from the wire.

The direction of the magnetic field depends on the direction of the electric current in the conductor.

The solenoid

The magnetic field lines of the individual coils in a solenoid add up to give a very strong, uniform field along the centre of the solenoid. However, the field lines cancel to give a weaker field outside the solenoid.

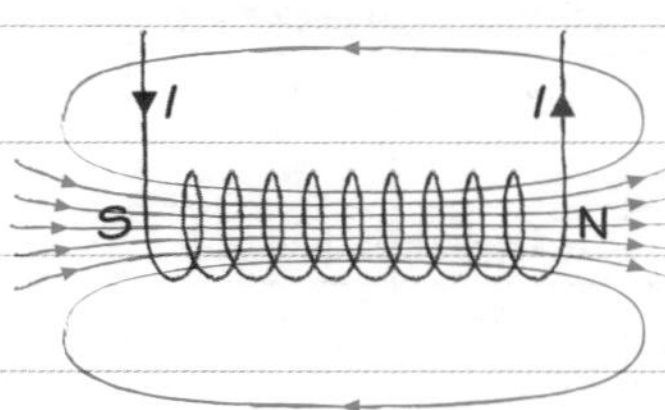

Worked example

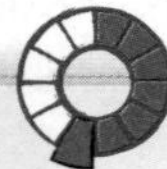

Explain how the strength of a magnetic field due to a current in a long straight wire changes when:

(a) the current increases **(1 mark)**

It will increase.

(b) the distance increases. **(1 mark)**

It will decrease.

Maths skills **Directly proportional** means that as **one variable increases** the other **variable increases** at **the same rate.** If the current doubles, the magnetic field strength will also double.

Maths skills **Inversely proportional** means that as **one variable increases** the other **decreases**. If the distance doubles, the field strength halves.

Now try this

1. Draw the magnetic field around a wire when the current is flowing downwards. **(2 marks)**
2. The current flowing in a long straight wire and the distance from the wire are both doubled. Explain what effect this will have on the magnetic field strength. **(3 marks)**
3. Suggest how the magnetic field strength inside a solenoid could be increased. **(3 marks)**

Current, magnetism and force

A current-carrying conductor, placed near a magnet, will experience a force due to the interaction of their magnetic fields.

Fleming's left-hand rule and the motor effect

An electric current flowing through a wire has a magnetic field. If you put the wire into the field of another magnet the two fields affect each other and the wire experiences a force. This is called the **motor effect**.

The maximum force on the wire occurs when the current is at right angles to the lines of the magnetic field. We can work out the direction of the force using **Fleming's left-hand rule**.

This rule uses 'conventional current' which flows from the + to the − of a cell. This is the opposite direction to the flow of electrons.

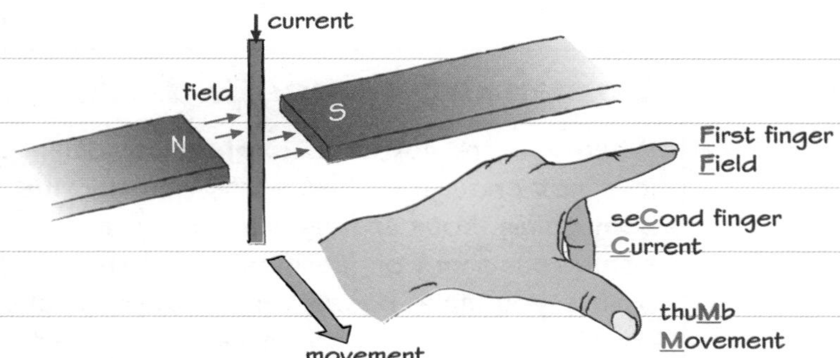

- There is **no force** if the **current** is **parallel** to the **field lines**.
- If the **direction** of either the **current** or the **magnetic field** is reversed the direction of the **force** is reversed.
- The size of the **force** can be **increased** by **increasing** the **strength** of the **magnetic field**, or **increasing** the **size** of the **current**.

Force on a current-carrying wire

Calculate the force on a wire using the equation:

force on a conductor at right angles to a magnetic field carrying a current (N) = magnetic flux density (T or N/A m) × current (A) × length (m)

$$F = B \times I \times l$$

Maths skills: Converting units

Magnetic flux densities are often given in millitesla, mT.

1 T = 1000 mT, so divide by 1000 to convert from mT to T.

For example,
3 mT = 3/1000 = 0.003 T

Worked example

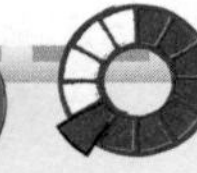

A 20 cm length of wire is placed in a 0.004 T magnetic field as shown. A current of 1.5 A flows through the wire.

Determine:

(a) the direction of movement of the wire **(2 marks)**

(b) the size of the force acting on the wire. **(2 marks)**

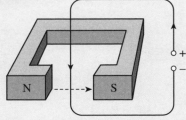

(a) Using Fleming's left-hand rule, the field acts from left to right (from N to S), the current is downwards, so the wire will move out of the plane of the paper towards you.

(b) The force acting on the wire is calculated using the equation

F = B × I × l, which leads to:

F = 0.004 T × 1.5 A × 0.2 m

F = 0.012 N

Now try this

1 Explain why a current-carrying wire experiences a force when placed near a magnet. **(2 marks)**

2 Calculate the force acting on a wire when B = 1.2 mT, I = 3.6 A and L = 50 cm. **(3 marks)**

3 For a current-carrying wire placed in the region of a magnetic field, explain how:

(a) the direction of the movement of the wire can be changed **(2 marks)**

(b) the speed of movement of the wire can be increased. **(3 marks)**

Extended response – Magnetism and the motor effect

There will be at least one 6-mark question on your exam paper. For these questions, you will need to think scientifically and structure your answer logically, showing how the points you make are related to each other.

You can revise the topics for this question, which is about **magnetism** and the **motor effect**, on pages 233 to 235.

Worked example

The diagram shows a conventional current flowing upwards in a current-carrying wire that has been placed in the region of a magnetic field.

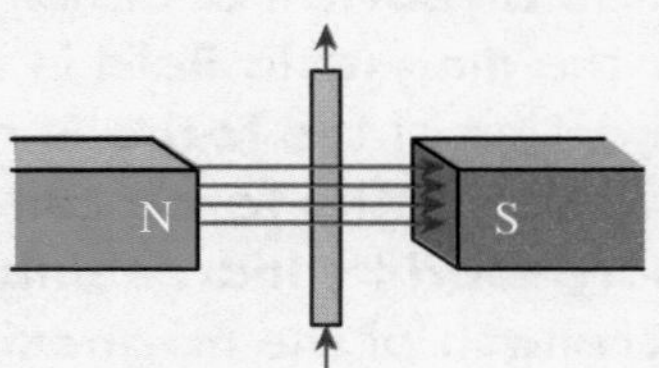

Explain what will happen to the wire when the size of the current that is flowing in the wire is increased. Your answer should refer to the size and direction of any forces acting on the wire.

(6 marks)

When a current flows in a long straight wire, a circular magnetic field is produced around the wire. This magnetic field will interact with the field between the magnetic poles and a force will act on the wire.

To start the answer, reference is made to the magnetic field around the wire interacting with the horizontal magnetic field that is acting between the magnetic poles.

Since the current is flowing upwards in the wire and the magnetic field between the poles is always from N to S, it is possible to use Fleming's left-hand rule to determine the direction of the force acting on the wire.

A short statement is made referring to the direction of the field as always being from N to S. This is always true.

The force acting on the wire will cause it to move into the plane of the paper (away from us). It will move faster away from us if the current is increased. It is also possible to describe how the size of the force will change if the current increases, based on the equation $\mathbf{F = B \times I \times l}$, assuming that the length of the wire and the magnetic flux density are kept constant.

This part of the answer refers to the direction of the force acting on the wire based on Fleming's left-hand rule. It also describes how the size of the force will change if the current increases based on the use of the equation $\mathbf{F = B \times I \times l}$.

Command word: Explain

When you are asked to **explain** something, it is not enough to just state or describe something. Your answer must contain some reasoning or justification of the points you make based on the appropriate physics. This needs to be communicated clearly using good spelling, punctuation and grammar.

Now try this

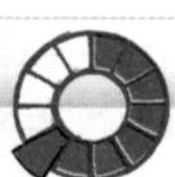

Magnetism can be used to attract a variety of materials of different masses. Explain how the strength of the magnetic field can be changed to make this possible. Your answer should refer to how the size of the magnetic field strength can be changed **with** and **without** the use of an **electric current**. **(6 marks)**

Electromagnetic induction and transformers

A voltage or current can be induced by having a conductor within the region of a changing magnetic field. Transformers make use of this for changing the size of an alternating voltage or current.

Inducing a current in a wire

If you **move** part of a loop of wire in a **magnetic field**, an electric current will flow in the wire. This is called **electromagnetic induction**, and the current is an **induced current.**

You can get the same effect by keeping the wire still and moving the magnet.

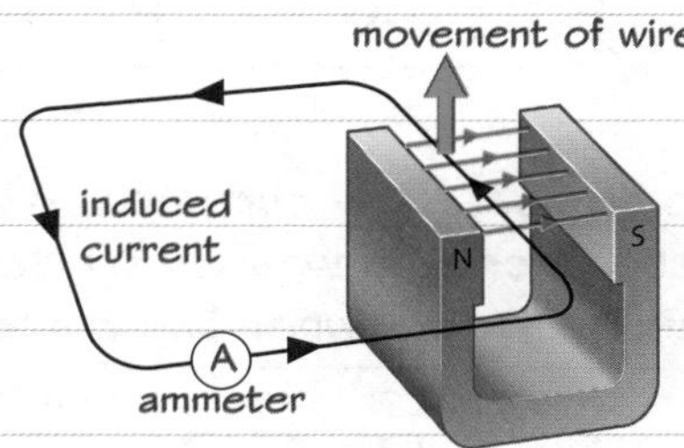
movement of wire
induced current
N
S
A
ammeter

Factors that affect the induced current

You can change the **direction** of the current by:

- changing the **direction of motion** of the wire
- changing the **direction** of the **magnetic field**.

You can increase the **size** of the current by:

- moving the wire **faster**
- using **stronger** magnets
- using **more loops** of wire, so there is more wire moving through the magnetic field.

Transformers

A transformer can be used to change the potential difference of an alternating electricity supply.

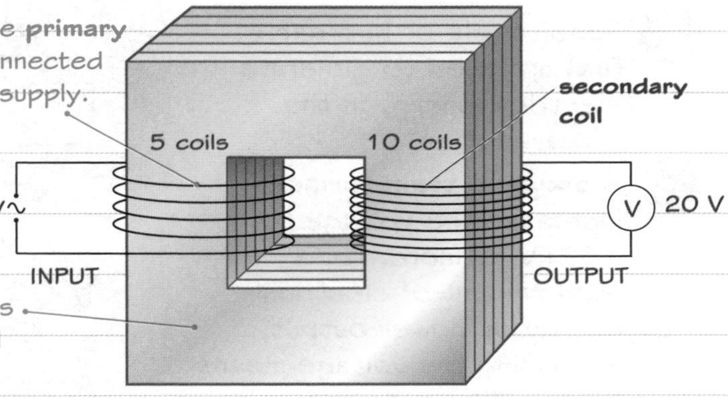
Remember that the **primary coil** is the one connected to the electricity supply.
secondary coil
5 coils
10 coils
10 V~
V
20 V
INPUT
OUTPUT
The core must be made of a magnetic material so that it can channel the magnetic lines of force from the primary coil to the secondary coil.

When an alternating current flows through the primary coil it produces a changing magnetic field in the iron core. This magnetic field induces an alternating potential difference across the secondary coil. The alternating potential difference in the secondary coil leads to an alternating current when the circuit is completed.

Worked example

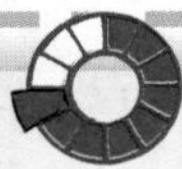

The current supplied to a transformer is 18 A at 12 V. The output potential difference is 144 V. Calculate the output current, assuming that the transformer is 100% efficient. **(4 marks)**

$V_p \times I_p = V_s \times I_s$

$12\text{V} \times 18\text{A} = 144\text{V} \times I_s$

$I_s = 12\text{V} \times \dfrac{18\text{A}}{144\text{V}} = 1.5\text{A}$

Now try this

1. State **two** ways in which you can change:
   (a) the direction of the induced current in a wire **(2 marks)**
   (b) the size of the induced current in a wire. **(2 marks)**
2. A transformer with an input potential difference of 30 kV has a power output on the secondary coil of 3 kW. Calculate the current in the primary coil. State the assumption that you make. **(3 marks)**
3. Explain why a transformer will work with an a.c. input but not a d.c. input. **(3 marks)**

Transmitting electricity

The **National Grid** is a system of wires that **transmits electricity** from power stations to where the electricity is needed and used.

The National Grid

Electricity is generated and transported to our homes, hospitals and factories by the National Grid. The National Grid is the **wires** and **transformers** that transmit the electricity; it does **not** include the **power stations** and the **consumers**.

To ensure that transmission is efficient, and that very little energy is lost as heat, the voltage and current values need to be chosen carefully at different stages.

Energy, current and voltage

Power, like energy, is always conserved. The equation that governs this is:

$$V_p \times I_p = V_s \times I_s$$

So we can transmit power at a low current and a high voltage (I × **V**) or we can transmit it at a high current and a low voltage (**I** × V).

$$P = I \times V \quad \text{or} \quad P = I \times V$$

We choose to use a low current and a high voltage because less energy is wasted to the surroundings as thermal energy stores when the current is low.

Transformers and the National Grid

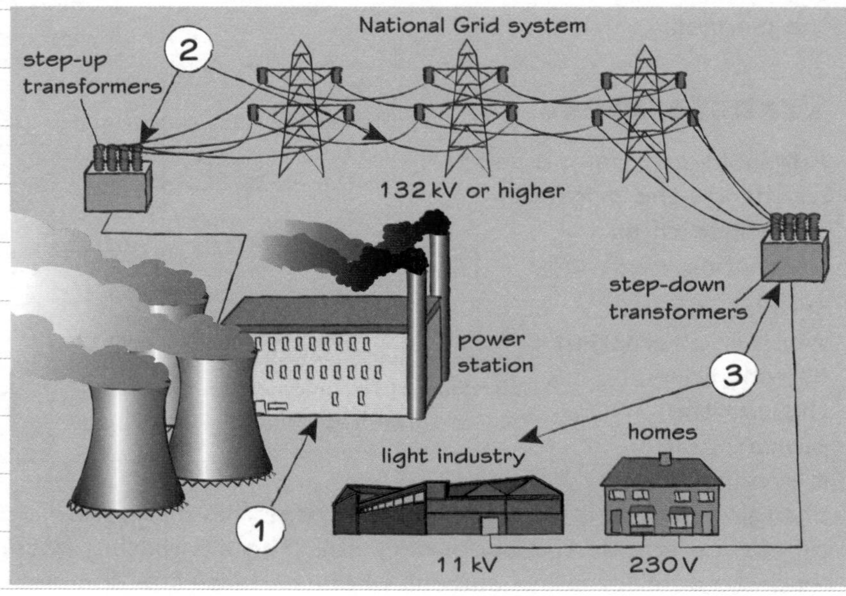

1. **Fossil fuels** or **nuclear fuel** are used to generate electrical energy in the power station.
2. A **step-up transformer** increases the voltage to 132 kV or more. For a fixed amount of electrical energy or power output, increasing the voltage means there will be a decrease in current. Electrical energy is transmitted at a high voltage and low current through the wires to reduce energy losses as heat in the wires. It also means thinner wires can be used, which reduces costs.
3. **Step-down transformers** decrease the voltages from the National Grid for safer use in our homes and industry. Reducing the voltage means there will be an increase in current.

Worked example

Explain why electrical power is transmitted at a high voltage and a low current on the National Grid. **(3 marks)**

Power is transmitted at a high voltage and a low current because when a high current is used more energy is wasted to the surroundings as thermal energy. Less of the input energy will be transmitted to the consumer as useful electrical energy.

Less current means thinner wires. Thinner wires means less wire, so less copper to purchase by the company. Low-resistance wires also reduce power losses.

Now try this

1 State what the National Grid:
(a) includes **(2 marks)**
(b) does not include. **(2 marks)**

2 Explain why transferring electrical energy at a high current is not efficient. **(3 marks)**

3 Describe the best way of transmitting 10^9 W of power over the National Grid. **(2 marks)**

Extended response – Electromagnetic induction

There will be at least one 6-mark question on your exam paper. For these questions, you will need to think scientifically and structure your answer logically, showing how the points you make are related to each other. You can revise the topics for this question, which is about **electromagnetic induction**, on pages 237–239.

Worked example

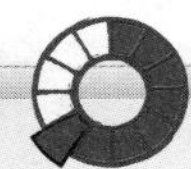

An alternating current can be induced in a conductor using a magnetic field.
Explain how this current is induced, and how the size of this induced current can be increased. **(6 marks)**

A current is induced in a conducting wire when it moves through a magnetic field. This happens because the wire moves through the lines of magnetic flux and cuts them. No current will be induced if the wire is stationary or moving parallel to magnetic field lines. If the direction of movement of the wire is reversed, the direction of the induced current is reversed. An alternating current is produced by regularly reversing the direction of movement.

The size of the induced current can be increased if the speed of movement of the wire, the magnetic flux density or the length of the wire present in the field are increased.

A transformer can also be used to increase an alternating current. The induced voltage will decrease if the number of turns in the secondary coil is less than the number of turns in the primary coil – this is called a step-down transformer. Since the voltage has decreased, then the current in the secondary coil must have increased in order to ensure that the principle of conservation of energy has not been broken.

Command word: Explain

When you are asked to **explain** something, it is not enough to just state or describe something. Your answer **must** contain some reasoning or justification of the points you make.

The answer starts with a clear explanation of how a current is induced. It is also made clear that the current is alternating because the wire changes direction. The nature of the energy transfers has been mentioned, with a reference that the movement of the wire leads to an electrical output.

Having explained how a current is induced, the factors that lead to an increase in the size of the induced current are then stated.

The answer then refers to the fact that a transformer can also lead to a larger current being induced in the secondary coil. Notice here that the student realises that it is a step-down transformer that leads to a greater current being induced since the decrease in voltage means there will be an increase in current to satisfy the equation $I_p \times V_p = I_s \times V_s$.

Electromagnetic induction can be seen as the motor effect in reverse. In the motor effect, a current flowing in a wire causes it to move when placed in the region of a magnetic field. Here, the energy change is the other way around – the movement of a wire through lines of magnetic flux leads to a current being induced in the wire, so the change is from kinetic to electrical.

Now try this

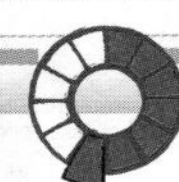

Transformers are used to step up and step down potential difference in the transmission of electricity across the UK. Explain how transformers work and why electricity is transmitted at a large voltage and a small current in the transmission lines of pylons. **(6 marks)**

Changes of state

Substances can change from one state of matter to another. The **change of state** depends on whether **energy is gained** or **lost** by the substance.

The three states of matter

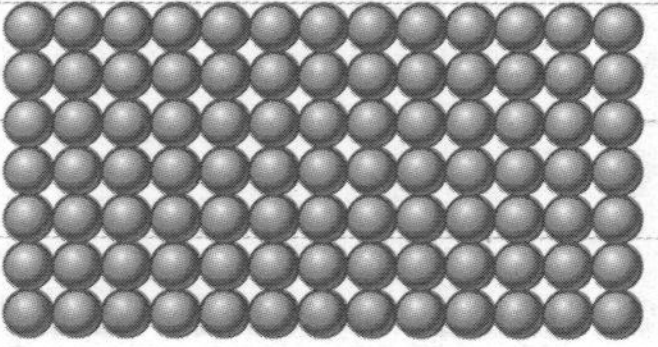

solid

In a solid the particles vibrate but they cannot move freely.

liquid

In a liquid the particles can move past each other and move around randomly.

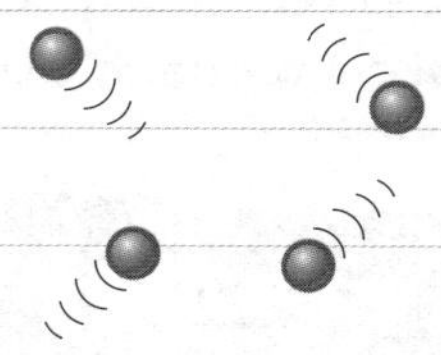

gas

In a gas the particles move around very fast and they move all the time. This is because they have a lot of **kinetic energy**.

Changing state

Changes of state are physical changes rather than chemical changes and they can be reversed.

The mass before the change is equal to the mass after the change. For example, when 500 kg of a solid melts it becomes 500 kg of a liquid.

When a body is heated or cooled, its temperature will change if its state of matter does not change. For example, if 750 ml of a liquid is heated, then the temperature will increase until it reaches its boiling point.

When a material changes state, it does so at a constant temperature. For example, when water boils, 1 kg of water at 100 °C will turn to 1 kg of steam at 100 °C. For more about this, see page 244.

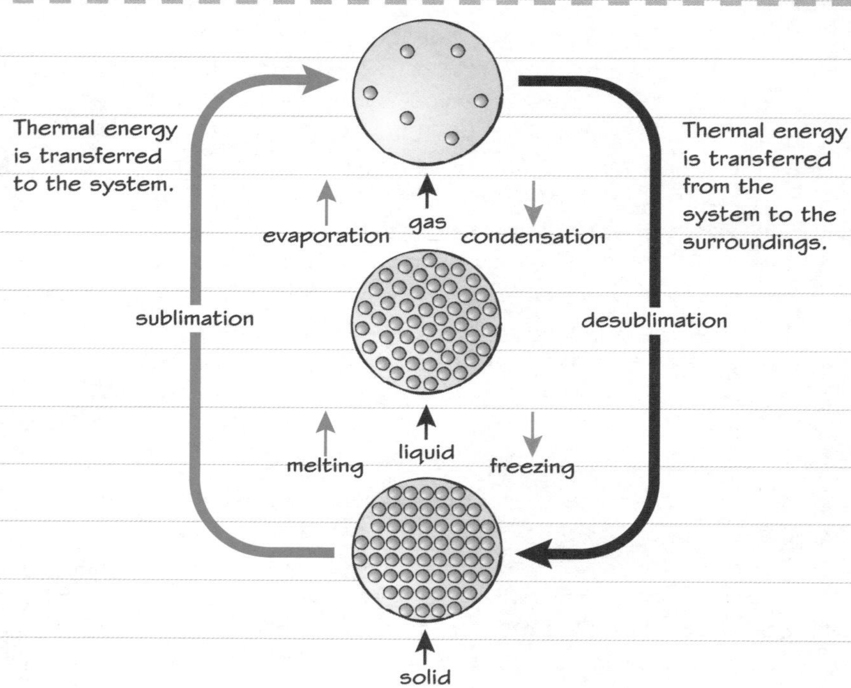

Worked example

Describe what happens when 2.0 kg of steam is allowed to cool against a glass plate, collected and then placed in a freezer. **(4 marks)**

2.0 kg of water will be produced when the steam turns to water at 100 °C. It will then cool as a liquid from 100 °C to room temperature. When placed in the freezer it will cool down from room temperature and turn to ice at 0 °C before continuing to cool to around −18 °C.

Now try this

1 Describe what happens when 1.5 kg of ice is left on a kitchen bench. **(3 marks)**

2 What effect does cooling a gas have on its pressure and volume? **(3 marks)**

3 Describe what happens when 200 kg of solid copper is heated to beyond its boiling point. **(5 marks)**

Density

The **density** of a material is a measure of the **amount of matter** that it contains **per unit volume.**

Density, mass and volume

Changing the amount of material will change both its mass and its volume. If the volume of a block of a particular material is doubled, its mass will also double.

The density of a material does not vary greatly for a given state of matter and relates to how closely packed the atoms or molecules are within the volume that it occupies.

The density of a material in the different states of matter varies greatly because of the arrangement of the particles in the state.

In a solid, the particles are closely packed together. The number of particles in a given volume is high.

In a liquid, the particles are usually less densely packed than in a solid. The number of particles in a given volume is less than for a solid. Solids are high density.

In a gas, the particles are spread out. The number of particles in a given volume is low. Gases are low density.

Calculating density

You can calculate density using the equation:

$$\text{density (kg/m}^3) = \frac{\text{mass (kg)}}{\text{volume (m}^3)}$$

$$\rho = \frac{m}{V}$$

LEARN IT! IT'S NOT ON THE EQUATIONS LIST

m
$\rho \times V$

Changing the volume of a substance will change its mass, but its density will be constant.

Other units for density can also be used, such as g/cm^3. The density of copper is $8.96\,g/cm^3$. This means that every $1\,cm^3$ of copper metal has a mass of 8.96 g.

Worked example

(a) A body has a mass of 34 000 kg and a volume of $18\,m^3$. Calculate its density. **(4 marks)**

$$\text{density} = \frac{\text{mass}}{\text{volume}} = \frac{34\,000\,\text{kg}}{18\,\text{m}^3} = 1900\,\text{kg/m}^3$$

(b) Calculate the mass of $0.2\,m^3$ of this material. **(4 marks)**

mass = density × volume
$= 1900\,\text{kg/m}^3 \times 0.2\,\text{m}^3 = 380\,\text{kg}$

Maths skills: Using correct units

When performing calculations involving density, mass and volume check that:

- you are using the correct equation to find density, mass or volume
- the units are consistent.

When the mass is in kg and the volume is in m^3 then the density will be in kg/m^3. Sometimes you may be dealing with g and cm^3, so check the units.

Worked example

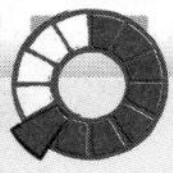

A solid has a density of $3.6\,g/cm^3$ and a mass of 320 g. Calculate its volume. **(4 marks)**

$$\text{volume} = \frac{\text{mass}}{\text{density}} = \frac{320\,\text{g}}{3.6\,\text{g/cm}^3} = 88.9\,\text{cm}^3$$

Maths skills: Checking units

Since volume = mass ÷ density then the units here will be g ÷ g/cm^3. This is the same as g × cm^3/g, which means the grams cancel and we are left with cm^3 – a unit of volume.

Now try this

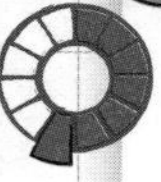

1 Calculate the density of a material that has a mass of 862 g and a volume of $765\,cm^3$. **(3 marks)**

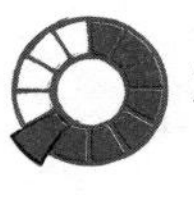

2 The density of aluminium is $2700\,kg/m^3$. Calculate the mass of a block of aluminium with a volume of $50\,000\,cm^3$. **(4 marks)**

Had a look ☐ Nearly there ☐ Nailed it! ☐

Investigating density

You can measure the volume and mass of solids and liquids to determine their densities.

Core practical

Aim

To determine the density of solids and liquids.

Apparatus

measuring cylinder, displacement can, electronic balance, ruler, various solids and liquids

Method 1 (for a solid)

1 Measure the mass of the solid using an electronic balance and record its mass in your results table.

2 Determine the volume of the solid. This can be done by measuring its dimensions and using a formula, or by using a displacement can to see how much liquid it displaces. Record this value in your results table.

Method 2 (for a liquid)

1 Find the mass of the liquid by placing the measuring cylinder on the scales, and then zeroing the scales with no liquid present in the measuring cylinder. Add the liquid to the desired level.

2 Record the mass of the liquid in g from the balance and its volume in cm^3 from the measuring cylinder.

Be careful to read the volume value correctly – at the bottom of the meniscus.

Results

Your results can be recorded in a table.

Material	Mass (g)	Volume (cm^3)	Density (g/cm^3)

3 Find the density of the solid by using the equation density = mass ÷ volume

Conclusion

The density of a solid and a liquid can be determined by finding their respective masses and volumes. Dividing the mass by the volume gives you the density of the solid or liquid.

There is more information about the relationship between mass, volume and density on page 241.

Be careful to use solids and liquids that are safe to use in this investigation. Mercury (a liquid metal) was used in these investigations until it was found to be carcinogenic.

When taking volume readings, make sure that you read the scales with your eye at the same level as the meniscus. Otherwise a parallax error will arise and your values will be incorrect.

Maths skills: Converting between units

Density can be given in g/cm^3 or kg/m^3.

Since 1 kg = 1000 g and $1\ m^3 = 1\,000\,000\ cm^3$:

- to convert from g/cm^3 to kg/m^3 multiply by 1000
- to convert from kg/m^3 to g/cm^3 divide by 1000.

For example, water has a density of $1\ g/cm^3$ or $1000\ kg/m^3$.

Now try this

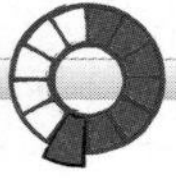

1 A measuring cylinder is placed on an electronic balance and its display is set to zero. The measuring cylinder is then filled with a liquid up to a volume of $256\ cm^3$. The mass shown is 454 g. Calculate the density of the liquid. **(3 marks)**

2 Explain how misreading the mass and volume values may occur and how this can lead to a higher than normal density value. **(4 marks)**

Energy and changes of state

You can calculate how much energy you need to **change the temperature** and **state** of a substance.

Specific heat capacity

Specific heat capacity is the **thermal energy** that must be **transferred** to change the **temperature** of **1 kg** of a material **by 1 °C**. Different materials have different specific heat capacities. Water has a value of 4200 J/kg °C, which means that 1 kg of water needs 4200 J of thermal energy to be transferred to raise its temperature by 1 °C.

You can calculate the thermal energy required using the equation: $\Delta Q = m \times c \times \Delta\theta$

change in thermal energy (J) = mass (kg) × specific heat capacity (J/kg °C) × change in temperature (°C)

Specific latent heat

Specific latent heat is the **energy** that must be transferred to change **1 kg** of a material from one **state of matter to another.** There are usually two values for specific latent heat:

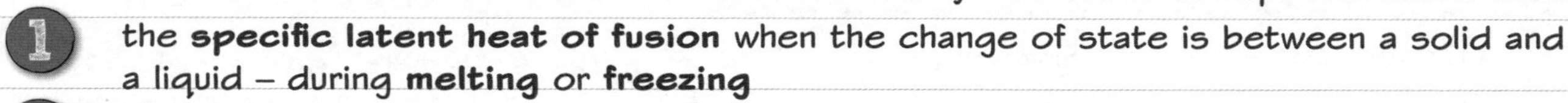

1. the **specific latent heat of fusion** when the change of state is between a solid and a liquid – during **melting** or **freezing**

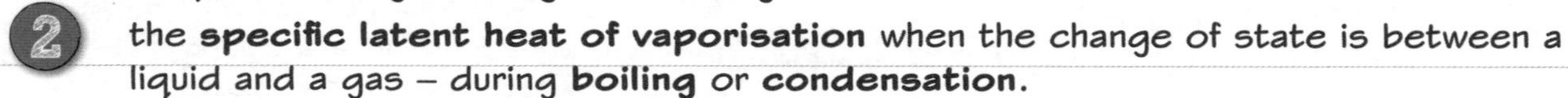

2. the **specific latent heat of vaporisation** when the change of state is between a liquid and a gas – during **boiling** or **condensation.**

You can calculate the thermal energy required using the equation:

thermal energy for a change of state (J) = mass (kg) × specific latent heat (J/kg)

$$Q = m \times L$$

Specific latent heat should **not** be confused with specific heat capacity. Calculations involving **specific latent heat** involve changes of state and **never involve a change in temperature**, since **changes of state always occur** at a **constant temperature.**

Worked example

(a) A mass of 800 g has a specific heat capacity of 900 J/kg °C and is heated from 20 °C to 80 °C. Calculate how much thermal energy was supplied. **(3 marks)**

$\Delta Q = m \times c \times \Delta\theta$

$0.8 \times 900 \times (80 - 20) = 43\,200$ J

(b) Calculate the energy required to convert 40 kg of ice at 0 °C to liquid water at 0 °C. The specific latent heat of fusion of water is 334 000 J/kg. **(3 marks)**

$Q = m \times L$

$Q = 40 \times 334\,000 = 13\,360\,000$ J

Worked example

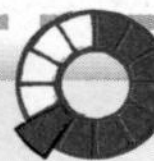

Ice at –40 °C is heated until it becomes steam at 110 °C. Sketch a graph to show the changes that take place during this process.

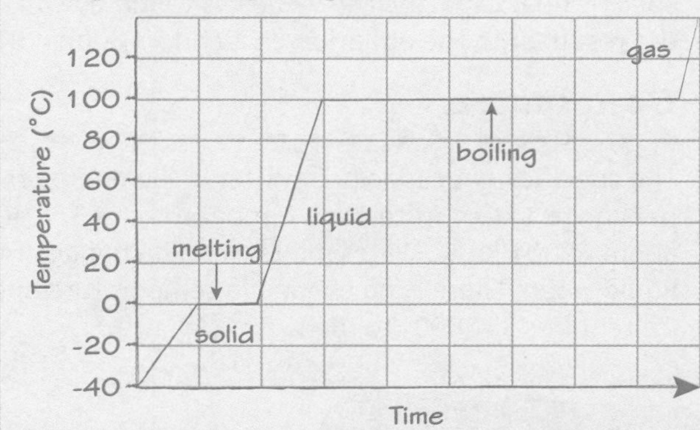

Now try this

1 Calculate the energy required to raise the temperature of 15 kg of water from 18 °C to 74 °C. **(3 marks)**

2 A mass of 1250 g of metal is connected to a heater rated as 2.8 A and 16 V and heated for 12 minutes. The temperature of the metal block increases by 26 °C. Calculate its specific heat capacity. **(4 marks)**

See page 229 for the equation to work out the energy transferred by the heater.

Thermal properties of water

You can determine the specific heat capacity of water using an electric heater. You can also observe how water behaves when a sample of ice is melted.

Core practical

Aim

To determine the specific heat capacity of water and to describe the behaviour of ice when melting.

Apparatus

water, ice, thermometers, electric heater, power supply, insulation, beakers, electronic balance

Method

1 Set the apparatus up as shown in the diagram.

2 Measure the mass (m) of the water using an electronic balance.

3 Record the potential difference (V) of the power supply and the current (I) through the heater.

4 Take temperature readings every 30 seconds for a time (t) until the water reaches the desired temperature. Record the initial and final temperatures of the water and find the change in temperature (ΔT).

thermometer

to power supply

electric heater

Results

Plot a graph of temperature against time. Calculate the specific heat capacity. The value c is found by substituting the results into the equation $c = (V \times I \times t) / (m \times \Delta T)$.

Conclusion

The specific heat capacity of water is the energy required to change 1 kg of water by a temperature of 1 °C. Its value is about 4200 J/kg °C. Water melts when it changes from ice to liquid water. There is no change in temperature and 1 kg of ice requires 334 000 J to melt.

There is more information about specific heat capacity on page 243, and more information about the energy transferred by a known current and potential difference on page 229.

Having insulation around the beaker will transfer less energy to the surroundings and give a more accurate value for the specific heat capacity of the water.

When recording the temperature readings, avoid parallax errors by reading the thermometer scale at eye level. Record the temperature values regularly, at equal intervals and ensure that the thermometer is in the middle of the liquid.

A graph of temperature against time for ice melting to water is shown here.

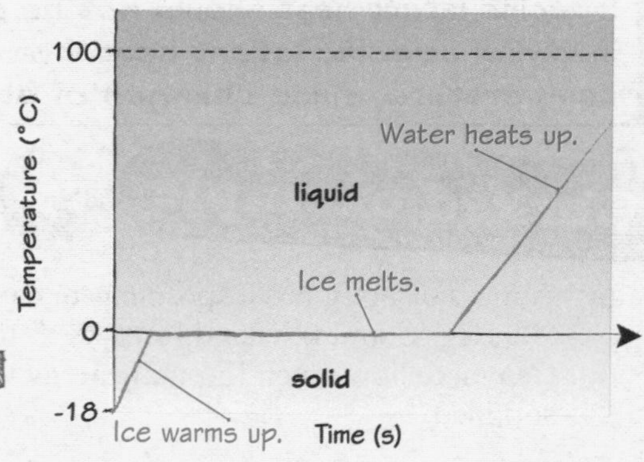

The word 'specific' here in this context is for 1 kg of mass. In other words, the values for specific heat capacity and latent heat are for when 1 kg either changes temperature by 1 °C or when it changes state.

Now try this

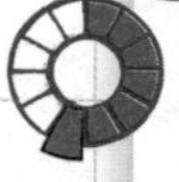

1 Explain why insulating the beaker leads to a more accurate value for the specific heat capacity of water. **(3 marks)**

2 Explain why melting occurs without there being any change in temperature. **(3 marks)**

3 Draw the apparatus that would allow you to obtain the temperature–time graph of melting ice. **(3 marks)**

Pressure and temperature

The **pressure** of a **gas** can be explained in terms of the motion of its particles.

The pressure of a gas

Gas pressure depends on the motion of the particles in the gas. Gas particles strike the walls of a container at many different angles. So the pressure of a gas produces a net force at right angles to the wall.

Pressure can be increased by:

- **increasing** the **temperature** of the gas
- **increasing** the **mass** of the gas
- **decreasing** the **volume** of the gas.

Pressure and temperature

The **pressure** of a **fixed mass** of **gas** at a **constant volume** depends on the **temperature** of the gas. When the temperature increases:

1. The gas molecules have a greater average kinetic energy.
2. The gas molecules move faster.
3. There are more collisions between the molecules and the walls of the container each second.
4. More force is exerted on the same area each second.
5. The pressure of the gas increases.

Absolute zero

As a gas is cooled the average speed of its particles falls and its volume gets smaller. At −273 °C the gas volume shrinks to zero. This temperature is called **absolute zero**. This is the zero temperature on the Kelvin scale. 0 K is equivalent to −273 °C.

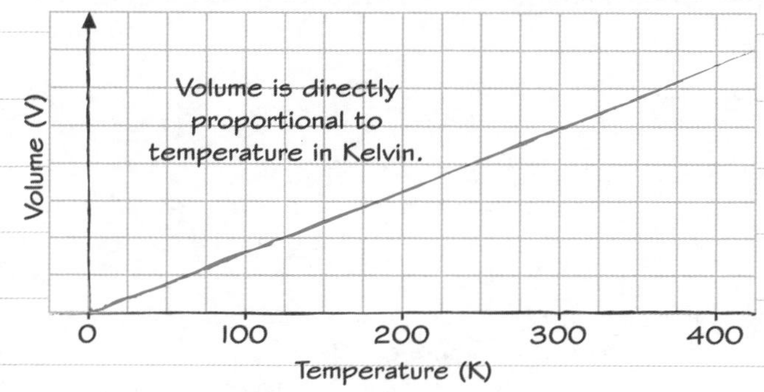

The temperature of a gas is directly proportional to the average kinetic energy of the gas molecules.

To convert between the Kelvin and Celsius scales use the appropriate formula:

- Celsius → Kelvin **add** 273 to the Celsius temperature
- Kelvin → Celsius **subtract** 273 from the Kelvin temperature

Worked example

Explain gas pressure using kinetic theory. **(3 marks)**

The particles (atoms or molecules) in a gas are continuously moving in a random way and colliding with the container walls. The force from these collisions produces pressure on the walls. On average the number and force of collisions is the same in all directions, so pressure is the same on all the walls of the container.

Worked example

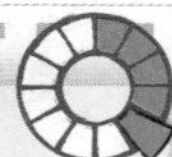

Convert:

(a) 150 °C to Kelvin **(1 mark)**

150 + 273 = 423 K

(b) 150 K to degrees Celsius. **(1 mark)**

150 − 273 = −123 °C

Now try this

1 (a) Convert 20 °C to a temperature on the Kelvin scale. **(1 mark)**
(b) Convert 300 K to its equivalent temperature on the Celsius scale. **(1 mark)**

2 The temperature of a fixed amount of gas is increased from −23 °C to 227 °C. Explain what happens to the average kinetic energy of the particles in this gas. **(3 marks)**

3 (a) State what happens to the average speed of gas particles as a gas is cooled. **(1 mark)**
(b) Explain the effect this has on the pressure of the gas in a rigid sealed container. **(2 marks)**

Extended response – Density

There will be at least one 6-mark question on your exam paper. For these questions, you will need to think scientifically and structure your answer logically, showing how the points you make are related to each other.

You can revise the topics for this question, which is about the **particle model**, on pages 240–245.

Worked example

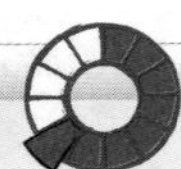

A team of investigators has been asked to determine whether rings being sold in a jeweller's shop are made from pure gold or whether they are fake.
Explain how the investigators could determine whether the rings were made from pure gold, which has a density of 19.3 g/cm^3.
Your answer should refer to a suitable experimental method that will allow you to determine the density of the material and any equations you would need to use. **(6 marks)**

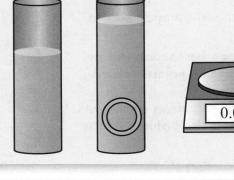

Density = mass ÷ volume; it is the mass per unit volume of a material. It is fixed for a substance in a given phase of matter.

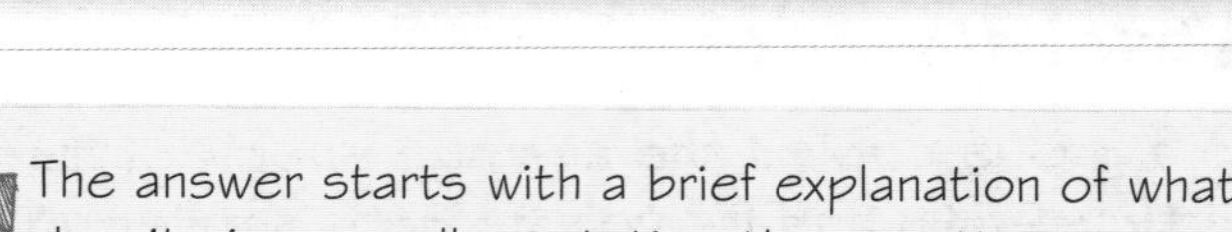

The answer starts with a brief explanation of what density is, as well as stating the equation.

To find the mass of a ring, we would put it on the electronic balance and get its mass in grams. To find the volume of the ring, we would lower it into a measuring cylinder of water and find the difference in the volume readings on the measuring cylinder scale. We do this by subtracting the volume reading before the ring has been submerged in the water from the volume reading after the ring has been submerged. This difference in volume is equal to the volume of the ring in cm^3. Dividing the value for the mass of the ring in g by the volume of the ring in cm^3 will get the value for the density of the ring in g/cm^3.

The method addresses how the mass and volume can be calculated. You could also refer to the use of a displacement can to determine the ring's volume.

If the value obtained for the density is very close to 19.3 g/cm^3 then the ring *might be* made from gold. If the density is very different from the value of 19.3 g/cm^3 then it is not pure gold.

The final part of the answer explains how the density value obtained tells you whether the material is gold or not – notice how the answer states that the ring *might be* gold, since you cannot say for definite as other materials may have this density.

When you structure your answer, make sure that you address the question in an order that allows you to deal with each of the points being asked for. Do not just list equations and facts. Communicate to the examiner that you understand how to find the density of the rings and how it will allow you to determine if the rings are real or fake based on the known value for the density of gold.

Now try this

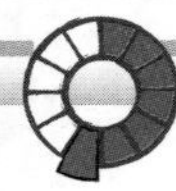

A gas contains many particles that are moving in a container of a fixed volume.
Explain how the pressure of a gas can be increased by changing its temperature.
Your answer should refer to energy, forces and the motion of the particles in the gas. **(6 marks)**

Elastic and inelastic distortion

The distortion of a material may be described as being **elastic** or **inelastic** and **requires more than one force.**

Bending, stretching and compressing

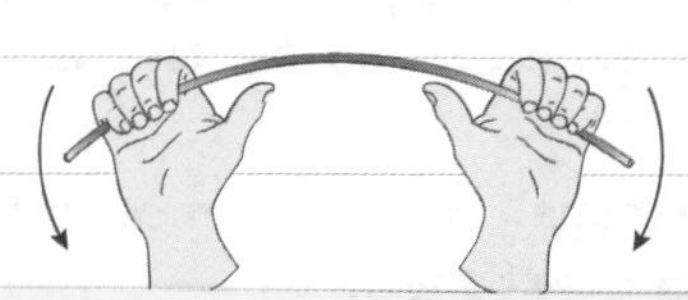

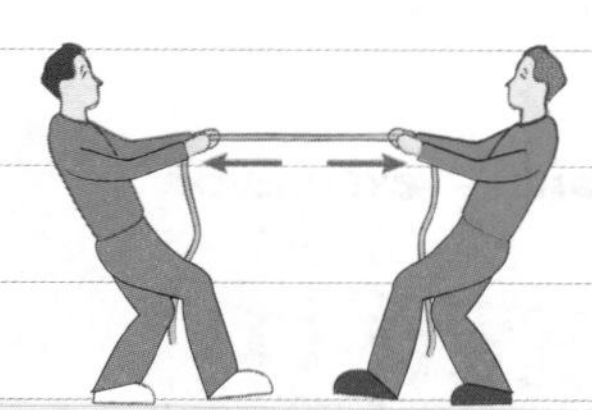

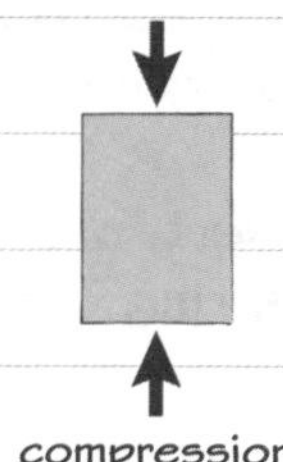

Bending requires two forces, one acting clockwise and one acting anticlockwise.

Stretching requires two forces of tension, acting away from each other.

Compression involves two equal forces acting towards one another.

Elastic distortion

Elastic distortion means that a material will return to its original shape when the deforming force is removed.

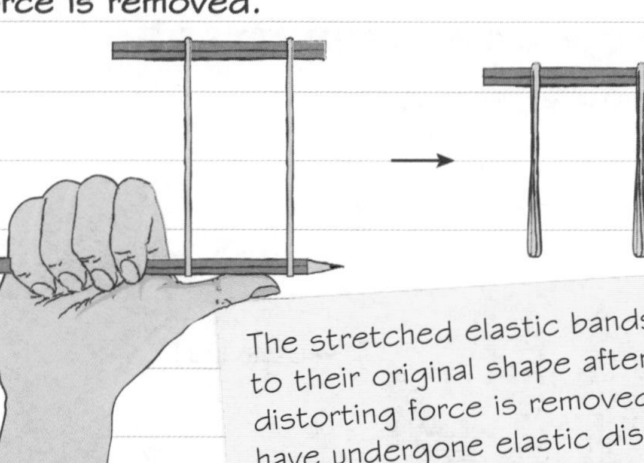

The stretched elastic bands return to their original shape after the distorting force is removed. They have undergone elastic distortion.

Inelastic distortion

Inelastic distortion means that a material will not return to its original shape when the deforming force is removed.

This spring has not returned to its original shape after being distorted, so it has undergone inelastic distortion.

Worked example

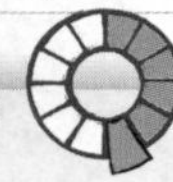

Describe what this graph shows for a spring that has been stretched. **(3 marks)**

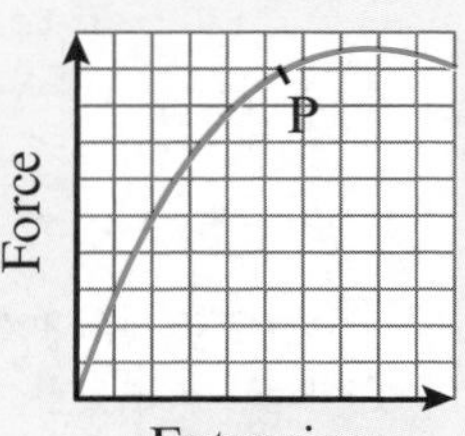

Initially, the relationship between force and extension is linear, so the behaviour of the spring is elastic for smaller forces.

Point P is the elastic limit. For forces applied after this, the relationship is no longer linear and the spring will not return to its original shape – it will exhibit inelastic distortion.

Now try this

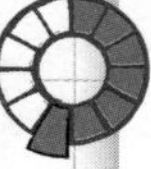

1 Explain the types of forces that cause:
(a) stretching **(2 marks)**
(b) compression. **(2 marks)**

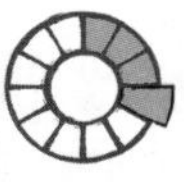

2 Give examples of materials that exhibit:
(a) elastic distortion **(2 marks)**
(b) inelastic distortion. **(2 marks)**

3 Explain why materials that are distorted often do not return to their original shape. **(3 marks)**

Springs

Forces can lead to the **distortion** of **elastic** objects, resulting in **energy** being stored.

Elastic distortion

A force applied to a spring can make it undergo linear elastic distortion. The equation that describes this is:

force exerted on a spring (N) = spring constant (N/m) × extension (m)

$F = k \times x$

LEARN IT! IT'S NOT ON THE EQUATIONS LIST

You can calculate the work done in stretching a spring using the equation:

energy transferred in stretching (J) = 0.5 × spring constant (N/m) × extension² (m²)

$E = \frac{1}{2} \times k \times x^2$

Force and linear extension

For elastic distortion, **extension** is **directly proportional** to the **force** exerted, **up to the elastic limit**.

The extension is the amount the spring stretches.

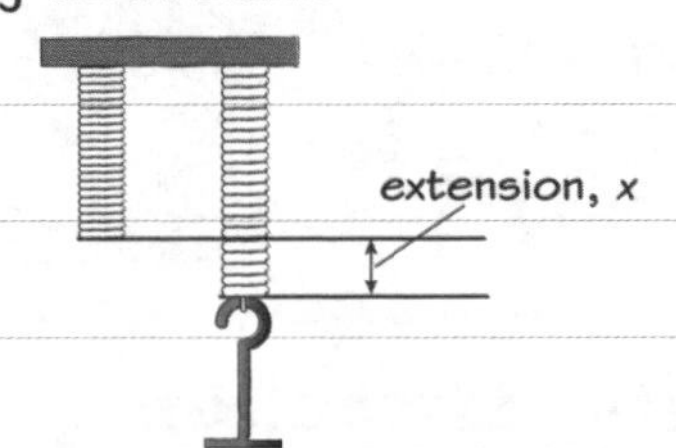
extension, x

Worked example

A spring is distorted elastically. It increases its length by 25 cm when a total weight of 12 N is added.

Calculate

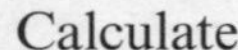

(a) the spring constant of the spring **(3 marks)**

$k = \frac{F}{x} = \frac{12\,\text{N}}{0.25\,\text{m}} = 48\,\text{N/m}$

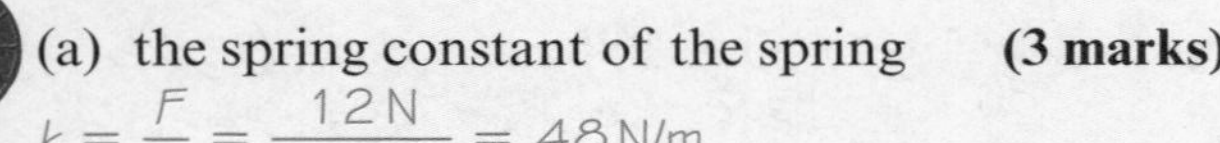

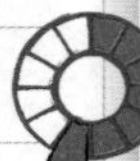

(b) the total energy stored in the spring. **(4 marks)**

energy stored = work done

$E = \frac{1}{2}k \times x^2$

$= \frac{1}{2} \times 48\,\text{N/m} \times (0.25\,\text{m})^2 = 1.5\,\text{J}$

When calculating the spring constant, work done or energy stored, ensure that force is in N and extension is in m.

Always use the extension. You may need to calculate this:

extension = total stretched length – original length

Force–extension graph

The gradient tells you the value of the spring constant, k. This is only true when the spring is showing elastic behaviour, i.e. when the line is straight.

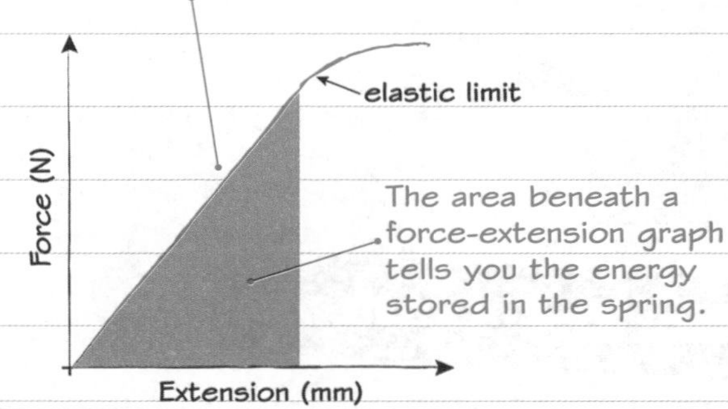
elastic limit
Force (N)
Extension (mm)
The area beneath a force-extension graph tells you the energy stored in the spring.

The area under the graph is the area of a triangle $= \frac{1}{2} \times F \times x$

The force is given by the equation $F = k \times x$

Substituting this into the equation gives

$E = \frac{1}{2} \times k \times x \times x = \frac{1}{2}k \times x^2$

Now try this

1 (a) Calculate the force that is required to extend a spring of spring constant 30 N/m by 20 cm. **(3 marks)**
(b) Calculate how much work is done on the spring under these conditions. **(3 marks)**

2 Explain how the stiffness of a spring is related to its spring constant. **(3 marks)**

3 Describe what happens to a spring for it to start displaying inelastic distortion. **(3 marks)**

Forces and springs

It is possible to determine the extension of a spring, and the work done, by applying different forces to the spring and then plotting a graph.

Core practical

Aim

To determine the extension and work done when applying forces to a spring.

Apparatus

spring, ruler, weights, retort stand, boss and clamp

Method

Arrange the apparatus as shown, with the ruler placed vertically alongside the spring and parallel to it. Add masses or weights and collect enough values to plot a graph of force against extension.

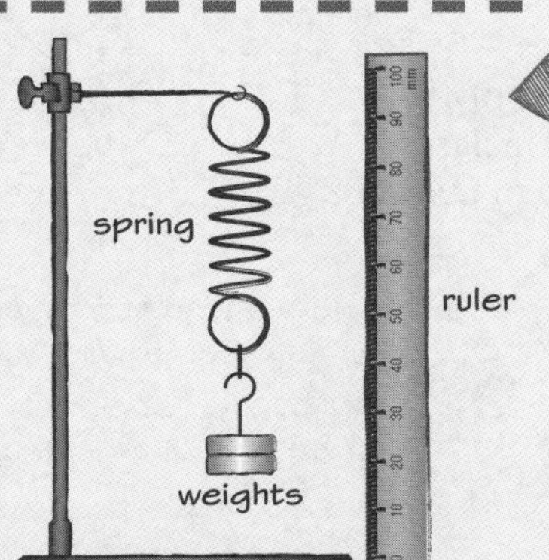

Results

Taking readings for force and extension allows a force–extension graph to be plotted:

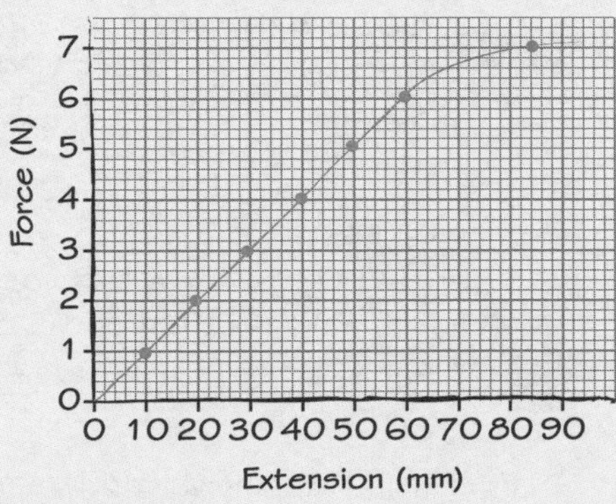

Conclusion

The gradient of the linear part of a force–extension graph gives you the value for the spring constant of the spring when it behaves elastically. The area beneath the graph equals the work done or the energy stored in the spring as elastic potential energy.

Wear eye protection when taking readings. A stretched spring stores a lot of elastic energy and this could damage your eyes if released.

When taking readings of the extension, read the values on the ruler at eye level to avoid parallax errors.

Weight and mass

Weight is a force, not a mass. A mass of 100 g is equivalent to a weight or force of 1 N.

Extension is equal to

new length (m) − original length (m)

Work done is equal to the energy stored in the spring.

The gradient of a force–extension graph gives you the spring constant.

The area beneath the line is the work done, or the energy stored as elastic potential energy, by the spring.

Maths skills You will need to convert extension values from mm to m before you can calculate an energy value, and state the result in joules, J. Forces also need to be stated in N, so any mass values recorded need to be converted from g to N.

Now try this

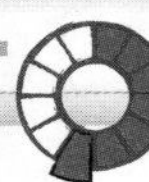

1 Calculate the energy stored by a spring of spring constant 0.8 N/m when extended by 48 cm. **(3 marks)**

2 Explain whether the spring constant and energy stored can be calculated for springs that are being compressed. **(3 marks)**

Had a look ☐ Nearly there ☐ Nailed it! ☐

Extended response – Forces and matter

There will be at least one 6-mark question on your exam paper. For these questions, you will need to think scientifically and structure your answer logically, showing how the points you make are related to each other.

You can revise the topics for this question, which is about **forces and matter**, on pages 247–249.

Worked example

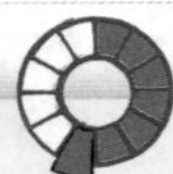

A student performs an investigation to compare the behaviour of two different springs. The force-extension graph for these springs is shown on the right.
Explain how the springs compare by using information from the graph.
Your answer should refer to the spring constants of the springs, the energy they store and the elastic or inelastic nature of their behaviour. **(6 marks)**

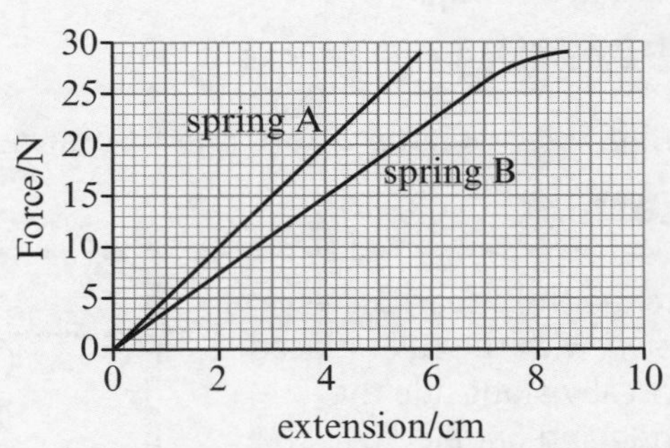

The force-extension graph shows two springs that have both been extended with a maximum load of 29 N. Spring A produces the greater gradient, so it has a higher spring constant than spring B. This means that spring A requires a greater force to extend it by the same amount. Spring A has a spring constant of 500 N/m whereas spring B has a spring constant of 375 N/m.

Spring A behaves elastically throughout as the gradient is constant throughout the experiment due to the extension being directly proportional to the force applied. Spring B behaves elastically up to a load force of about 25 N, at which point it starts to behave inelastically. This is shown by the reduction in gradient as the force leads to a greater and greater extension as the layers of atoms slide further apart. Upon removal of the load, spring A will return to its original shape but spring B will not – it will remain extended.

The energy stored by the springs is calculated by using the equation $E = \frac{1}{2} k x^2$. For the same extension, x, spring A will store a greater elastic potential energy due to its higher spring constant. For an extension of 4 cm, spring A will store 0.4 J whereas spring B will store only 0.3 J.

The answer starts with a quick explanation of what the graph shows The gradient tells you the spring constant. For spring A it is 25 N ÷ 5 cm = 5 N/cm and for spring B it is 22.5 ÷ 6 cm or 3.75 N/cm

Having discussed the spring constant values, the answer then focuses on the elastic and inelastic nature of the springs, using values from the graph and giving an explanation as to what will happen to the springs.

A gradient of 5 N/cm is equal to 500 N/m, since 100 times more force would be needed to cause an extension of 1 m compared to 1 cm.

Finally, the answer addresses the energy stored in each of the springs. Notice how the extension and the spring constant have to be converted into m and N/m in order to do the calculation.

4 cm = 0.04 m,
so $E = \frac{1}{2} \times 500 \times 0.04^2 = 0.4$ J (for spring A)
so $E = \frac{1}{2} \times 375 \times 0.04^2 = 0.3$ J (for spring B)

Now try this

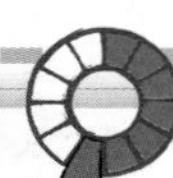

Describe the difference between elastic and inelastic distortion and explain how the spring constant of a spring and the elastic potential energy it stores can be determined when it is behaving elastically.
Your answer should refer to any graph, or equations, that may be useful. **(6 marks)**

Answers

Extended response questions

Answers to 6-mark questions are indicated with a star (*).
In your exam, your answers to 6-mark questions will be marked on how well you present and organise your response, not just on the scientific content. Your responses should contain most or all of the points given in the answers below, but you should also make sure that you show how the points link to each other, and structure your response in a clear and logical way.

Biology

1. Plant and animal cells

1 Muscle cells respire more **(1)**; they need more ATP/energy **(1)**; they carry out (aerobic) respiration/produce ATP **(1)**.

2 The plant is supported by the cell wall around each cell **(1)** and the vacuole in each cell when it is full **(1)**.

3 Not all plant cells photosynthesise, e.g. root hair cells. **(1)** Cells that do not photosynthesise do not need chloroplasts. **(1)**

2. Different kinds of cell

1 Similarity – any one of the following: has cell membrane/cytoplasm/DNA/ribosomes **(1)**

Difference – any one of the following: no nucleus/no nuclear membrane/no mitochondria/has plasmid/has cell wall **(1)**

2 It has a long thin root hair/protrusion **(1)** which increases the surface area for absorbing water/diffusion **(1)**.

3. Microscopes and magnification

magnification = 40 × 10 = ×400 **(1)**

image = 1.2 mm = 1200 µm so actual length of cell = 1200/400 = 3 µm **(1)**

4. Dealing with numbers

1 cell 0.1 millimetres (0.1 mm); chloroplast 2 micrometres (2 µm); protein molecule 10 nanometres (10 nm) **(2)**

2 (a) 1.3×10^{-4} m **(1)**

(b) 130 micrometres (130 µm) **(1)**

3 approx. 1 µm **(1)**

5. Core practical – Using a light microscope

(a) The description should cover the following points:

- Place the lowest power objective below the eyepiece.
- Place the slide on the microscope stage.
- Use the coarse focusing wheel to focus the image.
- If needed, move a higher power objective into place and focus using the fine focusing wheel.

(b) C

6. Core practical – Drawing labelled diagrams

Drawing of one red blood cell and one white blood cell using all rules described:

- clean sharp pencil lines
- drawn from photo
- correct relative sizes
- labelled to show nucleus and cytoplasm in white blood cell (no labelling expected on red blood cell)
- labels surrounding drawing with no overlapping label lines
- appropriate title with magnification **(3)**.

7. Enzymes

1 Part of the enzyme that the substrate fits into **(1)**; complementary shape to substrate/ right shape for substrate to fit in **(1)**

2 This means that enzymes are always at their optimum temperature **(1)** so that reactions occur at the fastest rate **(1)**.

8. Core practical – pH and enzyme activity

1 (a) rate of reaction = $\frac{1}{80}$ = 0.013 **(1)**

(b) It took the least time for the starch/ amylase mixture to react with iodine, so the rate of reaction was at its fastest **(1)**.

2 The values of pH used are much closer together than in the first experiment **(1)**, so the results should produce a more accurate value for the optimum pH **(1)**.

9. The importance of enzymes

1 The enzyme that digests lipids/lipase has a specific active site **(1)**; only lipids will fit in/ proteins will not fit in **(1)**.

2 ribosomes **(1)**

10. Getting in and out of cells

1 Osmosis is the net movement of water molecules from a region of higher water concentration to a region of lower water concentration **(1)** across a partially permeable membrane **(1)**.

2 (a) Neither diffusion nor osmosis need energy. **(1)**

(b) Diffusion does not need energy but active transport does need energy/ diffusion takes place down a concentration gradient, active transport takes place against a concentration gradient. **(1)**

3 The cells will still be able to absorb water **(1)**, by osmosis because that is a passive process **(1)**. The cells will not be able to absorb mineral ions **(1)**; mineral ions are absorbed by active transport and that needs energy from respiration **(1)**.

11. Core practical – Osmosis in potatoes

(a)

Solution concentration (mol dm^{-3})	Percentage change (%)
0.0	+22.0
0.2	+5.0
0.4	−8.0
0.6	−14.9
0.8	−24.1

(2)

(b) The solute concentration of the potato cells was approximately 0.3 mol dm^{-3}. **(1)** This is because at that concentration there would be no osmosis/net movement of water into or out of the potato cells, and so no gain or loss in mass. **(1)**

12. Extended response – Key concepts

*Answer could include the following points **(6)**:

Description of graph:

- As substrate concentration increases the activity of the enzyme increases.
- Reference to values at key points on the graph, e.g. activity fastest between about 75 and 85 mol/dm^3.
- Change in activity at different concentrations, as shown by steepness of curve.
- Activity is greatest where the graph line is steepest.
- Change in activity slows at higher substrate concentration.
- Curve flattens after about 125 mol/dm^3.

Explanation of graph:

- Starch is the substrate of amylase.
- At lower substrate concentration, there are many free active sites on enzyme molecules.
- At lower substrate concentration, adding more substrate means more substrate molecules can fit into free active sites and be broken down, so activity increases.
- At higher substrate concentration, most active sites of enzyme molecules are filled with substrate molecules.
- At higher substrate concentrations, adding more substrate does not increase activity because substrate molecules must wait until an active site becomes free.

13. Mitosis

They will be identical to the parent cell **(1)** with the same number of chromosomes **(1)**.

14. Cell growth and differentiation

1 It stops mitosis at prophase/before metaphase **(1)**, therefore new cells are not formed **(1)**.

2 a cell that is adapted to carry out a specific function **(1)**

3 It allows cells to become specialised **(1)** so they can carry out a function more effectively **(1)**.

15. Growth and percentile charts

1 Any one from: measure its change in height over time, measure its change in mass over time, measure the change in number and size of leaves over time (or similar). **(1)** *The key point to mention is change over time.* If the measurements increase over time, then the plant is growing. **(1)**

2 A percentile chart compares the growth (or an aspect of growth e.g. length, body mass) of an individual **(1)** against the growth (of the same aspect) of other individuals of the same sex and age **(1)**.

3 The increase in size of a balloon is not a permanent increase in size **(1)** while the increase in size of a child is a permanent increase in the number of cells in the child's body **(1)**.

16. Stem cells

1 Cutting the top off the shoot removes the meristem **(1)**, so the growing and dividing region is removed **(1)**.

2 Embryos have to be destroyed when the stem cells are removed **(1)**; some people think it is wrong to do this because they believe embryos have a right to life **(1)**.

3 (a) They are easier to extract OR They can produce more different kinds of cell. **(1)**

(b) Adult stem cells from the patient will be recognised **(1)** but embryonic stem cells will not and the body will reject them **(1)**.

17. Neurones

1 Sensory neurones carry electrical impulses from receptor cells to the central nervous system. **(1)**

Motor neurones carry nerve impulses from the central nervous system to effectors. **(1)**

Relay neurones link other neurones together and make up the nervous tissue of the central nervous system. **(1)**

2 The long axon and dendron of the sensory neurone means the cell can collect impulses from receptor cells and carry them through the body to the central nervous system/ spinal cord/brain. **(1)**

The myelin sheath insulates the neurone from surrounding neurones and helps the electrical impulse to travel faster. **(1)**

18. Responding to stimuli

1 receptor (in skin) → sensory neurone **(1)** → relay neurone (in spinal cord) **(1)** → motor neurone **(1)** → effector (muscle) **(1)**

2 Only one neurone produces neurotransmitter. **(1)**

There are only receptors for the neurotransmitter on the neurone receiving the impulse. **(1)**

19. Extended response – Cells and control

*Answer could include the following points **(6)**:

Benefits:

- Embryonic stem cells (ESCs) are able to turn into many types of cell.
- ESCs could be injected into the heart to develop into new specialised cells, or used to make new specialised cells in the lab before injecting them into the heart.
- ESCs good for treating heart damage because several different types of specialised cell may be damaged
- Replacing damaged cells will improve health of patient.

Risks:

- ESCs in the body may produce the wrong type of specialised cell in the wrong place.
- ethical problems with using ESCs because they involve destruction of embryos
- ESCs may cause cancer if cell division is uncontrolled.
- New specialised cells may be rejected by immune system as they are not originally from the body.
- To avoid rejection, patient would need to take drugs for life, which is expensive and may produce harmful side effects.

20. Meiosis

1 four cells produced **(1)**; haploid **(1)**; genetically different **(1)**

2 Meiosis reduces the chromosome number to half the diploid number **(1)**; during fertilisation two gametes fuse, so the zygote has the full diploid number **(1)**; if meiosis did not happen before fertilisation, the number of sets of chromosomes in the zygote would double each time **(1)**. *Another possible answer is that mixing of genes is important for variety in offspring.*

3 haploid = one set of chromosomes **(1)**; diploid = two sets of chromosomes/ chromosomes are in pairs **(1)**

4 B **(1)**

21. DNA

1 any suitable sentences that give appropriate definitions, such as: (a) A gene is a short section of DNA that codes for a specific protein. **(1)** (b) There are four bases in DNA: A, T, C and G. **(1)**

2 GCTA **(1)**; because A pairs with T and C pairs with G **(1)**

3 DNA forms a double helix structure **(1)** which consists of two strands coiled together **(1)** and held together by the weak hydrogen bonds **(1)** that link the complementary base pairs of A and T or C and G **(1)**.

22. Genetic terms

1 A chromosome is a long strand of DNA. **(1)** A gene is a small section of a chromosome/ DNA that gives the instructions for producing a particular characteristic. **(1)** An allele is an alternative form of a gene. **(1)**

2 (a) It is heterozygous **(1)** because it has one dominant and one recessive allele **(1)**.

(b) purple flowers **(1)**, because purple is dominant over white **(1)** *Remember that phenotype is what it looks like, so writing down the alleles would give you no marks for this question.*

23. Monohybrid inheritance

(a) genetic diagram or Punnett square showing this information (any appropriate letter used):

		parent genotype Bb	
	parent gametes	B	b
parent genotype bb	b	Bb brown	bb black
	b	Bb brown	bb black

(**1 mark** for setting out the parent genotypes and gametes correctly, **1 mark** for completing the offspring genotypes and phenotypes correctly)

The predicted outcome of phenotypes is 50% chance of brown and 50% chance of black (could also be presented as a ratio 1:1, or probability 1 in 2 for both colours). **(1)**

(b) The actual outcome is different to the predicted outcome because we would expect 2 black and 2 brown baby rabbits but we ended up with all black. **(1)** This is because at fertilisation it is chance which alleles are inherited. **(1)**

24. Family pedigrees

(a) heterozygous/Tt (other letters allowed) **(1)**

(b) genetic diagram or Punnett square showing offspring of Tt × tt **(1)**; 50% or 0.5 **(1)**

25. Sex determination

1 A **(1)**

2 chance is 50% **(1)**; diagram or Punnett square as shown on page 25 (**1 mark** for parent phenotype, **1 mark** for gametes, **1 mark** for possible outcomes)

3 There is always an X from the mother **(1)** because women are all XX and can only pass on an X chromosome **(1)**.

26. Variation and mutation

1 two from the following for **1 mark** each: change in DNA/gene; produces new allele; may produce a new/different protein **(2)**

2 Most body cells will not contain the new allele. **(1)** Most body cells do not divide so they will not produce new cells containing the new allele. **(1)**

27. The Human Genome Project

1 Advantage: person can lead a healthy lifestyle/named precaution e.g. do not smoke to reduce the chances of high blood pressure developing. **(1)**

Disadvantage: may depress person since there is no certainty high blood pressure will occur/ person could be discriminated against in some kinds of employment/ insurance. **(1)**

2 They have the same genes but different alleles **(1)**, therefore a different sequence of bases in their DNA **(1)**.

28. Extended response – Genetics

*Answer could include the following points **(6)**:

- Parents must be heterozygous/Rr
- because some offspring have smooth coat/rr.
- (Smooth-coated guinea pigs) inherit one allele from each parent.
- parental gametes shown as R and r
- genotypes of offspring as 1 RR:2 Rr:1 rr
- phenotypes correctly identified as RR and Rr rough-coated, rr smooth-coated.

29. Evolution

Mutation occurs conferring resistance. **(1)** Lice with mutant allele survive longer/non-resistant die. **(1)**

(Resistant) lice pass on allele to offspring. **(1)** Allele frequency of new allele/resistant allele increases. **(1)**

30. Human evolution

1 any two suitable features, such as: increase in brain size, increase in height, upright posture and ability to walk over long distances **(2)**

2 An increase in the complexity of the way the tools are made and in the range of tools produced **(1)** suggests that humans were developing in intelligence/skill OR suggests that humans developed their skills of tool-making as the skills were passed from generation to generation **(1)**.

31. Classification

1 3 domains are: Eubacteria, Eukaryota (including plants, animals, fungi, protists), Archaea (mainly bacteria living in hot/salty conditions) **(3)**

2 Research on genes/DNA **(1)** shows prokaryotes/bacteria need to be split into 2 groups **(1)**; some bacteria/archaea have genes that work more like eukaryotes **(1)**. **(max. 2)**

32. Selective breeding

1 Advantage: get crop plants/animals with higher yield/more meat/disease resistance etc. **(1)**

Disadvantage: you breed closely related organisms so they show less genetic diversity/may show genetic defects

OR

may inherit a good characteristic but also less desirable characteristics

OR

Many offspring do not inherit the desired characteristics. **(1)**

2 Natural selection occurs by chance and involves survival of organisms with a more advantageous phenotype. **(1)**

Selective breeding is carried out with a specific aim or result in mind. **(1)**

3 little genetic variation in the population **(1)**; so if a new disease appears, it might spread very easily **(1)**

33. Genetic engineering

1 transferring a gene from one organism to another/inserting new genes into an organism/modifying the genome of an organism **(1)** so the desired characteristic is produced in the organism/so that the organism has a new characteristic **(1)**

2 The gene for herbicide resistance is put into some plant cells. **(1)** The cells are treated/grown by tissue culture so they develop into new plants. **(1)**

3 (a) It makes it easier to see which mice have the human disease gene in their cells. **(1)**

(b) The gene for the disease is cut out of a human chromosome and joined to a glow gene from a jellyfish. **(1)** The combined genes are mixed with early mouse embryos so that the genes are taken into the cell nuclei and joined to the chromosomes. **(1)**

34. Stages in genetic engineering

1 human gene cut out of human DNA using restriction enzymes that leave sticky ends on the gene **(1)**; plasmid removed from bacterium and cut open using same restriction enzymes that produce matching sticky ends **(1)**; human gene and plasmid mixed with DNA ligase enzyme that joins the sticky ends together to make recombinant plasmid **(1)**; recombinant plasmid inserted into another bacterium where it makes the human insulin **(1)**

2 The bacteria can be grown in ideal conditions in large fermenters **(1)** so that they produce large quantities of insulin quickly and cheaply **(1)**.

35. Extended response – Genetic engineering

*Answer could include the following points **(6)**:

- Cut gene (for HGH) from human cell using restriction enzyme.
- Cut plasmid with same restriction enzyme.
- Join gene to plasmid using DNA ligase.
- Insert plasmid into bacteria.
- Grow bacteria in huge fermenters.
- Extract bacteria and purify growth hormone.

36. Health and disease

1 non-communicable because it is not caused by a pathogen; **(1)**

blood clots are not spread from one person to another **(1)**

2 spreads rapidly at first because people are susceptible; **(1)** people then develop immunity so fewer susceptible people **(1)**

37. Common infections

(a) HIV or Ebola

two suitable signs for each, e.g.

(HIV) no signs/symptoms for a long time; flu-like symptoms on first infection; many repeated infections as immune system stops working

OR

(Ebola) diarrhoea; muscle pain; extreme tiredness; internal and external bleeding **(2)**

(b) malaria

two suitable signs, e.g. fever; tiredness; muscle aches and pains; sweating and chills **(2)**

(c) Chalara

two suitable signs, e.g. leaf loss; bark lesions/damage **(2)**

38. How pathogens spread

The mosquito is a vector because it infects the human with the parasite/pathogen. **(1)**

Plasmodium/the protist is the pathogen because this infects the cells and causes the disease/symptoms. **(1)**

39. STIs

1 sexually transmitted infection **(1)**

spread by contact with sexual fluids/not using a condom during sexual intercourse **(1)**

2 This prevents contact with sexual fluid from his partner. **(1)**

40. Human defences

1 unbroken skin/mucus in breathing system **(1)**; lysozyme in mucus/saliva/tears OR hydrochloric acid in stomach **(1)**

2 This stops pathogens infecting the skin. **(1)**

The burn has damaged the outer layer of cells so pathogens can enter the body easily. **(1)**

41. The immune system

The pathogen has an antigen on its surface (that is unique to it). **(1)**

A lymphocyte with an antibody that fits the antigen is activated. **(1)** This lymphocyte divides repeatedly and secretes antibodies that destroy the pathogen. **(1)**

42. Immunisation

1 A vaccine contains a dead or harmless form of a pathogen or antigenic material that is used to immunise a person. **(1)** Immunisation means giving a vaccine to cause an immune response in the body. **(1)**

2 any two from: This gives the person another injection of antigens; increases the number of memory cells present; immune response greater; more antibodies made. **(2)**

43. Treating infections

1 Antibiotics only kill bacteria/do not destroy viruses. **(1)**

2 Antibiotic cannot bind to human ribosomes **(1)**; too big/cannot fit **(1)**.

44. New medicines

1 two from:

B may be cheaper;

A may have side effects;

patient may be allergic to A;

A may not be suitable for pregnant women;

A may not be compatible with other drugs the patient is taking **(2)**

2 The benefit of using animals as models is that the drugs can be tested for safety without harming humans. **(1)** A problem with using animals as models for humans is that the drugs may not work in the same way in the animal as they would do in a human. **(1)** *(Other problems may be acceptable, e.g. many people are against the use of animals for testing.)*

45. Non-communicable diseases

(a) so we can compare the different groups, **(1)** as different groups have different numbers of people in them **(1)**

(b) (Yes) (no mark); two from:

risk of cancer much higher in people with *p16* mutation;

use of figures from graph to support answer;

but only gives data for two kinds of cancer/ may not be true for all cancers **(2)**

46. Alcohol and smoking

(a) two from:

The more cigarettes smoked, the more likely a person is to die of coronary heart disease.

The older a person is the more likely they are to die of coronary heart disease.

The longer a person has been smoking, the more likely they are to die of coronary heart disease. **(2)**

(b) Smoking causes blood vessels to narrow, **(1)** increasing blood pressure (leading to CHD) **(1)**.

47. Malnutrition and obesity

1 The study is a very large study and so the results are valid. **(1)** The study is only of American men, so the conclusion may not apply to other groups of people. **(1)** The conclusion does not distinguish between fit and unfit men who are not overweight, and these two groups may have differences in their risk factors for health that also cause some of the differences between fit and unfit overweight people. **(1)**

2 She has a BMI of 31 and is obese. **(1)**

48. Cardiovascular disease

two from:

There is always a risk that the person might not recover after the operation;

surgery is expensive;

it may take time to recover after operation/risk of infection. **(2)**

49. Extended response – Health and disease

*Answer could include the following points **(6)**:

- Smoking cigarettes introduces poisonous substances such as tar and nicotine into the body that can cause cells to become cancerous leading to lung cancer.
- Alcohol is also poisonous to cells and could cause cells to become cancerous, leading to cancer of the liver, stomach, throat.
- Lack of exercise and poor diet can lead to obesity that can make people more susceptible to cancer.
- Stress can make people more susceptible to cancer.
- Poor diet, such as lack of roughage (wholemeal cereals, vegetables and fruit) can lead to bowel cancer.

50. Photosynthesis

1 Energy is taken in. **(1)**

Energy in the form of light is transferred to energy in the form of chemical (potential) energy in sugars. **(1)**

2 photosynthesise **(1)**; take in light energy **(1)**; make own food/sugars/organic molecules/ more cells **(1)**; source of food for animals/ humans **(1)**

51. Limiting factors

1 something that limits the rate of photosynthesis **(1)** when it is at a low level **(1)**

2 Photosynthesis produces oxygen. **(1)** The more rapidly oxygen is released, the faster the photosynthesis reactions must be happening. **(1)**

52. Core practical – Light intensity

1 Carbon dioxide is taken in by plants when they photosynthesise. **(1)** Removing carbon dioxide from the solution will increase its pH. **(1)** The rate of change in pH will indicate the rate at which carbon dioxide is removed from the solution. **(1)**

2 Temperature also affects the rate of photosynthesis. **(1)** If temperature is not controlled, then it will change with distance and so confuse the results. **(1)**

3 The bottle will act as a control **(1)** as it should show no change in pH as there will be no photosynthesis in the bottle **(1)**.

53. Specialised plant cells

1 contain many mitochondria **(1)**; supply energy from respiration/ATP for active transport **(1)** (of mineral ions)

2 **1 mark** for feature and **1 mark** for explanation, e.g.

pits in side walls; for sideways movement of water/mineral ions

no cell contents; so water flow not obstructed

no end walls; so water flow not obstructed

thick walls made of lignin; so cell does not collapse/strength. **(max. 6)**

54. Transpiration

1 movement of water through a plant **(1)** from roots to leaves **(1)**

2 Water enters root hair cell by osmosis;
moves through plant via xylem;
evaporates from leaf as water vapour;
via stomata;
draws water out of leaf cells/xylem. **(max. 4)**

3 On the surface the stomata could get blocked by water droplets **(1)**; this would reduce/ prevent gas exchange/transpiration **(1)**.

55. Translocation

1 from the phloem, **(1)** because dissolved sugars are transported around the plant in phloem **(1)**

2 Sucrose is made in the leaves after photosynthesis. **(1)** In the spring and summer most of the sugars will be transported in phloem to the growing parts of the plant, such as the tips of shoots and roots. **(1)** In the autumn, less sucrose will be transported to these growing parts and more will be transported to the potatoes for storage. **(1)**

3 xylem **(1)**; phloem **(1)**

56. Water uptake in plants

(a) $\pi r^2 d = 3.14 \times 0.5^2 \times 50 = 39.25\,mm^3$ **(1)**

rate = $39.25/5 = 7.85\,mm^3/min$ **(1)**

(b) higher rate/faster **(1)** because (increased temperature) increases rate of evaporation of water from leaf **(1)**

(c) all conditions kept the same except temperature **(1)**; one example is, e.g. air movement/light intensity **(1)**

57. Extended response – Plant structures and functions

*Answer could include the following points **(6)**:

- Bubbles were of oxygen gas.
- oxygen produced by photosynthesis
- rate of photosynthesis affected by temperature because increased temperature indicates molecules moving faster and more likely to react
- rate of photosynthesis affected by light intensity because higher light intensity means more energy is transferred from light to plant molecules so more reactions
- Temperature and light intensity both increase during the day as the Sun gets higher in the sky and decrease as the Sun gets lower towards evening.
- As temperature and light intensity increase, the rate of photosynthesis increases.
- As temperature and light intensity decrease, the rate of photosynthesis decreases.

58. Hormones

1 a chemical produced by an endocrine gland;

travels in blood;

affects target organ/cells **(max. 2)**

2 Hormones are chemical; nerve impulses are electrical/ionic.

Hormones travel in blood to all parts of the body; nerve impulses travel directly to the site of action.

Nerve impulses usually only affect one organ; hormones may affect several organs.

Nerve impulses short-lived; hormones usually last longer **(max. 2)**

59. Adrenalin and thyroxine

1 (a) heart/lungs/liver/blood vessels **(1)**

(b) speeds up heart rate/speeds up breathing rate/increases blood glucose/increases blood pressure/constricts blood vessels **(1)**

(Any 2 targets with appropriate effect for **2 marks**)

2 stops heart rate increasing/keeps heart rate low **(1)**

3 Thyroxine stimulates cellular respiration **(1)**; use up more glucose so eat more **(1)**; respiration releases heat therefore feel warmer **(1)**.

60. The menstrual cycle

1 If the egg is fertilised **(1)** the uterus lining is maintained **(1)**.

2 Contraceptive pills are very effective at preventing pregnancy, so that a couple can decide when they want to have children. **(1)** The pill has some positive side effects, such as reducing the risk of some cancers, and some negative effects, such as increasing the

risk of thrombosis/blood clot. **(1)** As long as a woman does not smoke heavily and is not greatly overweight, then the benefit of controlling fertility outweighs the risk of side effects. **(1)**

61. Control of the menstrual cycle

1 (a) in the ovaries **(1)**
  (b) in the pituitary gland **(1)**
2 fall in oestrogen and progesterone levels **(1)**
3 (a) thickening of uterus wall and triggering of LH release **(1)**
  (b) thickening of uterus wall **(1)**
  (c) stimulates growth and development of egg in ovary **(1)**
  (d) triggers ovulation **(1)**

62. Assisted Reproductive Therapy

1 The woman is given fertility drugs that contain FSH, which makes eggs mature in her ovaries. **(1)** The drugs also contain LH, which stimulates mature eggs to be released. **(1)** The released eggs are collected and fertilised with sperm outside her body. **(1)** When the embryos have developed into tiny balls of cells, one or two are placed in the woman's womb to develop until birth. **(1)**
2 Clomifene stimulates FSH and LH **(1)**; FSH causes follicles to mature **(1)**; LH stimulates ovulation **(1)**.

63. Blood glucose regulation

1 Gland is pancreas **(1)**, target organ is liver **(1)**.
2 A change in blood glucose causes mechanisms to act that bring about the opposite change **(1)**, so that blood glucose concentration is maintained within a small range **(1)**.
3 A

64. Diabetes

1 More insulin is needed if a big meal is taken rather than a small meal.
  More insulin is needed if a meal contains a lot of sugar/carbohydrates.
  Less insulin is needed if the person is exercising; because glucose used up in respiration. **(max. 3)**
2 There is evidence to suggest that obesity/being overweight is related to the risk of developing type 2 diabetes. **(1)** As the number of people who are obese increases this means that the number of people with type 2 diabetes is also likely to increase. **(1)** Controlling weight may help to prevent this. **(1)** However, it will not affect the number of people who develop type 1 diabetes. **(1)** *Although there is no evidence that the proportion of people with type 1 diabetes is changing much.*

65. Extended response – Control and coordination

*Answer could include the following points **(6)**:

- Blood glucose concentration rises after absorbing glucose from a digested meal.
- Blood glucose concentration falls as glucose is taken from blood by cells for respiration or storage.
- As blood glucose concentration rises, cells in pancreas respond by releasing the hormone insulin into the blood.
- Insulin causes cells to take up more glucose.
- Liver and muscle cells store extra glucose as glycogen.
- When blood glucose concentration falls too low, other cells in pancreas respond by releasing the hormone glucagon into the blood.
- Glucagon causes liver cells to break down glycogen to glucose and release glucose into blood.
- Blood glucose concentration is kept within limits, which is homeostasis.
- Control of blood glucose concentration shows negative feedback as a change in one direction causes a change in the opposite direction to happen.

66. Exchanging materials

1 lungs **(1)** *There are other possible answers, e.g. small intestine, kidney, but you will be expected to know about the lungs.*
2 Flatworm is very flat and thin;
  large surface area:volume ratio;
  every cell in the flatworm is close to the surface;
  can rely on diffusion. **(max. 3)**

67. Alveoli

1 one from: less oxygen/more carbon dioxide/more water **(1)**
2 more oxygen needed for respiration **(1)**; more carbon dioxide to breathe out **(1)**
3 maintains a concentration gradient **(1)**; larger surface area for exchange **(1)**

68. Blood

1 oxygen in red blood cells **(1)**; glucose in plasma **(1)** *You must say which substance is carried by each part of the blood. In an exam, you would not get marks for just saying 'red blood cells and plasma'.*
2 Some white blood cells surround and destroy pathogens. **(1)** Other white blood cells produce antibodies that destroy pathogens. **(1)**
3 Platelets respond to a break/wound in a blood vessel by triggering the blood clotting process. **(1)** The clot blocks the wound. **(1)** This prevents pathogens getting into the body. **(1)**

69. Blood vessels

1 (a) B
  (b) A
2 Every body cell needs oxygen and glucose for respiration, and to get rid of waste products such as carbon dioxide. **(1)** If this does not happen quickly enough then the cell may be damaged. **(1)** Capillaries are the blood vessels that exchange substances with the body cells. **(1)**. A short distance for diffusion between a body cell and a capillary helps to make diffusion rapid. **(1)**

70. The heart

1 They prevent the blood flowing backwards in the heart. **(1)**
2 The right ventricle pumps blood to the lungs, which are not far away. **(1)** The left ventricle pumps blood out around the rest of the body, including as far as the toes and fingers. **(1)** So the left ventricle needs to pump with greater force than the right ventricle. **(1)**

71. Aerobic respiration

1 the breakdown of glucose to release energy **(1)** using oxygen **(1)**
2 If pH or temperature vary too much, the rate of reaction will slow down or even stop. **(1)** This would rapidly lead to death because respiration provides the energy for essential processes in the body. **(1)**

72. Anaerobic respiration

1 It releases less oxygen per glucose molecule. **(1)**
  It does not release enough energy for rapid activity. **(1)**
2 There will be a high concentration of lactic acid in the muscles after the vigorous exercise **(1)**, so breathing is faster to take in more oxygen than normal. This is needed to oxidise the lactic acid to carbon dioxide and water. **(1)**

73. Rate of respiration

(a) rate of oxygen uptake (cm^3/min) = volume absorbed in 20 min ÷ 20 **(1)**
  5 °C: 0.03 cm^3/min
  10 °C: 0.05 cm^3/min
  15 °C: 0.085 cm^3/min
  20 °C: 0.12 cm^3/min **(1)**
(b) suitable scales **(1)**; axes labels with quantity and unit **(1)**; axes points plotted accurately **(1)**; line or curve of best fit drawn **(1)**
(c) As temperature increases, the rate of oxygen uptake of the organisms increases **(1)**, which means that their rate of respiration increases **(1)**.

74. Changes in heart rate

1 As exercise level increases/gets more intense **(1)** the heart rate increases **(1)**. OR As exercise level reduces/gets easier **(1)** the heart rate decreases **(1)**. *It is not enough to say 'heart rate increases'. You need to link how heart rate changes as exercise level changes to get both marks.*
2 4500/75 = 60 cm^3
  (**2 marks** for correct answer; **1 mark** for correct method with arithmetic error)
3 Cardiac output is heart rate multiplied by stroke volume. **(1)** A trained athlete will have a similar cardiac output because their stroke volume is much larger. **(1)**

75. Extended response – Exchange

*Answer could include the following points **(6)**:

- gases exchanged by diffusion in the lungs
- diffusion of oxygen from the air in the lungs into the blood
- diffusion of carbon dioxide from the blood into air in the lungs
- Alveoli create a very large surface area for exchange.
- each alveolus closely associated with blood capillary
- Rate of diffusion depends on area, so the larger the area the more rapid the exchange of gases.

- Large number of capillaries means large surface area for exchange.
- Thin walls of alveolus and blood capillary means short diffusion distance, which increases rate of exchange.
- Continuous blood flow and ventilation of lungs maintain steep concentration gradient which increases rate of diffusion.

76. Ecosystems and abiotic factors

large surface area:volume ratio, **(1)** so loses a lot of heat **(1)**

77. Biotic factors

1 so that the plants do not compete **(1)** for an environmental factor such as light/water/nutrients **(1)**

2 Predators of milk snakes may confuse them with the poisonous coral snakes **(1)** and avoid them because they know that coral snakes are poisonous **(1)**.

78. Parasitism and mutualism

1 A parasite is an organism that feeds on and causes harm to a host organism while living on or in the host. **(1)**

2 when two organisms both benefit from a close relationship **(1)**

3 The bacteria benefit from getting food from the plant and protection from the environment. **(1)** The plant benefits from getting nitrogen compounds/nitrates from the bacteria, which it needs to make proteins for healthy growth. **(1)**

79. Fieldwork techniques

1 to estimate population number/size **(1)**

2 when you want to study the effect of abiotic and biotic factors on where organisms live/to link a change in distribution with a change in a physical factor **(1)**

80. Core practical – Organisms and their environment

(a) graph with axes as for graph shown in Core practical; with primrose data correctly substituted for cowslip data **(2)** (**1 mark** for correct axes, **1 mark** for accurately plotted points joined by straight lines)

(b) Primroses are more common where there is some light **(1)**, but not where there is a lot of light in the open meadow **(1)**.

81. Human effects on ecosystems

(Growth of surface plants) blocks light so plants die **(1)**; increase in bacteria uses up oxygen **(1)**; fish and other animals die **(1)**.

82. Biodiversity

any suitable answer that includes explanation of how this change would lead to increased biodiversity, such as habitat for range of species **(max. 2)**

83. The carbon cycle

1 Decomposers release carbon back into the air as carbon dioxide from respiration. **(1)**

2 Respiration releases carbon from compounds in living organisms as carbon dioxide gas into the air. **(1)**
Photosynthesis takes carbon dioxide gas from the air and converts it into carbon compounds in plants. **(1)**
Combustion releases carbon from compounds in fossil fuels as carbon dioxide gas into the air. **(1)**

84. The water cycle

1 Water evaporates;
forms clouds;
condenses into precipitation/rain;
runs off land in rivers;
is taken up by plants;
filters through ground **(max. 5)**;
then more evaporation and the cycle continues.

2 less water available in rivers;
less precipitation;
less water in rivers/aquifers/drier soil **(3)**

85. The nitrogen cycle

1 Clover roots contain nitrogen-fixing bacteria which means fewer nitrates removed from soil **(1)**;
adds nutrients to the soil when clover plants ploughed in **(1)**.

2 The stubble contains proteins. **(1)** Decomposers break down these proteins and release the nitrogen compounds as ammonia into the soil. **(1)** This makes the ammonia available to nitrifying bacteria, which convert it to nitrates. **(1)** Plants that grow in this soil later will be able to take in the nitrates and use them to make proteins, which they need for healthy growth. **(1)**

86. Extended response – Ecosystems and material cycles

*Answer could include the following points **(6)**:

- farmed fish specially bred for rapid growth
- farmed fish treated with antibiotics and other medicines to keep them healthy
- farmed fish supplied with plenty of food
- Some of the food and medicines given to farmed fish get out into the environment.
- Mineral ions from food will encourage growth of water plants nearby which could change the environment.
- Medicines that kill organisms (e.g. parasites) could reduce biodiversity in the environment.
- Growing farmed fish reduces impact of fishing wild salmon so maintains biodiversity.
- Growing salmon provides food for growing human population.

Chemistry

87. Formulae

1 A molecule consists of two bromine atoms **(1)** chemically joined together **(1)**.

2 $CuCO_3$ **(1)**
No marks if 3 is shown as superscript or element symbols do not start with capital letters.

3 two from the following for **1 mark** each:

- It contains two oxygen atoms and two hydrogen atoms for every magnesium atom.
- It has three different elements/five atoms in total, chemically joined together.
- It contains one magnesium ion and two hydroxide ions.

88. Equations

1 (a) $2Mg + O_2 \rightarrow 2MgO$ **(1)**
(b) $N_2 + 3H_2 \rightarrow 2NH_3$ **(1)**
(c) $CH_4 + 2O_2 \rightarrow CO_2 + 2H_2O$ **(1)**

2 (a) $CuO(s) + 2HNO_3(aq) \rightarrow Cu(NO_3)_2(aq) + H_2O(l)$ (**1 mark** for balancing, **1 mark** for state symbols)
(b) $2Fe(s) + 3Cl_2(g) \rightarrow 2FeCl_3(s)$ (**1 mark** for balancing, **1 mark** for state symbols)

89. Ionic equations

1 $Pb^{2+} + 2Br^- \rightarrow PbBr_2$ **(1)**

2 (a) $BaCl_2(aq) + Na_2SO_4(aq) \rightarrow BaSO_4(s) + 2NaCl(aq)$ **(1)**
(b) $Ba^{2+}(aq) + SO_4^{2-}(aq) \rightarrow BaSO_4(s)$ **(1)**

90. Hazards, risks and precautions

1 one from: to warn about the dangers; to let people know about precautions to take; for people who cannot read **(1)**

2 chlorine: use a fume cupboard/open the windows **(1)**
hydrogen: avoid naked flames **(1)**

91. Atomic structure

1 central nucleus **(1)** containing protons and (usually) neutrons **(1)**
surrounded by electrons **(1)** in shells **(1)**

2 Atoms contain equal numbers of protons and electrons. **(1)**
Protons have a +1 charge and electrons have –1 charge (so charges cancel out). **(1)**

3 $(9.1094 \times 10^{-31})/(1.6726 \times 10^{-27})$ **(1)**
$= 5.446 \times 10^{-4}$ or $\frac{1}{1836}$ **(1)**

92. Isotopes

1 Each atom has the same number of protons and electrons **(1)**, 35 protons and electrons **(1)**, different numbers of neutrons **(1)**, 44 or 46 neutrons **(1)**.

2 $(35 \times 75.8) + (37 \times 24.2) = 2653 + 895.4 = 3548.4$ **(1)**

$A_r = \frac{3548.4}{100} = 35.484 = 35.5$ to 1 decimal place **(1)**

93. Mendeleev's table

1 in order of increasing atomic mass **(1)**, taking into account the properties of the elements and their compounds **(1)**

2 one from: It had gaps/there were pair reversals/some groups contained metals and non-metals. **(1)**

94. The periodic table

Both relative atomic masses are 59 when rounded to the nearest whole number. **(1)**

95. Electronic configurations

1 diagram with three circles **(1)** with two electrons in the first circle, eight in the second and seven in the outermost circle **(1)**, e.g.

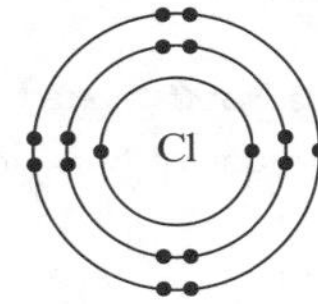

You can show electrons as dots or as crosses.

2 Number of occupied shells is the same as the period number **(1)**; the number of electrons in the outer shell is the same as the group number (except for the elements in group 0, which have full outer shells) **(1)**.

96. Ions

1 (a) Li^+ **(1)**

(b) Mg^{2+} **(1)**

(c) S^{2-} **(1)**

(d) Br^- **(1)**

2 a charged particle **(1)** formed when an atom or group of atoms loses or gains electrons **(1)**

3 (a) protons = 20 **(1)**

neutrons = (40 – 20) = 20 **(1)**

electrons = (20 – 2) = 18 **(1)**

(b) protons = 9 **(1)**

neutrons = (19 – 9) = 10 **(1)**

electrons = (9 + 1) = 10 **(1)**

97. Formulae of ionic compounds

(a) CaS **(1)**

(b) $FeCl_2$ **(1)**

(c) NH_4OH **(1)**

(d) $(NH_4)_2CO_3$ **(1)**

(e) Na_2SO_4 **(1)**

98. Properties of ionic compounds

(a) Its ions are not free to move/are held in fixed positions (in the lattice). **(1)**

(b) Heat it **(1)** until it melts/becomes molten **(1)**.

Remember: aluminium oxide is insoluble, so you cannot dissolve it in water.

(c) Aluminium oxide has a high melting point **(1)** so a lot of energy is needed **(1)**.

99. Covalent bonds

1 (strong) bond formed when a pair of electrons is shared **(1)** between two atoms **(1)**

2 dot-and-cross diagram for oxygen, e.g.

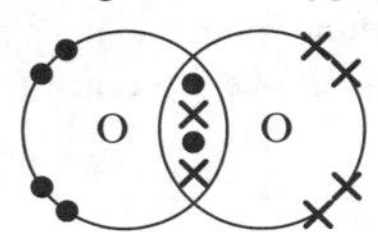

two overlapping circles with two dots and two crosses in the overlap **(1)**

two pairs of non-bonding electrons in each oxygen atom **(1)**

100. Simple molecular substances

1 There are weak intermolecular forces between ammonia molecules **(1)**, which are easily overcome/need little energy to break **(1)**.

2 The intermolecular forces between petrol molecules and water molecules are weaker **(1)** than those between petrol molecules **(1)** and between water molecules **(1)**.

101. Giant molecular substances

1 the following points for **1 mark** each to a maximum of **4 marks**:

similarities:

- only contain carbon atoms
- atoms covalently bonded to each other
- lattice structure.

differences:

- Diamond has four covalent bonds per atom *and* graphite has three.
- Diamond has no delocalised electrons *and* graphite does.
- Graphite has a layered structure (with weak intermolecular forces) *and* diamond does not.

2 Graphite has weak intermolecular forces **between its layers (1)** so the layers can slide over each other **(1)**.

102. Other large molecules

1 two from the following for **1 mark** each: almost transparent/flexible/conducts electricity

2 Buckminsterfullerene is not a giant molecular substance **(1)**, weak intermolecular forces are overcome on heating **(1)**, but there are many strong covalent bonds to break in diamond **(1)**.

103. Metals

1 (a) Layers of positive ions/atoms **(1)** can slide over each other **(1)**.

(b) Delocalised electrons/free electrons/sea of electrons **(1)** can move through the structure **(1)**.

2 (It should conduct electricity) because it will have delocalised electrons. **(1)**

104. Limitations of models

one limitation of each model:

- Empirical formula does not show: actual number of atoms/how the atoms are arranged/three-dimensional shape of molecule/bonding and non-bonding electrons/sizes of atoms relative to their bonds. **(1)**
- Molecular formula does not show: how the atoms are bonded/three-dimensional shape of molecule/bonding and non-bonding electrons/sizes of atoms relative to their bonds. **(1)**
- Structural formula does not show: three-dimensional shape of molecule/bonding and non-bonding electrons/sizes of atoms relative to their bonds. **(1)**
- Drawn formula does not show: three-dimensional shape of molecule/bonding and non-bonding electrons/sizes of atoms relative to their bonds. **(1)**
- Ball-and-stick model does not show: element symbols/bonding and non-bonding electrons/sizes of atoms relative to their bonds. **(1)**
- Space-filling model does not show: element symbols/bonding and non-bonding electrons/some atoms in complex models. **(1)**
- Dot-and-cross diagram does not show: three-dimensional shape of molecule/sizes of atoms relative to their bonds. **(1)**

105. Relative formula mass

(a) 18 **(1)**

(b) 44 **(1)**

(c) 40 **(1)**

(d) 154 **(1)**

(e) 134.5 **(1)**

(f) 142 **(1)**

(g) 78 **(1)**

(h) 234 **(1)**

106. Empirical formulae

(a) Mg: $\frac{0.36}{24} = 0.015$ and O: $\frac{0.24}{16} = 0.015$ **(1)**

Divide by 0.015

Mg: $\frac{0.015}{0.015} = 1$ and O: $\frac{0.015}{0.015} = 1$ **(1)**

Empirical formula is MgO. **(1)**

(b) Air/oxygen must be let in (to react with the magnesium). **(1)**

Magnesium/magnesium oxide must not be allowed to escape. **(1)**

107. Conservation of mass

1 M_r of $CH_4 = 12 + (4 \times 1) = 16$ **(1)**

M_r of $O_2 = (2 \times 16) = 32$ **(1)**

10 g of CH_4 reacts with $\left(2 \times \frac{32}{16}\right) \times 10$ g of O_2 **(1)**

= 40 g **(1)**

2 $2K + Cl_2 \rightarrow 2KCl$ **(1)**

A_r of K = 39 and M_r of KCl = 39 + 35.5 = 74.5 **(1)**

20 g of KCl is produced from $\left(\frac{39}{74.5}\right) \times 20$ g of K **(1)**

= 10.5 g or 10.47 g **(1)**

108. Reacting mass calculations

1 The reactant not in excess is the limiting reactant. **(1)** When all its particles have been used up, the reaction stops/no product forms. **(1)**

2 amount of $CO_2 = \frac{3.3}{44} = 0.075$ (mol) **(1)**

amount of $H_2 = \frac{0.30}{2} = 0.15$ (mol) **(1)**

ratio of CO_2:H_2 = 0.075:0.15 = 1:2 **(1)**

Right-hand side is: $\rightarrow CO_2 + 2H_2$

Equation is: $C + 2H_2O \rightarrow CO_2 + 2H_2$ **(1)**

109. Concentration of solution

1 (a) $\frac{0.40}{0.50} = 0.80$ g dm^{-3} **(1)**

(b) $100 \text{ cm}^3 = \frac{100}{1000} = 0.100 \text{ dm}^3$ **(1)**

$\frac{1.25}{0.100} = 12.5$ g dm^{-3} **(1)**

2 $150 \text{ cm}^3 = \frac{150}{1000} = 0.150 \text{ dm}^3$ **(1)**

mass = concentration × volume

mass = 40 × 0.150 = 6.0 g **(1)**

110. Avogadro's constant and moles

1 (a) $\frac{22.5}{18} = 1.25$ mol **(1)**

(b) 3 × 1.25 = 3.75 mol **(1)**

(c) $3.75 \times 6 \times 10^{23} = 2.25 \times 10^{24}$ **(1)**

2 (a) amount = $\frac{6.0}{12} = 0.50$ mol **(1)**

number = $0.50 \times 6.02 \times 10^{23} = 3.0 \times 10^{23}$ atoms **(1)**

(b) amount = $\frac{1.00 \times 10^{12}}{6.02 \times 10^{23}} = 1.66 \times 10^{-12}$ mol **(1)**

mass = $12 \times 1.66 \times 10^{-12} = 2.0 \times 10^{-11}$ g **(1)**

111. Extended response – Types of substance

*Answer could include the following points **(6)**:

- Mercury contains positively charged mercury ions.
- It is not in a regular lattice.
- There is a sea of delocalised electrons.
- The electrons are free to move and carry charge.
- Liquid zinc chloride contains oppositely charged particles.
- They are attracted to each other but are not in a regular lattice.
- The ions are free to move and carry charge.
- Paraffin oil molecules are attracted to each other by weak intermolecular forces.
- The molecules are free to move.
- The molecules are uncharged so paraffin oil does not conduct electricity.

112. States of matter

change of state: sublimation **(1)**

Forces of attraction (between particles) are overcome (as energy is transferred to the particles). **(1)**

Arrangement changes from regular to random. **(1)**

Arrangement changes from close together to far apart. **(1)**

Particles no longer just vibrate about fixed positions but can move quickly in all directions. **(1)**

113. Pure substances and mixtures

1 A pure substance contains only one element or compound **(1)**, but mineral water will be a mixture **(1)** of water and other substances **(1)**.

2 18-carat gold must be a mixture **(1)** because pure substances have a sharp melting point/mixtures melt over a range of temperatures **(1)**.

114. Distillation

(a) to cool the vapour **(1)**; to turn it from the gas state to the liquid state **(1)**

(b) (different) boiling points **(1)**

(c) hottest at the bottom and coldest at the top/gets cooler towards the top **(1)**

115. Filtration and crystallisation

Answer should include the following points for **1 mark** each to a maximum of **4 marks**:

- Add water to dissolve the copper chloride.
- Filter to remove the glass and dust (as a residue).
- Heat the filtrate to produce copper chloride crystals (by crystallisation).
- Pat the copper chloride crystals dry with filter paper/put them in a warm oven.

116. Paper chromatography

1 A 0.2 **(1)** E 0.9 **(1)**

2 The substance forms stronger attractive forces with the solvent/mobile phase **(1)** than with the paper/stationary phase **(1)**.

117. Core practical – Investigating inks

Answer could include the following points for **1 mark** each to a maximum of **2 marks**:

- Clamp the boiling tube.
- Use a blue flame with the air hole half open.
- Do not heat to dryness.
- Put the test tube in a beaker of cold water/iced water.

118. Drinking water

1 safe to drink **(1)**

2 Sedimentation removes **large** insoluble particles **(1)**; filtration removes **small** insoluble particles **(1)**; chlorination kills microbes/sterilises the water **(1)**.

119. Extended response – Separating mixtures

*Answer could include the following points **(6)**:

- Add water to the mixture.
- Stir the mixture.
- Sodium carbonate should dissolve but calcium carbonate should not.
- This is because their solubility is different/calcium carbonate is (almost) insoluble.
- Filter the mixture.
- Calcium carbonate stays behind as a residue.
- Wash the residue in the filter paper (to remove any sodium carbonate).
- Leave washed residue in a warm place to dry.
- Filtrate is sodium carbonate solution.
- Heat to evaporate (most of) the water.
- Leave solution to cool/crystallise sodium carbonate.
- Filter/decant to separate crystals.
- Leave crystals in a warm place to dry.

120. Acids and alkalis

1 Sodium chloride solution is neutral but hydrogen chloride solution is acidic. **(1)**

Sodium chloride does not release hydrogen ions when it dissolves but hydrogen chloride does **(1)**.

Hydrogen ions make the solution acidic. **(1)**

2 The indicator turns pink **(1)** because the solution is alkaline, containing hydroxide ions. **(1)**

3 **1 mark** for each correct row to **2 marks** maximum:

	Acidic solution	Alkaline solution
Red litmus	(stays) red	(turns) blue
Blue litmus	(turns) red	(stays) blue

121. Strong and weak acids

1 The concentration of hydrogen ions **(1)** is the same **(1)**.

2 0.073 g dm^{-3} **(1)**

122. Bases and alkalis

1 An alkali is a soluble base **(1)** but most bases are insoluble (so are not alkalis) **(1)**.

2 NaCl – sodium chloride **(1)**; $Cu(NO_3)_2$ – copper nitrate **(1)**; $MgCl_2$ – magnesium chloride **(1)**; $Al_2(SO_4)_3$ – aluminium sulfate **(1)**; $CaCl_2$ – calcium chloride **(1)**

123. Core practical – Neutralisation

(a) two precautions for precise measurements for **1 mark** each to give a total of **2 marks**, e.g.

- Use a measuring cylinder with a capacity similar to the volume being used/use a (volumetric) pipette.
- Use a balance of 2 decimal places (±0.01 g).
- Use a pH meter.
- Calibrate the pH meter.

(b) Add a spot of liquid to the universal indicator paper. **(1)**

Match the colour to the pH colour chart. **(1)**

Work carefully, e.g. dip a glass rod into the reaction mixture/put the indicator paper on a white tile/wait 30 s for colour to develop. **(1)**

(c) The pH will be 0–6 at the start **(1)** because the solution contains excess $H^+(aq)$ ions **(1)**.

The pH will be 7 at some point **(1)** because the acid will be neutralised by the base/alkali. **(1)**

The pH will be 8–14 at the end **(1)** because the solution contains excess $OH^-(aq)$ ions **(1)**.

124. Core practical – Salts from insoluble bases

(a) during the reaction: bubbling **(1)**; the white solid disappearing **(1)**

when calcium carbonate in excess: a white solid mixed with the liquid/cloudy mixture **(1)**

(b) **1 mark** for each of the following points to **6 marks** maximum:

- Put some dilute nitric acid into a beaker/suitable container.
- Add calcium carbonate.
- Add until bubbling stops/excess solid is seen.
- Filter (to remove excess solid).
- Heat the filtrate.
- This will remove some of the water.
- Leave to cool.
- Decant/pour away/filter to remove the excess liquid.
- Leave crystals in a warm place/dry with paper.

125. Salts from soluble bases

Filter the mixture **(1)** so the insoluble powder and indicator stay behind (and the salt solution passes through) **(1)**.

126. Making insoluble salts

(a) calcium carbonate **(1)**

(b) no precipitate **(1)**

(c) lead sulfate **(1)**

127. Extended response – Making salts

*Answer could include the following points **(6)**:

- suitable solutions: lead(II) nitrate and sodium chloride/potassium chloride/ammonium chloride
- Lead compounds are toxic.
- Wear eye protection.
- Wear gloves.
- $Pb(NO_3)_2(aq) + 2NaCl(aq) \rightarrow PbCl_2(s) + 2NaNO_3(aq)$
- $Pb(NO_3)_2(aq) + 2KCl(aq) \rightarrow PbCl_2(s) + 2KNO_3(aq)$
- $Pb(NO_3)_2(aq) + 2NH_4Cl(aq) \rightarrow PbCl_2(s) + 2NH_4NO_3(aq)$
- Mix the two solutions.
- Filter the mixture.
- Lead(II) chloride stays behind as a residue.
- Wash the residue in the filter paper.
- Use water/distilled water.
- This works because lead(II) chloride is insoluble.
- Leave washed residue in a warm place to dry.

128. Electrolysis

(a) potassium (at the negative electrode) **(1)**; iodine (at the positive electrode) **(1)**
Note that iodide is incorrect.

(b) negative electrode: $K^+ + e^- \rightarrow K$ **(1)**
positive electrode: $2I^- \rightarrow I_2 + 2e^-$ **(1)**

(c) Oxidation occurs at the positive electrode **(1)** because iodide ions lose electrons **(1)**.

129. Electrolysing solutions

(a) H^+ from water and acid **(1)**; OH^- from water **(1)**; SO_4^{2-} from acid **(1)**

(b) hydrogen from the cathode/negative electrode **(1)**; $2H^+(aq) + 2e^- \rightarrow H_2(g)$ **(1)**; oxygen at the anode/positive electrode **(1)**; $4OH^-(aq) \rightarrow 2H_2O(l) + O_2(g) + 4e^-$ **(1)**

130. Core practical – Investigating electrolysis

(a) Ethanol evaporates more quickly/has a lower boiling point than water. **(1)**

(b) **1 mark** for each of the following points to **2 marks** maximum:

- Some copper did not stick to the cathode/fell off during the experiment.
- Some copper fell off the cathode during washing/drying/weighing.
- Some metal fell off the anode during the experiment.
- The anode was not pure copper/contained other metals.

131. Extended response – Electrolysis

*Answer could include the following points **(6)**:

dissolving in molten potassium chloride:

- Ions are not free to move in solid lithium chloride.
- Ions must be free to move for electrolysis to happen.
- Ions will be free to move in liquid/molten potassium chloride.

cathode process:

- Lithium ions are attracted to the cathode.
- Lithium ions gain electrons and become lithium atoms.
- $Li^+ + e^- \rightarrow Li$
- This is reduction because the ions gain electrons.

anode process:

- Chloride ions are attracted to the anode.
- Chloride ions lose electrons and become chlorine atoms.
- Covalent bond forms between two chlorine atoms.
- There is a shared pair of electrons.
- Pairs of chlorine atoms form chlorine molecules.
- $2Cl^- \rightarrow Cl_2 + 2e^-$
- This is oxidation because the ions lose electrons.

132. The reactivity series

(a) magnesium/calcium **(1)**

(b) iron **(1)** and zinc **(1)**

133. Metal displacement reactions

(a) magnesium, zinc, iron, copper **(1)** This is in order of decreasing change in temperature. **(1)**

(b) 0.0 °C/no change in temperature **(1)** This is because silver cannot displace copper from copper sulfate solution/no reaction. **(1)**

134. Explaining metal reactivity

Aluminium is more reactive than iron **(1)** because it can displace iron from iron(III) oxide **(1)**. This is because it has a greater tendency to lose electrons/form cations. **(1)**

135. Metal ores

1 Copper is more resistant to corrosion **(1)** because it is lower in the reactivity series **(1)**.

2 Oxidation is a gain of oxygen. **(1)** Reduction is a loss of oxygen. **(1)**

3 Platinum is low in the reactivity series/below gold/very unreactive **(1)**, so it does not combine with other elements **(1)**.

136. Iron and aluminium

1 Tin is less reactive than carbon/carbon is more reactive than tin. **(1)**
Calcium is more reactive than carbon/carbon is less reactive than calcium **(1)**.
So carbon can reduce tin oxide to tin, but not calcium oxide to calcium. **(1)**

2 Electrolysis uses electricity. **(1)** This is expensive/more expensive than using carbon. **(1)**

137. Biological metal extraction

1 Plants absorb metal compounds through their roots. **(1)** The metal compounds become concentrated in the plants. **(1)** The plants are harvested and burned. **(1)** The ash contains (high concentrations of) metal compounds. **(1)**

2 advantage: low-grade ores can be used/mining not needed/happens naturally **(1)**
disadvantage: slow/further processes needed **(1)**

138. Recycling metals

1 two from the following for **1 mark** each: less waste rock produced/less carbon dioxide (or named pollutant) produced/less waste sent to landfill

2 Less energy is needed to melt aluminium than to melt steel. **(1)**
More energy is needed to extract aluminium by electrolysis than is needed to extract iron in the blast furnace. **(1)**

3 Recycling uses metals already extracted from ores **(1)**, so no ore/less ore must be mined to produce new metal **(1)**.

139. Life-cycle assessments

1 The four main stages in a life-cycle assessment are: obtaining raw materials, manufacture, use, disposal. **(1)**

2 Another reason for carrying out a life-cycle assessment is to identify a stage that could be improved on. **(1)**

140. Extended response – Reactivity of metals

*Answer could include the following points **(6)**:

expected order of reactivity:

- magnesium > zinc > copper

basic method:

- Add one of the solutions to three test tubes/other suitable container, e.g. spotting tile.
- Add a metal powder to each test tube.
- Observe a change in the solution and/or the metal powder, e.g. a colour change or temperature increase.
- Repeat with each of the other two solutions in turn.

expected results:

- An orange–brown coating of metal is obtained when magnesium or zinc is added to copper sulfate.
- The temperature increases when magnesium or zinc is added to copper sulfate.
- The blue colour of the solution fades when magnesium or zinc is added to copper sulfate.

- There is a black coating of metal when magnesium is added to zinc sulfate solution.
- There is no observed change with other combinations of metal and solution.

explanation:

- Magnesium and zinc are more reactive than copper/can displace copper from its compounds.
- Magnesium and zinc have a greater tendency than copper to lose electrons/are more readily oxidised/better reducing agents.
- $Mg + CuSO_4 \rightarrow MgSO_4 + Cu$
- $Zn + CuSO_4 \rightarrow ZnSO_4 + Cu$
- Magnesium is more reactive than zinc/can displace zinc from its compounds.
- Magnesium has a greater tendency than zinc to lose electrons/is more readily oxidised/is a better reducing agent.
- $Mg + ZnSO_4 \rightarrow MgSO_4 + Zn$
- Copper is less reactive than magnesium and zinc/cannot displace magnesium and zinc from their compounds.
- Zinc is less reactive than magnesium/cannot displace magnesium from its compounds.

141. The Haber process

(a) The symbol means that the reaction is reversible. **(1)**

(b) Add the substance to be tested to anhydrous copper sulfate. **(1)** If the copper sulfate turns blue, water is present. **(1)**

142. More about equilibria

(a) The equilibrium concentration is decreased **(1)** because the position of equilibrium moves to the left/in the direction of the greatest number of molecules of gas **(1)**.

(b) The equilibrium concentration is increased **(1)** because the position of equilibrium moves to the right/away from the endothermic reaction/in the direction of the exothermic reaction **(1)**.

(c) The equilibrium concentration is increased **(1)** because the position of equilibrium moves to the right/away from the substance increased in concentration **(1)**.

143. The alkali metals

(a) $2Cs + 2H_2O \rightarrow 2CsOH + H_2$

(**1 mark** for correct symbols and formulae; **1 mark** for correct balancing)

(b) two from the following for **1 mark** each:

- violent/very violent reaction explosion
- sparks/flames
- hydrogen/bubbles produced
- metal disappearing

(c) two from the following for **1 mark** each:

- good conductors of heat/electricity
- shiny when freshly cut
- soft
- relatively low melting points

144. The halogens

(a) The elements get darker going down the group. **(1)**

(b) melting point in the range 250 °C to 350 °C **(1)**, justification, e.g. extrapolated melting points from chlorine to iodine **(1)**

Accepted estimate for melting point is 302 °C.

145. Reactions of halogens

1 The outer shell in iodine is further from the nucleus/more shielded by inner electrons than it is in fluorine. **(1)**

Iodine gains electrons less easily than fluorine does. **(1)**

2 $2Al + 3I_2 \rightarrow 2AlI_3$

(**1 mark** for correct symbols and formulae; **1 mark** for correct balancing)

146. Halogen displacement reactions

(a) balanced equation:

$Br_2(aq) + 2KI(aq) \rightarrow 2KBr(aq) + I_2(g)$

(**1 mark** for correct symbols and formulae; **1 mark** for correct balancing)

ionic equation: $Br_2(aq) + 2I^-(aq) \rightarrow 2Br^-(aq) + I_2(aq)$

(**1 mark** for correct symbols and formulae; **1 mark** for correct balancing)

(b) Bromine atoms (in Br_2) gain electrons and are reduced to bromide ions. **(1)**

$Br_2 + 2e^- \rightarrow 2Br^-$ **(1)**

Iodide ions lose electrons and are oxidised to iodine. **(1)**

$2I^- \rightarrow I_2 + 2e^-$ **(1)**

(c) Astatine will not react with potassium iodide solution because:

- reactivity decreases down group 7 **(1)**
- astatine is less reactive than iodine **(1)**.

147. The noble gases

1 The densities of helium and neon are less than the density of air **(1)** but the densities of argon and krypton are greater than the density of air **(1)**.

2 It is inert/non-flammable. **(1)** It is denser than air so it sinks and excludes air. **(1)**

3 The electronic configuration is 2. **(1)** Its outer shell is full **(1)**, so it has no tendency to lose/gain/share electrons **(1)**.

148. Extended response – Groups

*Answer could include the following points **(6)**:

experiment:

- Put potassium chloride solution in three test tubes.
- Add a few drops of halogen solution to each one.
- Observe any changes.
- Record in a suitable table.
- Repeat with potassium bromide and potassium iodide.

precautions:

- Wear eye protection because solutions are irritants.
- Avoid contact with skin because solutions are irritants.
- Avoid breathing in vapours/keep lab well ventilated because vapours are toxic/harmful.

expected results:

- No changes with potassium chloride.
- Potassium bromide turns darker when chlorine solution is added.
- $Cl_2 + 2KBr \rightarrow 2KCl + Br_2$/bromine is produced.
- Potassium iodide turns darker when chlorine solution is added.
- $Cl_2 + 2KI \rightarrow 2KCl + I_2$/iodine is produced.
- Potassium iodide turns darker when bromine solution is added.
- $Br_2 + 2KI \rightarrow 2KBr + I_2$/iodine is produced.

using the results:

- Chlorine can displace bromine and iodine/oxidise bromide ions and iodide ions.
- Bromine can displace iodine/oxidise iodide ions.
- Iodine cannot displace chlorine or bromine/cannot oxidise chloride or bromide ions.
- Order of reactivity, most reactive to least reactive: chlorine, bromine, iodine.

149. Rates of reaction

(a) The rate of reaction increases **(1)**, the surface area:volume ratio of the marble increases **(1)** and so the frequency of collisions increases **(1)**.

(b) The rate of reaction decreases **(1)**, the acid particles become less crowded/there are fewer reactant particles in the same volume **(1)** and so the frequency of collisions decreases **(1)**.

(c) The rate of reaction increases **(1)**, the (acid) particles move around faster/have more energy **(1)**, more collisions have the necessary activation energy or higher energy **(1)** and so the frequency of successful collisions increases **(1)**.

150. Core practical – Investigating rates

three from the following for **1 mark** each:

- same volume of sodium thiosulfate solution
- same concentration of sodium thiosulfate solution
- same amount of cloudiness (measured using the disappearing cross)
- same volume of diluted acid.

These factors also affect the rate of reaction/measured rate of reaction. **(1)**

151. Exam skills – Rates of reaction

(a) 0 s to 45 s: $\frac{43}{45}$ **(1)** = 0.96 cm^3/s **(1)**

45 s to 90 s: $\frac{50 - 43}{45}$ **(1)** = 0.16 cm^3/s **(1)**

(b) Reactant particles are used up during the reaction **(1)**, so the concentration of hydrogen peroxide decreases **(1)**, the frequency of collisions decreases **(1)** and the rate of reaction decreases **(1)**.

(c) Measure and record the mass of manganese dioxide at the start **(1)**.

(afterwards) Filter **(1)**, dry **(1)**, measure and record the mass of manganese again. **(1)**

152. Heat energy changes

1 mark for each correct column (**0 marks** if two ticks in a column) to **4 marks**, e.g.

	Breaking bonds	Making bonds	Temperature of reaction mixture	
			Increases	Decreases
Exothermic process		✓	✓	
Endothermic process	✓			✓

153. Reaction profiles

1 Activation energy is the minimum energy **(1)** needed to start a reaction/to break the bonds in the reactant particles **(1)**.

2 (a) correct axes and shape of graph **(1)**; reactants and products identified **(1)**; activation energy identified **(1)**; energy change identified **(1)**, e.g.

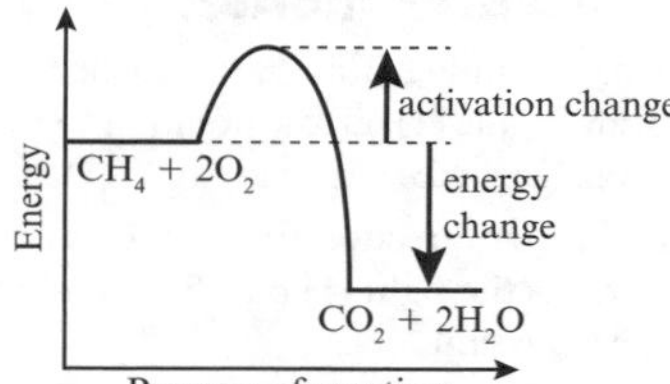

(b) correct axes and shape of graph **(1)**; reactants and products identified **(1)**; activation energy identified **(1)**; energy change identified **(1)**, e.g.

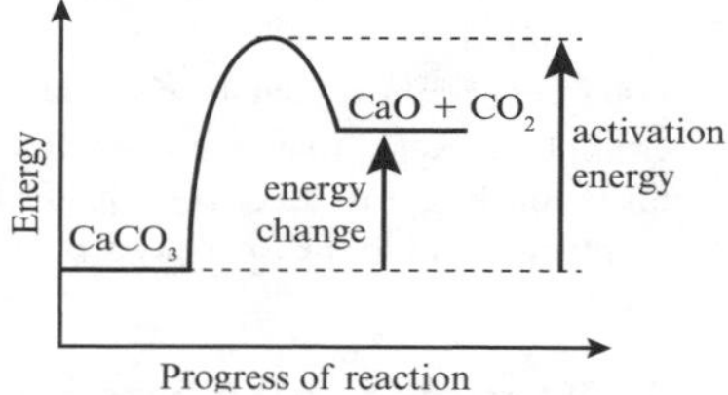

154. Calculating energy changes

1 (a) energy in to break bonds:
$4 \times (O–H) = (4 \times 464) = 1856$ kJ mol^{-1} **(1)**
energy out when bonds form:
$2 \times (H–H) + 1 \times (O{=}O) = (2 \times 436) + (1 \times 498)$ **(1)**
$= 872 + 498 = 1370$ kJ mol^{-1} **(1)**
energy change $= 1856 - 1370 = +486$ kJ mol^{-1} **(1)**

(b) Process is endothermic **(1)** because the energy change is positive/the energy taken in to break bonds is greater than the energy given out when bonds form **(1)**.

2 energy in to break bonds:
$4 \times (C–H) + 2 \times (O{=}O) = (4 \times 413) + (2 \times 498)$ **(1)**
$= 1652 + 996 = 2648$ kJ mol^{-1} **(1)**
energy out when bonds form:
$4 \times (O–H) + 2 \times (C{=}O) = (4 \times 464) + (2 \times 805)$ **(1)**
$= 1856 + 1610 = 3466$ kJ mol^{-1} **(1)**
energy change $= 2648 - 3466 = -818$ kJ mol^{-1} **(1)**

155. Crude oil

1 (a) A hydrocarbon is a compound of carbon and hydrogen **(1)** only **(1)**.
(b) covalent bond **(1)**

2 Crude oil takes millions of years to form/is made extremely slowly **(1)** or is no longer being made **(1)**.

3 chains **(1)**; rings **(1)**

4 (a) (petroleum) gases/petrol/kerosene/diesel oil/fuel oil **(1)**
not bitumen
(b) polymers/named polymer, e.g. poly(ethene) **(1)**

156. Fractional distillation

1 It is difficult to ignite/not very flammable **(1)**; it has a high viscosity/is very viscous/does not flow easily **(1)**.

2 Oil is evaporated **(1)** and passed into a column, which is hot at the bottom and cool at the top **(1)**. Hydrocarbons (rise), cool and condense at different heights. **(1)**

157. Alkanes

1 three from the following for **1 mark** each:
- same general formula
- molecular formulae of neighbouring members differ by CH_2
- gradual variation in physical properties
- similar chemical properties.

2 $C_{21}H_{44}$ **(1)**

3 –40°C (answer in range –35°C to –45°C) **(1)**

158. Incomplete combustion

1 There is a poor supply of air/oxygen **(1)**, so carbon in the fuel is only partially oxidised to carbon monoxide **(1)** or released as carbon particles/soot **(1)**.

2 It is odourless **(1)** and colourless **(1)**.

3 Soot shows that incomplete combustion is happening. **(1)** Carbon monoxide might be forming but soot does not prove this/carbon monoxide is likely to be forming as well (which could be confirmed using a carbon monoxide detector). **(1)**

159. Acid rain

1 two from the following for **1 mark** each:
- weathering of buildings/statues
- damage to trees
- harm to living things in river/lakes/soil

2 (a) Hydrocarbon fuels contain impurities of sulfur compounds. **(1)** The sulfur is oxidised to sulfur dioxide when the fuel is used. **(1)**
(b) Nitrogen and oxygen from the air **(1)** react together in the high temperatures inside an engine **(1)**.

160. Choosing fuels

1 Petrol contains hydrocarbons. **(1)** The carbon in these molecules is oxidised to carbon dioxide. **(1)** Hydrogen does not contain carbon/consists only of hydrogen. **(1)**

2 (a) Crude oil and natural gas are non-renewable. **(1)** They are being used faster than they are formed. **(1)**
(b) Carbon dioxide may be produced during the manufacture/transport of the fuel. **(1)** For example, fossil fuels are used (in power stations) to generate electricity/to react with steam to make hydrogen. **(1)**

161. Cracking

(a) C_2H_4 **(1)**
(b) This hydrocarbon can be used to make polymers. **(1)**

162. Extended response – Fuels

*Answer must include a supported judgement in favour of hydrogen, diesel oil, both or neither. It could include the following points **(6)**:

advantages of hydrogen:
- The only product is water.
- Hydrogen can be produced by the electrolysis of water.
- If water is the raw material, hydrogen could be a renewable resource.
- The electricity needed could be generated using renewable resources (sun, wind, tidal, biomass).
- No carbon dioxide is released when it is used.

disadvantages of hydrogen:
- It is expensive to produce.
- It is difficult to store.
- It may need to be stored under pressure/low temperatures.
- There are few filling stations (so the range is restricted).
- Hydrogen is usually produced from natural gas/coal.
- Carbon dioxide is released in these processes.

advantages of diesel oil:
- It is a liquid (rather than a gas) at room temperature.
- It is easy to store.
- There are many filling stations (so the range is not restricted).

disadvantages of diesel oil:
- Carbon dioxide is also produced.
- Carbon dioxide is a greenhouse gas/linked to global warming/climate change.
- It is produced from crude oil, which is a limited/non-renewable resource.
- It produces carbon particles/soot.
- It produces sulfur dioxide (unless sulfur impurities are removed from fuel before use).
- It produces oxides of nitrogen.
- Sulfur dioxide and oxides of nitrogen cause acid rain.

163. The early atmosphere

1 (a) Carbon dioxide dissolved in the oceans. **(1)**
(b) Plants produced oxygen **(1)** by photosynthesis **(1)**.

2 Scientists cannot be certain about the Earth's early atmosphere because no measurements were made then/no humans were on Earth then. **(1)**

164. Greenhouse effect

1 (a) carbon dioxide **(1)**, methane **(1)**
(b) carbon dioxide from burning fossil fuels **(1)**, methane from cattle/livestock/rice paddy field **(1)**

2 Greenhouse gases absorb heat radiated from the Earth. **(1)** The gases then release the heat (into the atmosphere). **(1)**

3 Fossil fuels give off carbon dioxide during combustion. **(1)** Increased consumption releases more carbon dioxide **(1)**, which is a greenhouse gas **(1)**, so the greenhouse effect increases **(1)**.

165. Extended response – Atmospheric science

*Answer could include the following points **(6)**:

processes that remove carbon dioxide:

- photosynthesis by plants/algae
- making oxygen and glucose/carbohydrates/ starch
- carbon dioxide dissolving in seawater

processes that release carbon dioxide:

- combustion of fossil fuels
- combustion of biomass/plants/trees
- respiration
- volcanic eruptions.

observed increase:

- rate of release greater than rate of removal
- increasing use of fossil fuels
- deforestation means less photosynthesis
- limit to the rate at which carbon dioxide can dissolve in the oceans.

Physics

166. Key concepts

1 (a) 3 **(1)** (b) 4 **(1)** (c) 5 **(1)**

2 (a) 12 × 60 × 60 **(1)** = 43 200 s **(1)**

(b) 64 000 m/3600 s **(1)** = 17.8 m/s **(1)**

3 58.3 MW = 58 300 000 W **(1)** = 5.83×10^7 W **(1)**

4 186 000 miles/s = 186 000 × 1609 **(1)** = 299 274 000 m/s **(1)** = 2.993×10^8 m/s **(1)**

167. Scalars and vectors

1 (a) C **(1)** (b) D **(1)**

2 (a) 4 m/s **(1)** (b) –4 m/s **(1)**

3 The satellite travels at a constant speed as it covers an equal distance per second **(1)**, but its direction is constantly changing as it moves in a circle **(1)** so its velocity must be changing as velocity is a vector quantity with both size and direction **(1)**.

168. Speed, distance and time

1 speed = 120 m ÷ 8 s **(1)**, speed = 15 **(1)** m/s **(1)**

2 gradient of zero at origin **(1)**, increasing gradient of graph shown **(1)**, distance and time shown as correct *y*- and *x*-axes, respectively **(1)**

3 112.7 km = 112 700 m **(1)**; 1 hour = 60 × 60 = 3600 s **(1)**; 112 700 m ÷ 3600 s = 31.3 m/s **(1)**

169. Equations of motion

1 a = (8 m/s – 2 m/s) ÷ 5 s **(1)** = 6 m/s ÷ 5 s = 1.2 **(1)** m/s² **(1)**

2 $v^2 - 0^2 = 2 \times 1.6\,\text{m/s}^2 \times 1800\,\text{m}$ **(1)** = 5760 m²/s; $v = \sqrt{5760}$ = 75.9 **(1)** m/s **(1)**

3 $v^2 - u^2 = 2ax$, $(18\,\text{m/s})^2 - (5\,\text{m/s})^2 = 2 \times 1.2\,\text{m/s}^2 \times x$ **(1)**; $x = (18^2 - 5^2) \div 2.4$ **(1)** = 124.6 **(1)** m **(1)**

170. Velocity/time graphs

1 (a) Gradient of a velocity–time graph gives acceleration. **(1)**

(b) Area beneath a velocity–time graph gives distance travelled **(1)**.

2 (a) correct axes shown: velocity on *y*-axis and time on *x*-axis **(1)**, horizontal line at 8 m/s shown for 12 s **(1)**, positive linear gradient of 1.5 shown for the next 6 seconds **(1)**, final velocity of 17 m/s shown after 18 s **(1)**

(b) distance travelled = area under graph **(1)**, distance travelled in first 8 s = 8 m/s × 12 s = 96 m **(1)**, distance travelled in second part of journey = 8 m/s × 6 s + $\frac{1}{2}$ × 9 m/s × 6 s = 48 m + 27 m = 75 m **(1)**, total distance = 96 m + 75 m = 171 m **(1)**

171. Determining speed

1 time taken **(1)** for the object to travel over a known distance **(1)**

2 Calculate the speed of the ball through the first light gate using speed = distance ÷ time. **(1)** Calculate the speed of the ball through the second light gate. **(1)** Work out the change in speed and the time difference for the two gates. **(1)** Calculate the acceleration using the equation acceleration = change in speed ÷ change in time. **(1)**

3 Answer should include reference to: measurement of a height, h, through which a ball will fall vertically **(1)**; the time taken for the ball to fall through the height, h **(1)**; the use of the equation $g = 2h/t^2$ **(1)** to find g; repeat values being taken **(1)** and an average value being calculated and compared to g; the falling object must accelerate and not reach terminal velocity during the fall **(1)**.

172. Newton's first law

1 balanced or equal and opposite **(1)** OR no forces acting **(1)**

2 The forces acting on the moving body are in different/opposite directions **(1)** of different sizes **(1)**, or a diagram drawn to show this. These forces could be antiparallel or even at an angle, such as at 90° to the initial movement. A change in direction will occur if the forces oppose the original motion or act in a way to change the direction.

3 diagrams showing: (a) equal forces of 100 N **(1)** and opposite **(1)** or no forces shown; (b) equal forces of 100 N **(1)** and opposite forces **(1)** shown; (c) two forces of 100 N **(1)** both acting to the left **(1)**

173. Newton's second law

1 F = 1.2 kg × 8 m/s² **(1)** = 9.6 **(1)** N **(1)**

2 $m = F \div a$ = 18.8 N ÷ 0.8 m/s² **(1)** = 23.5 **(1)** kg **(1)**

3 80 g = 0.08 kg **(1)**, 0.6 kN = 600 N **(1)**, $a = F \div m$ = 600 N ÷ 0.08 kg **(1)** = 7500 **(1)** m/s² **(1)**

174. Weight and mass

1 $g = W \div m$ **(1)** = 54 N ÷ 18 kg **(1)** = 3 N/kg **(1)**

2 Mass is 5 kg **(1)** as mass is not affected by gravitational field strength; weight = 5 kg × 25 N/kg **(1)** = 125 N **(1)**

175. Core practical – Force and acceleration

1 More accurate/repeat readings **(1)** can be taken with little or no human error **(1)** and calculations can be performed quickly by the data logger to provide speeds and accelerations, etc. **(1)**.

2 more accurate mass **(1)**, more accurate slope/ gradient **(1)**, reduce friction, more readings for mass **(1)**, repeat readings to determine the precision **(1)**

3 constant mass throughout/other variables controlled **(1)**, change slope by 10° at a time **(1)**, same starting point along slope **(1)**, record enough values for change in velocity and change in times between two light gates **(1)**, repeat readings to determine the precision **(1)**

176. Circular motion

(a) arrow pointing straight up labelled thrust from engine **(1)**, arrow pointing to centre of circle labelled centripetal force **(1)**

(b) It changes direction from straight up to towards the right **(1)** but the magnitude stays the same **(1)**.

177. Momentum and force

1 momentum = mass × velocity **(1)** = 25 kg × 6 m/s **(1)** = 150 kg m/s **(1)**

2 F = 1580 kg (16 m/s – 7 m/s) ÷ 0.8 s **(1)** = 17 775 **(1)** N **(1)**

3 Use safety features such as seat belts and crumple zones **(1)** to increase the time taken for the car to decelerate/change in momentum **(1)** so the force needed decreases **(1)**.

178. Newton's third law

1 For two objects involved in a collision, the forces involved are equal in magnitude but opposite in direction **(1)**, and as the time of impact is the same for both bodies, the impulses and the momentum must be equal and opposite **(1)**.

2 momentum before = 12 kg × 8 m/s = 96 kg m/s **(1)**, momentum afterwards = 96 kg m/s **(1)**, velocity afterwards = 96 kg m/s ÷ 20 kg **(1)** = 4.8 m/s **(1)**

179. Human reaction time

1 (a) Human reaction time is the time between a stimulus occurring and a response. **(1)** (b) any three of: tiredness **(1)**, alcohol **(1)**, drugs **(1)**, distractions **(1)**, age **(1)**

2 Reaction time is proportional to the square root of the distance an object falls. **(1)** The square root of 4 is 2, so four times the distance means twice the reaction time. **(1)**

180. Stopping distances

1 (a) any two for **one mark** each: speed, tiredness, alcohol, drugs, distractions, reaction time

(b) any two for **one mark** each: speed, icy or wet roads, worn tyres or brakes, mass of car, any reference to friction being less than normal

2 less friction/force between tyres and road **(1)** so increased braking distance **(1)**, so a greater stopping distance **(1)**

181. Extended response – Motion and forces

*Answer could include the following points **(6)**:

- reaction times and thinking distance. As reaction time increases so does thinking distance. Reaction time is the time between seeing the hazard and applying a force to the brakes.
- speed of car – affecting thinking and braking distance. More speed means a greater distance is covered when thinking and braking compared with a lower speed.
- alcohol, drugs or tiredness affecting reaction times and thinking distance – both will increase if drivers are under the influence of these since they will be less alert.
- condition of tyres – bald tyres means greater braking distance as there will be less grip.
- condition of road (icy or wet) resulting in less grip
- mass of car/number of passengers increases braking distance at a given speed due to greater inertial mass/constant force needed over a longer time to stop.

182. Energy stores and transfers

1 Energy store diagram should include the chemical energy, kinetic energy and gravitational potential energy stores and the thermal energy store shown as the wasted form. At each part of the journey of the car uphill, the total of these four stores should be constant. The diagram should show

chemical store (decreases) → (mechanical) (energy transfer)

→ kinetic store (increases)

\+ gravitational potential store (increases)

\+ thermal store (increases)

(4)

2 Diagram should show:

chemical store → kinetic store + gravitational potential store → gravitational store + thermal store

The initial energy store is chemical energy from food. **(1)** The energy transfer is mechanical. **(1)** The chemical store is transferred to an increasing kinetic store due to its motion and an increasing gravitational potential energy store **(1)** due to the increase in height. Finally, all of the chemical store has been transferred to gravitational store with some being wasted as a thermal store **(1)** and dissipated to the surroundings **(1)**.

183. Efficient heat transfer

1 efficiency = (14 J ÷ 20 J) × 100% **(1)** = 0.7 or 70% **(1)**

2 A kettle transfers most of the electrical energy to thermal energy for heating the water in the kettle **(1)** but some thermal energy is always transferred to the surroundings **(1)** instead of being transferred to the water, so it must be less than 100% efficient **(1)**.

184. Energy resources

1 (a) Diagram should show gpe → kinetic → electrical → (thermal energy as wasted) **(3)**

(b) Diagram should show chemical → heat or ... chemical → electrical or ... chemical → kinetic **(3)**

2 biofuels advantages: reliable, can be replaced, can be readily grown and converted to fuels for use in cars, can be stored, renewable, cheap **(any 1 for 1 mark)**; biofuels disadvantages: gives off CO_2/greenhouse gases when burned, not renewable if not regrown **(any 1 for 1 mark)**; solar advantages: renewable, almost infinite energy supply, no cost for the solar energy **(any 1 for 1 mark)**; solar disadvantages: solar cells are expensive to make and buy, unreliable (when not sunny or at night), difficult to store the energy in electrical form, production of cells in manufacturing will pollute **(any 1 for 1 mark)**

185. Patterns of energy use

1 two of: threats to food supplies for those who need them **(1)**, global warming and floods due to using fossil fuels **(1)**, running out of non-renewable resources **(1)**

2 Wood has been readily available **(1)** for thousands of years **(1)** and has not had to be discovered, mined or have any extra technology developed (e.g. nuclear reactors/ power stations) in order for it to be used to produce thermal energy **(1)**.

186. Potential and kinetic energy

1 $\Delta GPE = m \times g \times \Delta h$ = 5 kg × 10 N/kg × 18 m **(1)** = 900 **(1)** J **(1)**

2 30 km/h = 30 000 m ÷ 3600 s = $8.\dot{3}$ m/s **(1)**, KE = $\frac{1}{2}$ × 80 kg × $8.\dot{3}$ m/s **(1)** = 2778 **(1)** J **(1)**

3 KE = ΔGPE, $\frac{1}{2} \times m \times v^2 = m \times g \times \Delta h$ **(1)**, so $v^2 = 2 \times g \times \Delta h$ **(1)** 2 × 10 N/kg × 34 m **(1)**, $v = \sqrt{680}$ = 26 m/s **(1)**

187. Extended response – Conservation of energy

*Answer could include the following points (for all **six marks**, two advantages and two disadvantages with good supporting explanation should be supplied):

- advantages of wind – renewable, no fuel needed, no carbon dioxide gas emissions
- disadvantages – unsightly, unreliable
- advantages of coal – efficient, reliable
- disadvantages – non-renewable, carbon dioxide and sulfur dioxide gas emissions

188. Waves

1 (a)

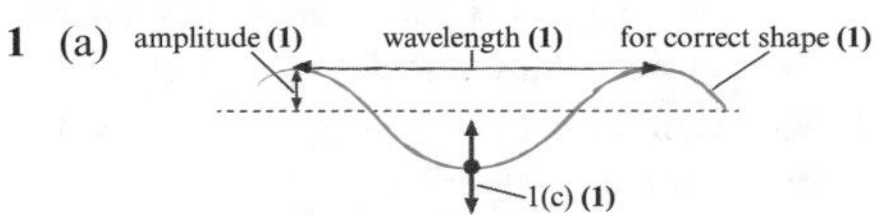

(b) ← or → **(1)** (c) arrow shown above

2 2.2 mm **(1)** *Any answer between 2.1 and 2.4 is acceptable.*

189. Wave equations

1 λ = 330 m/s ÷ 100 Hz **(1)** = 3.3 **(1)** m **(1)**

2 1 minute = 60 seconds **(1)**; x = 25 m/s × 60 s **(1)** = 1500 **(1)** m **(1)**

190. Measuring wave velocity

1 wave speed = frequency × wavelength = 4 Hz × 0.08 m **(1)** = 0.32 **(1)** m/s **(1)**

2 Speed of sound in air is found by measuring a distance for the wave to travel and a time over which the distance is covered. **(1)** Having a large distance **(1)** and a large time **(1)** for these measurements using equipment with a high degree of accuracy **(1)** will lead to a low percentage error.

191. Waves and boundaries

1 They can be reflected, refracted, transmitted or absorbed **(2 marks for all four, 1 mark for three)**.

2 Speed will change **(1)**, direction may change (unless travelling along the normal) **(1)** and wavelength of the wave will change **(1)**.

3 Light rays from the part of the pencil beneath the water change direction when they move from water to air. **(1)** When the eye detects the light rays, the brain traces them back as a straight line. **(1)** So they appear to come from a point above the bottom of the pencil, so the pencil appears bent. **(1)**

192. Core practical – Waves in fluids

1 error in wavelength measured for the wave when distance being measured **(1)**, error in frequency value of the wave from generating source **(1)**, error when taking any time values with a stopwatch in order to find for the wave speed **(1)**

2 More wavelengths gives a greater distance to measure **(1)** with a measuring device of a constant accuracy/resolution **(1)** so wavelength value will be more accurate since percentage error in value obtained will be less **(1)**; example given, e.g. a metre ruler being used to measure a wavelength of 10 cm with a 1 cm scale gives a 10% error, whereas measuring 10 wavelengths would mean a 1% error (or similar) **(1)**.

193. Extended response – Waves

*Answer could include the following points **(6)**:

- Longitudinal waves are produced when the vibration is parallel to the direction of energy transfer.
- Transverse waves are produced when the vibrations are at right angles to the direction of energy transfer.
- Transverse waves will travel across the surface of the water, e.g. once a stone has been dropped into the water.
- Longitudinal waves are produced under the surface in the main body of water, e.g. from a sound.
- Wave speed can be determined using speed = distance/time, using echoes.
- Wave speed can be determined by using wave speed = frequency × wavelength, using an oscilloscope.

194. Electromagnetic spectrum

1 (a) radio waves **(1)**, (b) gamma-rays **(1)**

2 4×10^{-7} m: $f = c \div \lambda = 3 \times 10^8$ m/s ÷ 4×10^{-7} m **(1)** = 7.5×10^{14} Hz **(1)**; 7×10^{-7} m: f = 3×10^8 m/s ÷ 7×10^{-7} m = 4.3×10^{14} Hz **(1)**

3 (a) frequency = $4 \times 10^{18} \div 60 = 6.7 \times 10^{16}$ Hz **(1)**; $\lambda = v \div f = 3 \times 10^8$ m/s $\div 6.7 \times 10^{16}$ Hz **(1)** = 4.5×10^{-9} m **(1)**

(b) X-rays **(1)**

195. Core practical – Investigating refraction

1 its speed **(1)**, its direction **(1)**

2 Angles of refraction are less than their corresponding angles of incidence **(1)**; angles of refraction are greater than the corresponding angles of refraction for those in the glass block **(1)**.

3 (a) Frequency does not change **(1)** because the frequency is determined by the oscillating source that produces the wave and remains constant regardless of any change in speed or wavelength **(1)**. (b) Wavelength decreases **(1)** because the wave speed decreases and the frequency remains constant **(1)**.

196. Wave behaviour

1 be reflected, be refracted, be transmitted, be absorbed **(2 marks for all four, 1 mark for two three)**

2 They are refracted by the ionosphere **(1)** and reflected back to receivers on the surface of the Earth **(1)**.

197. Dangers and uses

1 X-rays are useful to doctors as they allow them to detect broken bones **(1)** but they can damage cells **(1)**, which could cause them to mutate/lead to cancer **(1)**.

2 Gamma-rays can pass through the body and be detected by, e.g., a gamma camera, allowing a doctor to see if cancer is present. **(1)** Gamma-rays may cause cancer, but this is a risk worth taking if there is a possibility that the patient has a malignant cancer which may lead to death if not detected. **(1)**

3 Microwaves are absorbed by water molecules in the food **(1)** and their vibrations cause them to heat the food and cook it **(1)**. Infrared cooks food by heating the surface of the food only **(1)** before the heat then conducts **(1)** into the food, cooking it over a longer period of time **(1)**.

198. Changes and radiation

1 (a) Electrons move to higher energy levels **(1)** with the difference in energy between the levels equal to the energy of the electromagnetic radiation **(1)**.

(b) An electron falls down from a higher to a lower energy level **(1)** with a photon emitted having energy equal to the energy difference between the two levels **(1)**.

2 The energy difference between energy levels in the shells or orbits of electrons is much lower **(1)** compared with energy differences in the nuclei of atoms **(1)** when nuclear changes occur. The energy released in transitions between the energy levels of electrons is lower and leads to lower-energy visible light being emitted **(1)**, whereas the enormous energy changes involved during nuclear energy transitions results in much higher-energy gamma-rays being emitted **(1)**.

199. Extended response – Light and the electromagnetic spectrum

*Answer could include the following points **(6)**:

- radio waves: including broadcasting, communications and satellite transmissions
- microwaves: including cooking, communications and satellite transmissions
- infrared: including cooking, thermal imaging, short-range communications, optical fibres, television remote controls and security systems
- visible light: including vision, photography and illumination
- ultraviolet: including security marking, fluorescent lamps, detecting forged bank notes and disinfecting water
- X-rays: including observing the internal structure of objects, airport security scanners and medical X-rays
- gamma-rays: including sterilising food and medical equipment, and the detection of cancer and its treatment

200. Structure of the atom

The diameter of the atom is around 100 000 times greater than the diameter of a nucleus. **(1)** The diameter of the nucleus on the poster will be around 60 cm ÷ 100 000 **(1)** or a diameter of about 6×10^{-4} **(1)** cm **(1)**.

201. Atoms and isotopes

1 $^{11}_{5}$B **(1)**

2 Both contain 7 protons **(1)**, both contain 7 electrons **(1)**, nitrogen-15 contains 8 neutrons, one more than nitrogen-14 **(1)**.

202. Atoms, electrons and ions

1 (a) nucleus **(1)** (b) nucleus **(1)** (c) in orbits or shells around the nucleus **(1)**

2 Atoms of an element contain the same number of protons as electrons **(1)** and have a neutral charge overall **(1)**, whereas ions of an element have the same number of protons but can gain electrons so that they have a negative charge (more electrons than protons) **(1)** or lose electrons so that they have a positive charge (more protons than electrons) **(1)**.

3 The charged particle is deflected in an electric field **(1)**, e.g. a positive ion moves towards a negative charge or plate **(1)** OR direction of motion in a magnetic field **(1)**, e.g. a positive charge will move clockwise and a negative ion anticlockwise in the same magnetic field **(1)**.

203. Ionising radiation

1 They have the greatest mass **(1)** and the greatest charge **(1)** so can remove electrons more easily from the shells of atoms **(1)**.

2 You cannot predict **(1)** when a nucleus will decay by emitting radiation **(1)**.

204. Background radiation

1 D **(1)**

2 (a) The sector for radon would be much smaller. **(1)**

(b) The sector for radon would be bigger. **(1)**

3 Radon gas emits alpha particles which are highly ionising **(1)** and cause damage inside the body **(1)** but have too small a range in air to reach the body, so will not cause harm **(1)**.

205. Measuring radioactivity

1 Alpha particles are more ionising than gamma-rays **(1)** so more electrons are released per collision **(1)** so it is easier to detect from the current produced **(1)**.

2 If the film behind the plastic and lead is darkened then radiation has reached it. **(1)** If the radiation behind the plastic and lead has darkened then gamma-rays have been detected **(1)** and the person is at risk as these are dangerous outside the body and can lead to cellular mutations/cancer/death **(1)**.

206. Models of the atom

1 Plum pudding model with negative plums in a positive 'dough' **(1)**; Rutherford atom model with a small positive nucleus compared to overall atomic diameter **(1)**; Bohr model with electrons orbiting a positive nucleus in definite, discrete energy levels **(1)**

2 Positive alpha particles were fired at nucleus. **(1)** Most went through undeflected **(1)**, so most of the atom must be empty space **(1)** and there is a small nucleus as only a very few alpha particles bounced back **(1)**.

3 In the Bohr model electrons orbit the nucleus in very definite, discrete energy levels. **(1)** These energy levels are the same values for atoms of a given element. **(1)** Electronic transitions between energy levels in atoms of a gas lead to certain frequencies of electromagnetic radiation being emitted or absorbed in the spectra of light from stars. **(1)** The characteristic lines identify atoms of a particular element. **(1)**

207. Beta decay

1 (a) A neutron turns into a proton **(1)** and an electron is emitted **(1)**.

(b) A proton turns into a neutron **(1)** and a positron is emitted **(1)**.

2 A beta particle is an electron that is emitted from an unstable nucleus **(1)** and an electron in a stable atom is found in shells orbiting the nucleus **(1)**.

3 $^{66}_{28}$Ni **(1)** → $^{66}_{29}$Cu **(1)** + $^{0}_{-1}$e **(1)**

208. Radioactive decay

1 $^{232}_{90}$Th → $^{228}_{88}$Ra + $^{4}_{2}$He **(2)**

2 Final mass number will be $A - 5$ **(1)**, nucleus loses two protons and three neutrons **(1)**; atomic number of Z does not change **(1)**, number of protons decreases by 2 from alpha decay, then increases by 2 from two beta-minus decays **(1)**.

209. Half-life

1 120 Bq to 30 Bq is one quarter of the activity or two half-lives in 4 hours **(1)**, so one half-life is half of four hours or 2 hours **(1)**.

2 After 15 minutes: $\frac{1}{2}$; after 30 minutes: $\frac{1}{4}$ **(1)**; after 45 minutes: $\frac{1}{8}$ **(1)**; after 60 minutes: $\frac{1}{16}$ **(1)**.

210. Dangers of radiation

1 two of: can cause cells to become ionised, which leads to chemical reactions taking places in cells **(1)** causing them to die **(1)** or mutate **(1)**, which can lead to cancer **(1)**

2 Ionising radiation knocks electrons from shells **(1)** making them more reactive and

allowing change to DNA to occur **(1)**, whereas non-ionising radiation cannot do this as there is not enough energy to knock electrons from shells **(1)**.

3 measure radiation using a semiconductor crystal as opposed to film **(1)**, record specific exposure **(1)**, do not need to be developed so can provide information immediately **(1)**

211. Contamination and irradiation

1 (a) Irradiation is the exposure of the body to ionising radiation **(1)** from a source that is outside the body **(1)**.

(b) Contamination is exposure to ionising radiation from a source that is taken into the body **(1)** by eating, breathing in, drinking, injection or via contact with ionising radiation on the surface of the skin **(1)**.

2 Alpha particles are highly ionising due to their charge and mass **(1)** and will cause damage inside the body as they have a short range **(1)** and will come into contact with cells. Outside the body they are unlikely to reach the body as they have a range of only 5 cm in air and so cannot easily reach the body to cause any ionisation in cells **(1)**.

3 Some sources of background radiation do not come into contact with the body such as cosmic rays from the Sun and medical X-rays **(1)**, but other sources such as radon gas and radioisotopes in food are taken into the body **(1)**.

212. Extended response – Radioactivity

*Answer could include the following points **(6)**:

- plum pudding model with electrons inside a positive dough
- Rutherford model with small, positive nucleus which is concentrated in a tiny space around which electrons orbit
- Bohr model with electrons in defined energy shells
- experimental methods for refining models (e.g. a description of how Rutherford's alpha particle experiment was conducted)
- evidence collected from experiments to show new, better models
- Evidence from experiments shows a better understanding and this new knowledge replaces older knowledge since the new model is better.
- Experiments need to be conducted to provide evidence based on calculations or theories that have been developed. Only when the experiment provides evidence for a new, better model is it accepted by the scientific community.

213. Work, energy and power

1 work = $F \times d$ = 350 N × 30 m **(1)** = 1050 **(1)** J **(1)**

2 energy = $P \times t$ = 1800 W × 8 × 60 s **(1)** = 864 000 **(1)** J **(1)**

3 power = $E \div t$ = 3600 W ÷ 60 s **(1)** = 60 **(1)** W **(1)**

4 $E = P \times t$, $\Delta E = m \times c \times \Delta T$ **(1)**, 4000 W × (46 × 60 + 40) s **(1)** = 36 kg × 4200 J/kg °C × ΔT **(1)**, ΔT = 74 **(1)** °C **(1)**

214. Extended response – Energy and forces

*Answer could include the following points **(6)**:

- Refer to power as the rate of energy transfer (or similar).
- Refer to gain in gravitational potential energy as $mg\Delta h$ where Δh is the vertical distance climbed by the runner and the climber.
- Δh is a control variable.
- time measurement needed
- Refer to the change in gravitational potential energy for both athletes over a period of time.
- Compare their power values for running or climbing using the equation: power = energy/time or $P = mg\Delta h/t$.
- The athlete with the greatest value for $mg\Delta h/t$ (or who can transfer the most energy per second) is the most powerful.

215. Interacting forces

1 Gravity is always an attractive force, whereas magnetism and the electrostatic force can be attractive or repulsive. **(1)**

2 Friction and drag always work to oppose motion/slow down a moving object. **(1)**

3 Contact forces are any two of: drag **(1)**, normal **(1)** and upthrust **(1)**. Non-contact force is the force of gravity acting on the ship. **(1)**

216. Free-body force diagrams

1 correct length shown with correct scale **(1)**, correct angle drawn to the horizontal using a protractor **(1)**, correct horizontal value found of value close to 57.5 N **(1)**, correct vertical component shown of value close to 48.2 N **(1)**

2 force of gravity acting vertically downwards **(1)**, smaller force of drag acting upwards **(1)**, both forces labelled **(1)**

217. Resultant forces

1 (a) two forces of 20 N acting parallel and upwards **(1)**, (b) two forces of 20 N acting anti-parallel **(1)**

2 It is accelerating **(1)** downwards **(1)**.

3 correct scale drawing of the two vectors **(1 each)**, resultant correctly drawn and measured as 13 N **(1)**, direction of resultant force 67.4° to the horizontal, acting left **(1)**

218. Extended response – Forces and their effects

*Answer could include the following points **(6)**:

- Contact forces involve surfaces touching or being in contact.
- Non-contact forces involve bodies interacting at a distance and not being in contact.
- Examples of contact forces include friction, upthrust, reaction force.
- Examples of non-contact forces are gravitational force, magnetic force, electrostatic force.
- Contact forces such as reaction force and upthrust, if balanced, will cause a body to remain still or float.
- Contact forces, if balanced, can cause a body to move at a constant speed.
- Non-contact force of gravity will cause a body to accelerate towards Earth if there is a resultant force.
- Non-contact force of gravity will cause a body to move at a constant speed if the force of gravity is balanced by opposing frictional or drag force.
- Contact force of friction opposes the motion of a body and tries to slow it down.

other examples of balanced/unbalanced forces used as examples for magnetism or electrostatic force

219. Circuit symbols

1 filament lamp **(1)**, motor **(1)**, LED **(1)**

2 ammeters in series **(1)**, voltmeters in parallel **(1)**

3 circuit to contain a cell **(1)**, lamp **(1)**, LDR **(1)**

220. Series and parallel circuits

1 series: current same at all points **(1)**, potential differences across components add to give potential difference across cell **(1)**. Parallel: current splits at a junction **(1)**, potential difference across each branch is same as potential difference across cell **(1)**.

2 Current will stop flowing in series if one bulb breaks **(1)**, so parallel arrangement so that lights stay on **(1)** if one or more bulbs break **(1)**.

3 More bulbs in parallel means a higher current **(1)** is drawn from the cell, so finite amount of stored chemical energy **(1)** decreases **(1)** more rapidly **(1)**.

221. Current and charge

1 charge: coulomb **(1)**; current: ampere **(1)**

2 $Q = I \times t$ = 0.25 A × (60 × 60) s **(1)** = 900 **(1)** C **(1)**

3 $Q = I \times t$ so $t = Q \div I$ **(1)** = 3 × 10^4 C ÷ 0.25 A **(1)** = 1.2 × 10^5 **(1)** s **(1)**

222. Energy and charge

1 $E = Q \times V$ = 24 C × 6 V **(1)** = 144 **(1)** J **(1)**

2 $V = E \div Q$ **(1)**, so since unit of energy is J and charge is C **(1)**, we get 1 V = 1 J/C **(1)**.

3 using $E = QV$ and $Q = It$, $E = VIt$ **(1)**, 500 J = 18 V × 240 s × I **(1)**, I = 0.12 **(1)** A **(1)**

223. Ohm's law

1 $V = I \times R$ = 3.2 A × 18 Ω **(1)** = 57.6 **(1)** V **(1)**

2 $R = V \div I$ = 28 V ÷ 0.4 A **(1)** = 70 **(1)** Ω **(1)**

224. Resistors

$I = V \div R$ = 12 V ÷ 120 Ω **(1)** = 0.1 **(1)** A **(1)**

225. I–V graphs

(a) graph of S-shaped curve through origin of I against V for filament lamp **(1)**, labelled axes **(1)**.

(b) As the current goes up, the temperature goes up **(1)** and as the temperature goes up the resistance goes up **(1)** and the gradient goes down **(1)**.

226. Core practical – Electrical circuits

1 Current will flow through the components **(1)** in either arrangement, with the shape of the I–V graph being determined by the structure **(1)** of the component and whether it obeys Ohm's law or not **(1)**, not on how it is arranged in a circuit, since heat will still be dissipated in each circuit **(1)**.

2 As the p.d. increases across the filament lamp, the current through the filament lamp also increases **(1)** and as current increases, the energy is dissipated as heat increases leading to more vibrations of the metal ions in the filament **(1)**, which increases the resistance of the lamp **(1)** so the current increases by less and less per unit increase in p.d. **(1)** leading to a graph of decreasing gradient as the p.d. gets bigger and bigger **(1)**.

227. The LDR and the thermistor

1 (a) temperature **(1)**, (b) light intensity **(1)**

2 Provide an example of a circuit where temperature **(1)** and light intensity **(1)** vary their resistances **(1)** to turn on an output device **(1)**. The marks can only be awarded if the operation of the circuit is explained in terms of changing light levels and temperature. An example would be for a circuit that turns a light/heater on or off in a greenhouse, for example.

228. Current heating effect

1 any three devices, e.g. oven, iron, heater, hair dryer, toaster, grill **(1 mark for each)**

2 Large current **(1)** transfers much thermal energy **(1)** which may lead to a fire **(1)**.

3 Greater current can flow **(1)**, power proportional to current squared **(1)**, more energy dissipated as heat **(1)**.

229. Energy and power

1 $I = P \div V = 3000\,\text{W} \div 230\,\text{V}$ **(1)** = 13 **(1)** A **(1)**

2 $P = E \div t = 100\,000\,\text{J} \div 180\,\text{s} = 556\,\text{W}$ **(1)**, $P = I \times V$, so I = 556 W ÷ 230 V = 2.4 A **(1)**, $R = P \div I^2 = 556\,\text{W} \div 2.4^2 = 97$ **(1)** Ω **(1)**

230. a.c. and d.c. circuits

1 (a) any three devices, e.g. remote controls, torches, mobile phones, calculators, watches, etc. **(1 mark each)**.

(b) any three mains-operated devices, e.g. electric oven, microwave oven, fridge, freezer, tumble dryer, hairdryer, TV, etc. **(1 mark each)**

2 (a) They both transfer energy. **(1)**

(b) Potential difference in a.c. alternates **(1)** but in d.c. it is constant **(1)**.

3 The potential difference of a.c. can be changed using transformers **(1)** so that energy losses while transmitting energy over the National Grid **(1)** are minimised **(1)**.

231. Mains electricity and the plug

1 (a) brown **(1)** (b) blue **(1)** (c) yellow and green **(1)**

2 advantage of fuse: cheaper **(1)**; advantage of circuit breaker – one of: more sensitive to current **(1)**, does not need to be replaced (reset by pressing a switch) **(1)**, more reliable **(1)**, responds faster **(1)**

3 Live wire touches metal case **(1)**, earth wire forms circuit with live wire and pulls a large current through the fuse **(1)**, large current melts fuse and isolates appliance **(1)**, device now at 0 V and no longer dangerous to user **(1)**.

232. Extended response – Electricity and circuits

*Answer could include the following points **(6)**:

- Fuse will melt if the current entering an appliance is too big.
- Earth wire pulls a large current through a fuse and melts the fuse wire if the live wire comes into contact with the metal casing.
- Mention that fuse/earth wire are safety devices.
- Mention fuse is connected to live wire so that device will be at 0 V when the fuse blows.
- Fuse/earth wire isolate the appliance so that device case cannot become live and so protects the user from electrocution/fires occurring due to overheating.
- The fuse rating should be close to, but above, the current taken by the appliance so as to avoid any risk of the device overheating/catching fire due to the current becoming too high and producing too much thermal energy.

233. Magnets and magnetic fields

1 (a) similar to the diagram for the bar magnet **(1)** but with fewer field lines **(1)**; (b) similar to the diagram for the uniform field **(1)** but with more field lines **(1)**

2 The permanent magnet always induces **(1)** the opposite pole **(1)** next to its pole causing it to attract **(1)**.

3 The permanent magnetic field of the magnets needs to interact **(1)** with the magnetic field around the coil once the current has started to flow in it **(1)**.

234. Current and magnetism

1 similar to diagram on page 234 **(1)**, but with the arrows in the opposite direction **(1)**

2 Current doubles so field doubles **(1)**, distance doubles so field halves **(1)**, so overall there is no change **(1)**.

3 Increase the size of the current flowing **(1)**, increase the number of turns of wire per metre on the coil **(1)**, insert a magnetic material as the core **(1)**.

235. Current, magnetism and force

1 A magnetic field is set up around a wire when a current flows through it **(1)** and this will interact with the field from the permanent magnet and experience a force **(1)**.

2 $F = B \times I \times L = 1.2 \times 10^{-3}\,\text{T} \times 3.6\,\text{A} \times 0.5\,\text{m}$ **(1)** = 2.16×10^{-3} **(1)** N **(1)**

3 (a) Change direction of the current flowing in the wire **(1)**, change the direction of the magnetic field **(1)**.

(b) Increase the size of the current flowing in the wire **(1)**, increase the strength of the magnetic field/magnetic flux density **(1)**.

236. Extended response – Magnetism and the motor effect

*Answer could include the following points **(6)**:

- The closer a magnet is to an object, the stronger the magnetic field that it will experience, so the greater the mass it can attract.
- A greater electric current will cause the strength of the magnetic field around an electromagnet to be greater.
- More turns on the coil or a coiled wire instead of a straight wire increases the strength of the magnetic field from the electromagnetic field, so a greater mass can be picked up or attracted.
- Using a soft iron core will increase the strength of the magnetic field around a solenoid/electromagnet so more mass can be attracted.
- Mention that certain materials where magnetism can be induced will be attracted, whereas others will not – iron, cobalt, nickel, steel will be, but others will not be.
- Mention that certain materials that are themselves permanent magnets will be attracted based on their magnetic field strengths, even when no current flows through an electromagnet.

237. Electromagnetic induction and transformers

1 (a) Reverse the direction of motion of the wire **(1)**, reverse the direction of the magnetic field **(1)**.

(b) Increase the speed of movement of the wire **(1)**. increase the size of the magnetic field strength/magnetic flux density **(1)**.

2 assuming 100% efficiency **(1)**, power in primary = 3000 W, so $I_p = P \div V_p = 3000\,\text{W} \div 30\,000\,\text{V}$ **(1)** = 0.1 A **(1)**

3 For electromagnetic induction to occur constantly, there needs to be a constantly changing magnetic field. **(1)** This will only happen with a.c. because it is constantly changing direction **(1)** whereas d.c. does not change direction **(1)**.

238. Transmitting electricity

1 (a) any two of transmission pylons, wires, transformers **(2)**

(b) power stations **(1)**, consumers **(1)**

2 High current means a large amount of heat is generated **(1)** which is lost to the surroundings **(1)** which means that less energy reaches the consumer as electrical energy **(1)**.

3 It would be best to transfer 10^9 W at a low current and a high voltage **(1)** because less energy is dissipated to the surroundings as a thermal store **(1)** when the current is low. (Reference to $P = I^2R$ in terms of power loss/energy loss per second being proportional to I^2)

239. Extended response – Electromagnetic induction

*Answer could include the following points **(6)**:

- Transformer has primary and secondary coil and works with a.c. input only.

- Transformer has a soft iron core which allows electromagnetic induction to induce voltage in secondary coil based on a.c. input in primary coil.
- The voltage will be stepped up/increased if the number of turns on the secondary coil is greater than on the primary coil.
- The current is stepped down when the voltage is stepped up to conserve power and not violate the principle of conservation of energy.
- Voltage is stepped up by transformers when it leaves the power station for transmission across the UK on wires.
- Voltage is stepped down to safer levels when electricity is supplied to homes, hospitals and factories.
- Lower current means thinner wires means less expensive to buy means lower cost to the consumer.

240. Changes of state

1 Thermal energy is transferred from the warmer room to the ice **(1)** and causes the ice to melt **(1)** and become 1.5 kg of liquid water **(1)**.

2 As the gas cools, the particles slow down and occupy a smaller volume **(1)** and make fewer collisions with the container they are in **(1)** so the pressure decreases too **(1)**.

3 Thermal energy is supplied to the copper metal, bonds are broken and it turns from a solid to a liquid at a constant temperature **(1)** once it reaches its melting point **(1)**. Further heating of the liquid copper causes its temperature **(1)** to rise until it reaches its boiling point, at which point it turns from a liquid to a gas at a constant temperature **(1)**. If the gaseous copper atoms continue to be heated then the temperature of the gas will increase **(1)**.

241. Density

1 density = mass ÷ volume = 862 g ÷ 765 cm^3 **(1)** = 1.1 **(1)** g/cm^3 **(1)**

2 50 000 cm^3 = 5000 ÷ 1 000 000 m^3 = 0.05 m^3 **(1)**, mass = density × volume = 0.05 m^3 × 2700 kg/m^3 **(1)** = 135 **(1)** kg **(1)**

242. Core practical – Investigating density

1 density = mass ÷ volume = 454 g ÷ 256 cm^3 **(1)** = 1.77 **(1)** g/cm^3 **(1)**

2 Mass reading can be misread by human error or by parallax error or by systematic error **(1)**; volume can be misread by parallax error or reading the top of the meniscus for water **(1)**; a value that is too high for the mass **(1)** or too low for the volume will lead to a value that is too high for the density **(1)**.

243. Energy and changes of state

1 ΔQ = 15 kg × 4200 J/kg °C × (74 °C – 18 °C) **(1)** = 3 528 000 **(1)** J **(1)**

2 $I \times V \times t = m \times c \times \Delta T$ **(1)**, c = 16 V × 2.8 A × (12 × 60) s ÷ (1.25 kg × 26 °C) **(1)** = 992 **(1)** J/kg °C **(1)**

244. Core practical – Thermal properties of water

1 Less thermal energy is lost to the surroundings **(1)** and more is used to heat the water **(1)** so the value obtained is closer to the true value required to cause a temperature change of 1 °C for that mass of liquid **(1)**; OR State that losing more to the surroundings **(1)** requires more energy than necessary to be supplied for unit temperature change **(1)** of unit mass **(1)**.

2 Energy supplied **(1)** does not lead to an increase in the kinetic energy of the particles **(1)**, just in their arrangement as they change phase from a solid to a liquid **(1)**.

3 Diagram should show ice in a funnel over a beaker, with an electric heater in the ice **(1)** and a thermometer (which should be away from the heater) **(1)** and a stopwatch **(1)**.

245. Pressure and temperature

1 (a) 20 + 273 = 293 K **(1)**

(b) 300 – 273 = 27 °C **(1)**

2 –23 °C + 273 K = 250 K **(1)** and 227 °C + 273 K = 500 K **(1)**; average KE of gas particles is proportional to temperature in Kelvin, so the average KE doubles **(1)**.

3 (a) The average speed is slower. **(1)**

(b) The pressure is reduced **(1)** because there are fewer collisions/the particles do not hit the walls as hard **(1)**.

246. Extended response – Density

*Answer could include the following points **(6)**:

- Pressure is caused by moving particles colliding with the sides of a container.
- Motion of the particles is random.
- Pressure is force per unit area.
- Increasing the temperature of a gas at constant volume means that the particles have greater kinetic energy and speed and so collide more frequently with the area, so greater pressure.
- The internal energy of a gas is entirely kinetic.
- When work is done on the gas the kinetic energy/speed of the particles increases.
- Faster moving particles collide with the walls/area more frequently, leading to an increase in pressure.

247. Elastic and inelastic distortion

1 (a) two equal forces **(1)** of tension acting in opposite directions **(1)**

(b) two equal forces **(1)** acting towards one another **(1)**

2 (a) any two suitable examples, e.g. springs **(1)**, elastic bands **(1)**, rubber **(1)**, skin **(1)**

(b) any two suitable examples, e.g. springs beyond elastic limit **(1)**, putty **(1)**, Plasticine **(1)**, bread dough **(1)**, wet chewing gum **(1)**

3 Forces applied lead to extension of material beyond elastic limit **(1)**, which means that forces of attraction between planes of atoms or ions **(1)** are no longer strong enough for the materials to return to the original structure and so they remain permanently distorted **(1)**.

248. Springs

1 (a) $F = k \times x$ = 30 N/m × 0.2 m **(1)** = 6 **(1)** N **(1)**

(b) work done = $\frac{1}{2}k \times x^2 = \frac{1}{2}$ × 30 N/m × (0.2 m)^2 **(1)** = 0.6 **(1)** J **(1)**

2 The greater the spring constant, the stiffer the spring **(1)** since a greater spring constant means more force is needed **(1)** to provide the same extension compared with a spring of lower spring constant **(1)**.

3 The force applied takes the spring beyond its elastic limit **(1)** and so the material will remain permanently deformed **(1)** and not return to its original shape **(1)**.

249. Core practical – Forces and springs

1 energy stored = $\frac{1}{2}kx^2 = \frac{1}{2}$ × 0.8 N/m × (0.48 m)^2 **(1)** = 0.09 **(1)** J **(1)**

2 Yes **(1)**, as energy is still being stored and there is a compression rather than an extension **(1)**, so the size of the compression/length used in the equations to determine the spring constant or the energy stored would be less than the original length **(1)**.

250. Extended response – Forces and matter

*Answer could include the following points **(6)**:

- Distortion is when a force causes a body to extend or change shape.
- The extension is directly proportional to the load/force applied for elastic distortion.
- The body/spring/material will return to its original shape when the force/load is removed.
- Inelastic distortion is when the body is extended beyond its elastic limit.
- The body will then not return to its original shape but will remain permanently extended.
- The spring constant for a spring is found from the gradient of the force–extension graph provided that the linear region only is used.
- The spring constant may be found from $k = F \div x$
- Energy stored is found from $E = \frac{1}{2}kx^2$ or from the area beneath the line.

The Periodic Table of the Elements

Key

relative atomic mass **atomic symbol** name atomic (proton) number

1	2											3	4	5	6	7	0
							1 **H** hydrogen 1										4 **He** helium 2
7 **Li** lithium 3	9 **Be** beryllium 4											11 **B** boron 5	12 **C** carbon 6	14 **N** nitrogen 7	16 **O** oxygen 8	19 **F** fluorine 9	20 **Ne** neon 10
23 **Na** sodium 11	24 **Mg** magnesium 12											27 **Al** aluminium 13	28 **Si** silicon 14	31 **P** phosphorus 15	32 **S** sulfur 16	35.5 **Cl** chlorine 17	40 **Ar** argon 18
39 **K** potassium 19	40 **Ca** calcium 20	45 **Sc** scandium 21	48 **Ti** titanium 22	51 **V** vanadium 23	52 **Cr** chromium 24	55 **Mn** manganese 25	56 **Fe** iron 26	59 **Co** cobalt 27	59 **Ni** nickel 28	63.5 **Cu** copper 29	65 **Zn** zinc 30	70 **Ga** gallium 31	73 **Ge** germanium 32	75 **As** arsenic 33	79 **Se** selenium 34	80 **Br** bromine 35	84 **Kr** krypton 36
85 **Rb** ribidium 37	88 **Sr** strontium 38	89 **Y** yttrium 39	91 **Zr** zirconium 40	93 **Nb** niobium 41	96 **Mo** molybdenum 42	[98] **Tc** technetium 43	101 **Ru** ruthenium 44	103 **Rh** rhodium 45	106 **Pd** palladium 46	108 **Ag** silver 47	112 **Cd** cadmium 49	115 **In** indium 49	119 **Sn** tin 50	122 **Sb** antimony 51	128 **Te** tellurium 52	127 **I** iodine 53	131 **Xe** xenon 54
133 **Cs** caesium 55	137 **Ba** barium 56	139 **La*** lanthanum 57	178 **Hf** hafnium 72	181 **Ta** tantalum 73	184 **W** tungsten 74	186 **Re** rhenium 75	190 **Os** osmium 76	192 **Ir** iridium 77	195 **Pt** platinum 78	197 **Au** gold 79	201 **Hg** mercury 80	204 **Tl** thallium 81	207 **Pb** lead 82	209 **Bi** bismuth 83	[209] **Po** polonium 84	[210] **At** astatine 85	[222] **Rn** radon 86
[223] **Fr** francium 87	[226] **Ra** radium 88	[227] **Ac*** actinium 89	[261] **Rf** rutherfordium 104	[262] **Db** dubnium 105	[266] **Sg** seaborgium 106	[264] **Bh** bohrium 107	[277] **Hs** hassium 108	[268] **Mt** meitnerium 109	[271] **Ds** darmstadtium 110	[272] **Rg** roentgenium 111	Elements with atomic numbers 112–116 have been reported but not fully authenticated						

*The lanthanoids (atomic numbers 58–71) and the actinoids (atomic numbers 90–103) have been omitted.

The relative atomic masses of copper and chlorine have been rounded to the nearest whole number.

Combined Science Equations List

In your exam, you will be provided with the following list of equations. Make sure you are clear which equations will be given to you in the exam. You will need to learn the equations that aren't on the equations list.

(final velocity)^2 – (initial velocity)^2 = 2 × acceleration × distance $v^2 - u^2 = 2 \times a \times x$
force = change in momentum ÷ time $F = \frac{(mv - mu)}{t}$
energy transferred = current × potential difference × time $E = I \times V \times t$
force on a conductor at right angles to a magnetic field carrying a current = magnetic flux density × current × length $F = B \times I \times l$
For transformers with 100% efficiency, potential difference across primary coil × current in primary coil = potential difference across secondary coil × current in secondary coil $V_p \times I_p = V_s \times I_s$
change in thermal energy = mass × specific heat capacity × change in temperature $\Delta Q = m \times c \times \Delta\theta$
thermal energy for a change of state = mass × specific latent heat $Q = m \times L$
energy transferred in stretching = 0.5 × spring constant × (extension)^2 $E = \frac{1}{2} \times k \times x^2$

Your own notes

Your own notes

Your own notes

Your own notes

Your own notes

Your own notes

Published by Pearson Education Limited, 80 Strand, London, WC2R 0RL.

www.pearsonschoolsandfecolleges.co.uk

Copies of official specifications for all Edexcel qualifications may be found on the website: www.edexcel.com

Produced and typeset by Phoenix Photosetting
Cover illustration by Miriam Sturdee

First published 2017

20 19 18 17
10 9 8 7 6 5 4 3 2 1

British Library Cataloguing in Publication Data
A catalogue record for this book is available from the British Library

ISBN 978 1 292 13163 4

Printed in Slovakia by Neografia

Acknowledgements
Content by Penny Johnson, Sue Kearsey, Stephen Winrow-Campbell and Steve Woolley is included.

The publishers are grateful to Nigel Saunders for his help and advice with this book.

The publisher would like to thank the following for their kind permission to reproduce their photographs:
(Key: b-bottom; c-centre; l-left; r-right; t-top)
Bridgeman Art Library Ltd: Musee de Picardie, Amiens, France 30l, Musee des Antiquities Nationales, St. Germain-en-laye, France 30r; **NASA:** JPL-Caltech / R. Hurt (SSC) 191; **Science Photo Library Ltd:** Herve Conge, ISM 6l, Steve Gschmeissner 6br; **Shutterstock.com:** Malota 26

A note from the publisher
In order to ensure that this resource offers high-quality support for the associated Pearson qualification, it has been through a review process by the awarding body. This process confirms that this resource fully covers the teaching and learning content of the specification or part of a specification at which it is aimed. It also confirms that it demonstrates an appropriate balance between the development of subject skills, knowledge and understanding, in addition to preparation for assessment.

Endorsement does not cover any guidance on assessment activities or processes (e.g. practice questions or advice on how to answer assessment questions), included in the resource nor does it prescribe any particular approach to the teaching or delivery of a related course.

While the publishers have made every attempt to ensure that advice on the qualification and its assessment is accurate, the official specification and associated assessment guidance materials are the only authoritative source of information and should always be referred to for definitive guidance.

Pearson examiners have not contributed to any sections in this resource relevant to examination papers for which they have responsibility.

Examiners will not use endorsed resources as a source of material for any assessment set by Pearson.

Endorsement of a resource does not mean that the resource is required to achieve this Pearson qualification, nor does it mean that it is the only suitable material available to support the qualification, and any resource lists produced by the awarding body shall include this and other appropriate resources.